Numerical Methods for Initial Value Problems in Physics

Francisco S. Guzmán

Numerical Methods for Initial Value Problems in Physics

Springer

Francisco S. Guzmán
Instituto de Física y Matemáticas
Universidad Michoacana de San Nicolás de Hidalgo
Morelia, Mexico

ISBN 978-3-031-33555-6   ISBN 978-3-031-33556-3 (eBook)
https://doi.org/10.1007/978-3-031-33556-3

© The Editor(s) (if applicable) and The Author(s), under exclusive license to Springer Nature Switzerland AG 2023

This work is subject to copyright. All rights are solely and exclusively licensed by the Publisher, whether the whole or part of the material is concerned, specifically the rights of translation, reprinting, reuse of illustrations, recitation, broadcasting, reproduction on microfilms or in any other physical way, and transmission or information storage and retrieval, electronic adaptation, computer software, or by similar or dissimilar methodology now known or hereafter developed.

The use of general descriptive names, registered names, trademarks, service marks, etc. in this publication does not imply, even in the absence of a specific statement, that such names are exempt from the relevant protective laws and regulations and therefore free for general use.

The publisher, the authors, and the editors are safe to assume that the advice and information in this book are believed to be true and accurate at the date of publication. Neither the publisher nor the authors or the editors give a warranty, expressed or implied, with respect to the material contained herein or for any errors or omissions that may have been made. The publisher remains neutral with regard to jurisdictional claims in published maps and institutional affiliations.

This Springer imprint is published by the registered company Springer Nature Switzerland AG
The registered company address is: Gewerbestrasse 11, 6330 Cham, Switzerland

Dedicated to Tana, Paco, Lisa, and Susi

Preface

The book contains the construction, implementation, and application of commonly used methods in the solution of Initial Value Problems (IVPs).

Along the years, within the groups I have worked with, we have seen the need to teach the theory and practice of numerical methods to graduate students, who are very strong in science but not so strong in programing or computer science. Personalized tutorials and small group workshops have been the common solution, where it is possible to revise the theory and implement programs hands-on. This book is intended to provide these two ingredients, theory and immediate implementation.

The chapters are organized from simple to difficult, starting with IVPs involving Ordinary Differential Equations (ODEs), followed by problems with Partial Differential Equations (PDEs) on a domain with 1+1 dimensions and finally problems with PDEs in 3+1 dimensions.

Examples and applications of IVPs associated to ODEs intend to cover various fields of science, including simple models of population, chaos, celestial mechanics, and astrophysics. Models and applications are expected to seed the motivation to continue along state of the art problems. The examples concerning PDEs concentrate on Wave, Schrödinger, Diffusion, and Fluid Dynamics equations, which are common to a variety of disciplines, for example electromagnetism, diffusion problems in chemistry or ecology, quantum mechanics or atomic gases, and all sorts of problems involving simple fluids in astrophysics or the laboratory.

At the end of each chapter, there is a brief description of how the methods can improve, along with one or two projects that can be developed with the methods and codes described. The type of projects would reflect my inclination toward space physics, explained by my academic origins centered in Numerical Relativity.

Of special importance nowadays, when Machine Learning (ML) and Artificial Intelligence (AI) methods shine, useful to solve inverse problems using causal inference, it is essential to keep in mind that direct problems are essential to train and validate ML and AI programs. This book contains a collection of direct problems, along with the methods and codes to solve them, that can be connected to interesting inverse problems.

The numerical methods in the book were selected because it is possible to solve a considerable variety of problems of various sorts. More specifically, the IVPs are solved on a discrete version of the domain of solution, and therefore the methods of solution are based on Finite Differences and Finite Volumes approaches.

The implementation of the methods described in the text is explained with `fortran 90` language, which is nearly a literal formula translation. Moreover, fortran 90 is nearly as easy to program as python and nearly as efficient as `c` or `c++` in most of architecture-compiler combinations. In summary, `fortran 90` allows to straightforwardly translate the ideas in the text into a program, and that is the reason why the codes included to reproduce the results in the book use this language. As supplementary material, we also include the c++ version of the codes for the readers more used to this language. For the analysis, the output is plain ascii that can we viewed with most data visualizers, and we provide instructions for `gnuplot`.

Hopefully this text, written at the action field, will help the reader to start solving state of the art problems.

Morelia, Mexico   Francisco S. Guzmán

Acknowledgments

I thank to all the students I've had the pleasure to supervise during their research. The ideas we developed together to explain, construct, and grasp the implementation of numerical methods are in one way or another reflected in this book.

Undergraduate students: Antonio Rendón Romero, Javier Vargas Arias, Alejandro Cruz Osorio, Néstor Ortiz Madrigal, Jesús M. Rueda Becerril, Adriana González Juárez, Iván R. Avilés Calderón, Venecia Chávez Medina, Itzayana Izquierdo Guzmán, Karla Sofía Zavala Alvarez, Miguel Angel Ceja Morales, María Otilia Segura Patiño, Ricardo Tovar Duarte, Alejandro Romero Amezcua, Paris Alejandro Dávalos Bravo, Flavio Rosales Infante, Ana Laura Colmenero César, Curicaveri Palomares Chávez. *Master in Science students:* Jacobo Israel Palafox González, Alejandro Cruz Osorio, Manuel David Morales Altamirano, Miguel Gracia Linares, José Juan González Avilés, Francisco Javier Rivera Paleo, Ricardo Ochoa Armenta, Venecia Chávez Medina, Itzayana Izquierdo Guzmán, Luis Felipe Mendoza Mendoza, Ricardo Tovar Duarte. *Doctoral students:* Argelia Bernal Bautista, Fabio Duvan Lora Clavijo, Alejandro Cruz Osorio, José Juan González Avilés, Miguel Gracia Linares, Francisco Javier Rivera Paleo, Ricardo Ochoa Armenta, Itzayana Izquierdo Guzmán, Iván Alvarez Ríos.

Special thanks to Francisco Guzmán Cajica, who developed the templates for the c++ codes that later on I specialized to solve each of the examples in the text. Flavio Rosales Infante, who programmed a message transmitter on top of a chaotic signal, proposed as a project also in Chap. 2. Karla Sofía Zavala Alvarez, whose results inspired the project on gravitational waves also in Chap. 2. Argelia Bernal Bautista, with whom we started to study the dynamics of the now well-known bosonic dark matter, reflected in projects of Chap. 3. Fabio Duvan Lora Clavijo and Alejandro Cruz Osorio, with who we started the relativistic plasma dynamics project CAFE, the first tuned implementation of High Resolution Shock Capturing methods in the group, which inspires Chap. 5; they also invested energy in the solution of the wave equation in constant mean curvature foliations of the space-time, which appear as a project in Chap. 4. José Juan González Avilés, who developed the Newtonian version of CAFE and started the Solar Physics research in the group, somehow honored with projects of Chap. 6, and whose research was awarded with the Weizmann prize to the best Doctoral Thesis in exact sciences by the Mexican Academy of Sciences.

I finally want to thank my mentors Tonatiuh Matos, Miguel Alcubierre, and Edward Seidel, to whom I owe infinities, and those incredible officemates I once learned a lot from, Ian Hawke, Jonathan Thornburg, Denis Pollney, Scott Hawley, and Peter Diener.

Contents

1 Introduction 1
   1.1   Errors and Precision of Numbers 2
   1.2   Rescaling Equations 4
   1.3   Other Methods 5
   1.4   Organization of the Book.. 5

2 Ordinary Differential Equations 7
   2.1   The General Problem 8
   2.2   Workhorse Example 1: Newton Cooling Law 8
   2.3   Euler Method 10
   2.4   Backward Euler Method.. 17
   2.5   Average Euler Method.. 19
   2.6   Workhorse Example 2: Harmonic Oscillator 20
   2.7   Error Theory, Convergence, and Self-Convergence 29
   2.8   Runge-Kutta Methods 31
   2.9   Workhorse Example 3: Damped and Forced Harmonic Oscillator .. 39
   2.10 Extrapolation.. 44
   2.11 Applications... 46
       2.11.1   Predator-Prey Model 47
       2.11.2   SIR Model for the Spread of Diseases..................... 49
       2.11.3   Lorenz System and Chaos.................................. 53
       2.11.4   Simple Pendulum 57
       2.11.5   Damped Pendulum 60
       2.11.6   Polytropic Star Models with Spherical Symmetry 63
       2.11.7   Polytropic Relativistic TOV Star Models with
                    Spherical Symmetry 69
       2.11.8   The String: An Eigenvalue Problem 74
       2.11.9   Test Particle Orbiting a Central Object 80
       2.11.10 Two Body Problem 86
   2.12 How to Improve the Methods in This Chapter 91
   2.13 Projects... 92

		2.13.1	Synchronization of Chaos and Message Encryption	92
		2.13.2	Gravitational Waves Emitted By a Binary System Using the Quadrupolar Formula	94
	References			97
3	**Simple Methods for Initial Value Problems Involving PDEs**			**99**
	3.1	Discretization of an IVP		100
	3.2	1 + 1 Wave Equation: The Paradigm of Hyperbolic Equations		108
		3.2.1	Dissipation and Dispersion	121
	3.3	1 + 1 Wave Equation Using Implicit Methods		122
	3.4	Diffusion Equation: The Paradigm of Parabolic Equations		132
		3.4.1	Simple Discretization Method	133
		3.4.2	Implicit Crank-Nicolson Method	135
		3.4.3	Dissipation and Dispersion Again	139
	3.5	1 + 1 Schrödinger Equation		139
		3.5.1	Particle in a 1D Box	140
		3.5.2	Particle in a Harmonic Oscillator Potential	141
		3.5.3	Implicit Crank-Nicolson Method Applied to Schrödinger Equation	143
	3.6	How to Improve the Methods in this Chapter		151
	3.7	Projects		151
		3.7.1	Bosonic Dark Matter	152
		3.7.2	Newtonian Boson Stars	152
	References			153
4	**Method of Lines for Initial Value Problems Involving PDEs**			**155**
	4.1	Method of Lines		155
	4.2	Time Integrators		156
	4.3	Advection Equation		159
	4.4	1+1 Wave Equation		163
	4.5	Wave Equation in a General 1+1 Minkowski Space-Time: Example of Characteristic Analysis		170
	4.6	1+1 Schrödinger Equation Using the Method of Lines		183
		4.6.1	Particle in a 1D Box	184
		4.6.2	Particle in a Harmonic Oscillator Potential	186
	4.7	Application: Solution of the Wave Equation on Top of a Schwarzschild Black Hole		188
	4.8	Application: Thermal Diffusion of Earth's Crust		196
	4.9	How to Improve the Methods in This Chapter		199
	4.10	Projects		200
		4.10.1	Solution of the Wave Equation in the Whole Space-Time	200
		4.10.2	Nonlinear Absorption of a Scalar Field	201
		4.10.3	Black Hole Shrinking	201
		4.10.4	Bose Condensates in Optical Lattices	202
	References			202

Contents   xiii

5 Finite Volume Methods 205
   5.1 Characteristics in Systems of Equations 206
        5.1.1 Constant Coefficient Case 207
        5.1.2 Variable Coefficient Case 208
        5.1.3 Finite Volume Method in 1+1 Dimensions 211
   5.2 1+1 Euler Equations 213
   5.3 Application: Spherically Symmetric Hydrodynamical
        Solar Wind 226
   5.4 Application: 1+1 Relativistic Euler Equations 233
   5.5 Application: Spherical Accretion of a Fluid onto a
        Schwarzschild Black Hole 241
   5.6 How to Improve the Methods in This Chapter 255
   5.7 Projects 255
        5.7.1 Evolution of a Relativistic TOV Star 256
        5.7.2 Nonlinear Michel Accretion 257
   References 257

6 Initial Value Problems in 3+1 and 2+1 Dimensions 259
   6.1 General Problem 259
        6.1.1 Expressions for Partial Derivatives 260
        6.1.2 Programming 261
   6.2 The 3+1 Wave Equation Using a Simple Discretization 263
        6.2.1 The Exact Solution 263
        6.2.2 Numerical Solution Using a Simple Discretization 264
        6.2.3 Example with Sources 268
   6.3 The 3+1 Wave Equation Using the Method of Lines 271
        6.3.1 The Time-Symmetric Gaussian Pulse 272
        6.3.2 Example with a Plane Wave 274
   6.4 The 3+1 Schrödinger Equation Using the ADI Scheme 277
        6.4.1 Two Exact Solutions 279
        6.4.2 Particle in a Harmonic Oscillator Potential 282
   6.5 3D Hydrodynamics 285
        6.5.1 2D Spherical Blast Wave 295
        6.5.2 3D Spherical Blast Wave 297
        6.5.3 2D Kelvin-Helmholtz Instability 297
   6.6 Diffusion Equation in 2+1 Dimensions 300
   6.7 Projects 305
        6.7.1 Stationary Solar Wind and CMEs 305
        6.7.2 Relativistic Hydrodynamics in 3+1 Dimensions 308
        6.7.3 Magnetohydrodynamics 309
   6.8 How to Improve the Methods in This Chapter 310
   References 310

7	**Appendix A: Stability of Evolution Schemes**...............................	313
8	**Appendix B: Codes**	319
	8.1 Summary of Codes	319
	8.2 Selected Codes..	322

Index... 357

Chapter 1
Introduction

Abstract The solution of Initial Value Problems (IVPs) in various branches of science has become essential in recent years. Models of evolution are used in most disciplines, social sciences, economy, biology, chemistry, geophysics, space sciences, and physics in general to mention a few. These problems are defined in terms of a Differential Equations (DEs) that describe the evolution of properties of a given system, for example, population densities, risk estimates, spread of species and diseases, diffusion of substances in mixtures, prediction of atmospheric currents, space weather processes, quantum and classical problems, and more.

The solution of Initial Value Problems in various branches of science has become essential in recent years. Models of evolution are used in most disciplines, social sciences, economy, biology, chemistry, geophysics, space sciences, and physics in general to mention a few. These problems are defined in terms of a Differential Equations that describe the evolution of properties of a given system, for example, population densities, risk estimates, spread of species and diseases, diffusion of substances in mixtures, prediction of atmospheric currents, space weather processes, quantum and classical problems, and more.

This type of problem is defined with one or various Differential Equations (DEs), Ordinary (ODEs) in the time domain or partial (PDEs) on a spatial-temporal domain. The objective is the construction of the solution to the equations for given initial and boundary conditions.

Existence and uniqueness of solutions to these problems are guaranteed provided some mathematical conditions of the equations related to a given Initial Value Problem (IVP), for example, initial conditions and the parameters in the equations. Once these properties are known one can search for a solution, either in closed or numerical form. The closed form is a formula that expresses the solution. In contrast, numerical solutions are approximations to a possibly existing closed solution, in most cases calculated in a limited domain. In this sense numerical solutions are modest in comparison with exact closed solutions.

Also in many cases, an IVP is formulated in an unbounded domain along spatial and temporal directions, while most numerical solutions can only be calculated in a limited chunk of space during a finite time window. The calculation of numerical solutions demand from us the appropriate spatial and temporal domain selection, where the solution can be interesting.

One commonly frustrating aspect of numerical solutions is that while closed solutions are contained in mathematical expressions, numerical solutions in the best case are presented as a collection of data and sometimes within plots. Even so, numerical solutions are valuable because they can be constructed for differential equations whose exact solutions are unknown.

The name IVP indicates that the solution is needed for a scenario whose initial conditions are known. Initial time is a space-like boundary where boundary conditions are imposed to start an evolution, whose result will be the solution in the whole domain that includes time-like boundaries. Implicitly the solution of an IVP is a process of evolution, evolution of the initial conditions imposed on the unknown functions of the DEs.

Evolution. The idea of evolution of initial conditions helps to imagine how the solution is constructed in time. Biological evolution is the best evidence that information is transmitted from one generation to the next. In the same manner, the evolution of initial data transmits information along the time domain the values of an unknown in the future with evolution rules governed not by natural selection but by the DEs that model a process. The numerical solution of IVPs is constructed during a discrete time domain, like biological evolution from generation to generation, in this case step by step in terms of the solution at the previous time.

The type of IVPs we use to illustrate the methods involving PDEs define a space-time domain of solution. We solve **hyperbolic** equations, specifically wave-like equations and locally hyperbolic problems associated to hydrodynamics, as well as **parabolic** type equations, that include diffusion, reaction-diffusion processes and Schrödinger equation.

1.1 Errors and Precision of Numbers

Nature of Approximate Solutions. The construction of numerical solutions, independently of the numerical method to be used, intrinsically carry an approximation to the functions or derivative operators in the DEs. An error is to be expected from origin, by concept of approximation. In calculations done with a computer, errors are eventually unavoidable, while the closed exact solution of an IVP is represented with a formula that one can evaluate with arbitrary precision.

Various types of errors are present in the numerical solution of an IVP. Essential to all combinations of compiler-architecture is the **round-off error**. Since numbers can be represented only up to a certain accuracy, numbers are rounded. This error is defined as the difference between the number we would like to use in a calculation and the one used; for a simplistic example, assume we would like to use the number

1.66 and the computer represents its value by 1.7, rounded to the closest version, and then the round-off error is −0.04.

Modern computers though can handle a bigger number of digits and therefore the round off error in the representation of a number is smaller. For example the number 1/3, calculated using numbers 1.0 and 3.0 using single, double, and quadruple precision, is respectively 0.333333343, 0.33333333333333331, and 0.333333333333333333333333333333333317, on a given 64 bits computer and using the gfortran compiler.

Arithmetic operations between these numbers propagate the errors, and the numerical solution of IVPs is based in the considerable calculation of arithmetic operations with rounded numbers which damages the quality of numeric approximations. Considering the experience in the representation of the number above, the propagation of the round-off error expands, for example, the number $\frac{1}{7} \cdot \frac{8}{3}$ is represented in various ways depending on the precision used to define the numbers 1,7,8,3. Respectively, the product is 0.380952388, 0.38095238095238093, and 0.380952380952380952380952380952380934 when using single, double, and quadruple precision. In the latter case the cyclic pattern of the number is better seen, while in the first two cases truncation prevents it.

Round-off errors are intrinsic to representations with different precision; they propagate along with arithmetic operations and are unavoidable. The best one can do is to use the higher precision allowed by the compiler-architecture combination used and maintain these errors below the **errors due to the numerical methods** used to implement a numerical solution. In the worst-case scenario, a balance between hardware capacity and precision is desirable, and this is the reason why in the book we assume that the compiler-processor combination allows the representation of real numbers with **double precision**.

The errors due to numerical methods are **truncation errors**. Numerical methods can only construct approximate solutions of IVPs, and the source of this kind of error is that the problem solved is only and approximated version of the original one. Different methods approximate better the equations of an IVP, and the better the approximation, the smaller the error. In Chapter 1 we describe in detail error estimates due to truncation errors and how they propagate in time. **Truncation error** serves to determine the consistency and convergence of numerical solutions and is key to calibrate their validity.

Notice finally that during the evolution of given initial conditions, the successive arithmetic operations involving rounded numbers, that on top carry truncation errors due to numerical methods, that errors propagate along the spatial direction at a fixed time, and that later on they are transmitted over time. Despite all these inconveniences, it is possible to construct reliable numerical solutions to IVP as will be clear with worked examples.

1.2 Rescaling Equations

It is possible to reduce the truncation errors when some coefficients in equations can be absorbed in the variables and leave the equations with coefficients easy to represent. One typical example is *Schrödinger equation*:

$$i\hbar \frac{\partial \hat{\Psi}}{\partial \hat{t}} = -\frac{\hbar^2}{2m}\frac{\partial^2 \hat{\Psi}}{\partial \hat{x}^2} + \hat{V}\hat{\Psi}, \qquad (1.1)$$

where $\hat{\Psi} = \hat{\Psi}(\hat{x}, \hat{t})$ is the wave function, $\hbar$ is Planck's constant, and m is the mass of the particle to be described, which is subject to the effects of the potential $\hat{V}$. Rescaling variables and functions with the definitions

$$\Psi = \frac{\hbar}{m}\hat{\Psi}, \quad x = \frac{m}{\hbar}\hat{x}, \quad t = \frac{m}{\hbar}\hat{t}, \quad V = \frac{\hat{V}}{m}, \qquad (1.2)$$

Equation (1.1) is transformed into

$$i\frac{\partial \Psi}{\partial t} = -\frac{1}{2}\frac{\partial^2 \Psi}{\partial x^2} + V\Psi, \qquad (1.3)$$

for $\Psi = \Psi(x, t)$, which is a more comfortable equation to deal with. It also helps reducing the sources of round-off error, for example, if the particle is the electron, then $m = 9.1093837 \times 10^{-31}$kg $= 0.51099895$MeV$\cdot$c^{-2}, while Planck's constant is $\hbar = 1.054571817... \times 10^{-34}J\cdot$s $= 6.582119569... \times 10^{-22}$ MeV$\cdot$s. In either of the two unit systems, it is simpler to use the coefficient $\frac{1}{2}$ in the first term of the scaled equation than that of the original equation $\frac{\hbar^2}{2m}$. Once the equation has been numerically solved for Eq. (1.3), one can always rescale units of physical quantities to those in the original equation by inverting relations (1.2).

Another important example is the *Wave Equation*:

$$\frac{1}{v^2}\frac{\partial^2 \phi}{\partial \hat{t}^2} - \frac{\partial^2 \phi}{\partial x^2} = 0$$

where $\phi = \phi(x, \hat{t})$ and v is the wave velocity. In the text we absorb the velocity by redefining time $t = v\hat{t}$, which leads to the equation $\frac{\partial^2 \phi}{\partial t^2} - \frac{\partial^2 \phi}{\partial x^2} = 0$ for $\phi = \phi(x, t)$. This version of the equation also simplifies the construction of solutions, and the solution of the original equation can be recovered by rescaling the time coordinate back.

This strategy is common to most examples in the book.

1.3 Other Methods

The title of the book intends to waive on the type of numerical methods used to solve IVPs. The methods used in each problem are the very basic ones. They all are constructed using simple and intuitive ideas and illustrate how they can be implemented. The examples and codes in the text provide a basic guide to program more sophisticated methods. These methods belong to a particular class, based on the definition of a discrete domain D_d, which is a subset of the domain D of solution in the continuum, and then a numerical solution is constructed on D_d. This applies to cases involving ODEs and PDEs.

In the case of PDEs, linear equations, including wave-like and diffusion equations, use the vertex centered Finite Differences discretization, whereas quasi-linear equations, specifically related to hydrodynamical problems, use the cell-centered Finite Volume discretization. More sophisticated improvements are suggested at the end of each chapter that improve accuracy or stability of solutions; however, the methods in the text belong to one of these two basic classes.

This means that we have left aside important methods also used in the solution of IVPs, particularly Spectral Methods. The basic idea of *spectral methods* consists in assuming that the functions involved in the DEs are expressed as a linear combination of orthogonal polynomials. The result is that the DEs transform into a set of equations for the coefficients of the expansion. Depending on the boundary conditions, some bases of polynomials are more used than other, for example, a cubic domain commonly uses the Chebyshev polynomials as the basis, whereas the Fourier basis is adequate for problems in a periodic domain.

This is an elegant, useful, and accurate method to solve DEs, however is not as basic as the methods in this book and uses a completely different idea as to be somehow described alongside with the methods in this book. These methods deserve their own space.

1.4 Organization of the Book

Theory of the numerical methods used is complemented with pieces of code that illustrate the implementation of the methods on the fly. For this `fortran 90` is used within the text because of its similarity of pseudocode.

We start in Chap. 2 with the solution of IVPs related to ODEs, where elementary numerical methods are constructed, followed by more accurate and typically used methods. A variety of applications are used to illustrate the potential of solving ODEs has in various scenarios. Chapter 3 describes the solution if IVPs related to PDEs in a space-time domain with one spatial dimension that we call problems in 1+1 dimensions, using Finite Differences approximations and simple evolution methods, whereas Chap. 4 contains the solution of problems also in 1+1, this time using the Method of Lines. In Chap. 5 we also solve 1+1 problems this time using

Finite Volume discretization, suitable to solve evolution equations whose solutions develop discontinuities. Finally in Chap. 6 we describe the strategy to solve 2+1 and 3+1 problems, problems in a domain with two and three spatial dimensions.

Finally, the codes in `fortran 90` and `c++`, usable to reproduce the results of the examples in the text are enclosed in appendices and supplementary material. Codes for exercises and challenges will remain undercover.

Chapter 2
Ordinary Differential Equations

Abstract In this chapter we illustrate the implementation of the simplest numerical methods that help solving Initial Value Problems associated to Ordinary Differential Equations. The chapter starts with the description of the general problem and its definition on a discrete domain. It continues with the construction of basic and intuitive numerical methods, namely, Euler type of methods. Later on, second-order accurate Runge-Kutta methods are constructed. Illustration of these methods starts with problems involving only one equation, the Newton Cooling law, which has a well-behaved attractor solution, followed by the Harmonic Oscillator, written as a system of two coupled first-order equations, and after that we tackle problems of systems consisting of nonlinear coupled equations. Error estimates of numerical solutions, as well as convergence and self-convergence tests are essential to our approach, and their implementation is described in detail. In the final section, we present a number of applications that range from purely dynamical systems to problems of celestial mechanics.

Keywords Ordinary differential equations · Basic methods · Error theory · Convergence

In this chapter we illustrate the implementation of the simplest numerical methods that help solving Initial Value Problems associated to Ordinary Differential Equations. The chapter starts with the description of the general problem and its definition on a discrete domain. It continues with the construction of basic and intuitive numerical methods, namely, Euler type of methods. Later on, second-order accurate Runge-Kutta methods are constructed. Illustration of these methods starts with problems involving only one equation, the Newton Cooling law, which has a well-behaved attractor solution, followed by the Harmonic Oscillator, written as a system of two coupled first-order equations, and after that we tackle problems of systems consisting of nonlinear coupled equations. Error estimates of numerical solutions, as well as convergence and self-convergence tests are essential to our approach, and their implementation is described in detail. In the final section, we present a number of applications that range from purely dynamical systems to problems of celestial mechanics.

2.1 The General Problem

Consider the generic Initial Value Problem associated to an Ordinary Differential Equation defined by

$$\frac{du}{dt} = f(t, u) \qquad u = u(t) \qquad (2.1)$$

$$D = [0, t_f] \qquad Domain$$

$$u(t = 0) = u_0 \qquad Initial\ Condition$$

where f is an arbitrary function of the independent variable t and the unknown function u.

In this chapter we describe the method to construct numerical solutions of this type of problem in a discrete domain. The method has two basic steps:

A. Construct a discrete version D_d of the domain D.
B. Construct the numerical solution on D_d.

Step A. In all the examples of this chapter, as well as for the description of theoretical concepts, we use a numerical domain D_d uniformly discretized as follows. For a given positive integer N, one defines $D_d \subset D$, as the set of points $t_i \in D$, such that $t_i = i\Delta t$, with $i = 0, ..., N$, where $\Delta t = t_f/N$ is the **numerical resolution**. That is, D_d is a set of $N + 1$ real numbers that separate N cells of size Δt. Consider keeping in mind the difference between the number of points of D_d and the number of cells implicitly defined.

The **numerical solution** of the problem (2.1) is the set of values $u_i = u(t_i)$ that at each point of D_d satisfies the equation approximately.

Step B. Construct the solution using different numerical methods, with different accuracy and other properties discussed later on, like convergence rate and stability.

We start by describing the basics of the simplest methods before constructing the most elaborate ones. For that we need a workhorse Initial Value Problem (IVP) associated to an Ordinary Differential Equation (ODE) like (2.1).

2.2 Workhorse Example 1: Newton Cooling Law

There is a workhorse example with an equation that is used to illustrate the numerical solution of IVPs, the **Newton Cooling Law**. This law models the time dependence of temperature of a substance in a thermal bath, for example, the

2.2 Workhorse Example 1: Newton Cooling Law

temperature of the coffee in a cup which is in contact with the air at room temperature. The expression of this law is

$$\frac{dT}{dt} = \kappa(T_a - T) \qquad (2.2)$$

where T is the substance temperature, T_a the ambient temperature, and κ is the cooling or heating rate of the substance in the room. This equation can be solved in a closed form by separation of variables as follows:

$$\frac{dT}{T_a - T} = \kappa dt \quad \Rightarrow$$

$$\ln(T - T_a) = -\kappa t + A \quad \Rightarrow$$

$$T - T_a = e^{-\kappa t + A},$$

with A a constant of integration. Now, assuming the initial temperature of the substance is $T(t = 0) = T_0$, the solution reads

$$T = T_a + (T_0 - T_a)e^{-\kappa t}. \qquad (2.3)$$

This solution converges exponentially in time to an asymptotic value which coincides with the room temperature. This attractor behavior of the solution is a friendly property that helps to illustrate how intuitive and simple numerical methods work. Examples of the solution (2.3) are shown in Fig. 2.1 for two different initial temperatures T_0, ambient temperature $T_a = 25C$ and $\kappa = 0.1\,\text{s}^{-1}$.

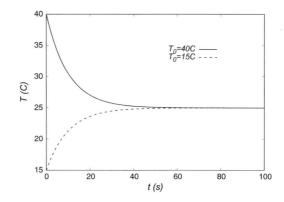

Fig. 2.1 Exact solution of Newton Cooling Law for two different initial temperatures $T_0 = 15,\ 40$ degrees, say Celsius (C), ambient temperature $T_a = 25C$ degrees, and constant $\kappa = 0.1\,\text{s}^{-1}$. Notice the temperature for the two solutions converges exponentially in time toward the value $T = T_a$

2.3 Euler Method

The simplest and more intuitive method is Euler method. This method is based on the definition of derivative from Calculus. Consider a zoom of the discrete domain near $t_i \in D_d$ as shown in Fig. 2.2, and assume Δt is small. Then an approximate value of the derivative of the function u at t_{i-1} can be estimated as

$$\left.\frac{du}{dt}\right|_{t_{i-1}} \simeq \frac{u_i - u_{i-1}}{t_i - t_{i-1}} = \frac{u_i - u_{i-1}}{\Delta t}. \tag{2.4}$$

Notice that it is only an approximation since we have omitted the limit when $\Delta t \to 0$ from the definition of derivative. Let us now assume we know the value of the solution u_{i-1} at point t_{i-1}, and then it is possible to evaluate any function f at (t_{i-1}, u_{i-1}). Using the above expression in the original definition of the general IVP (2.1), one has the approximation:

$$\frac{u_i - u_{i-1}}{\Delta t} \simeq f(t_{i-1}, u_{i-1}).$$

This is a **discrete version** of the equation of the IVP (2.1), and u_i is a solution of the problem at t_i in terms of t_{i-1}, u_{i-1} given by

$$u_i \simeq u_{i-1} + f(t_{i-1}, u_{i-1})\Delta t. \tag{2.5}$$

This expression can also be constructed considering the equation of a line that passes through the point (t_{i-1}, u_{i-1}) with slope $f(t_{i-1}, u_{i-1})$, in agreement with Fig. 2.2.

As simple as it is, this formula allows the construction of the numerical solution of a problem like (2.1), provided an initial condition, that in the discrete domain D_d would correspond to $u(t_0) = u_0$. Specifically, in the discrete domain, knowing u_0

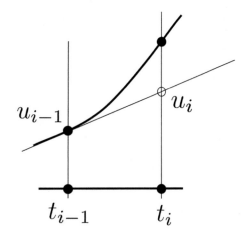

Fig. 2.2 Region near t_i showing the concept of approximate derivative. The thick curve represents a function defined in the continuum, whereas the straight line represents the tangent line to the curve $u(t)$ at t_{i-1}. This plot indicates the line only as an approximation to the exact curve. The white point indicates the approximate numerical solution u_i at t_i

2.3 Euler Method

one can use (2.5) to construct $u_1 = u(t_1)$. With this value use (2.5) again to construct $u_2 = u(t_2)$ and so forth until $u_N = u(t_N)$. The whole set of values $u_0, \ldots, u_N$ is a **numerical solution** of the problem in D_d.

More formally, expression (2.5) can be seen as a Taylor series expansion around t_{i-1}, which explicitly would read

$$u_i = u_{i-1} + (t_i - t_{i-1})\frac{du}{dt}\bigg|_{t_{i-1}} + \mathcal{O}(\Delta t^2)$$

$$= u_{i-1} + \Delta t \frac{du}{dt}\bigg|_{t_{i-1}} + \mathcal{O}(\Delta t^2)$$

$$= u_{i-1} + f(t_{i-1}, u_{i-1})\Delta t + \mathcal{O}(\Delta t^2).$$

The difference between this expression and (2.5) is that in the later, we have an estimate of the error. The **error** term $\mathcal{O}(\Delta t^2)$ is the difference between the exact solution at t_i and the numerical solution u_i. Notice that in the expansion above, one assumes u_{i-1} to be the exact solution at t_{i-1}, because the black dot belongs to the curve in Fig. 2.2. This is the reason why this error is called the **local error** of the numerical solution.

One can estimate the error of the method starting from the initial condition at t_0 which is an exact value, until t_N, the local error of each step accumulates N times and adds to the **global error** of the numerical solution.

By applying (2.5) successively from u_0 to u_1, from u_1 to u_2, and so on, until u_N, the accumulated error is proportional to $N \, \mathcal{O}(\Delta t^2)$, and since by definition $\Delta t = (t_N - t_0)/N$, that is $N = (t_N - t_0)/\Delta t$, the accumulated error is of order $\mathcal{O}(\Delta t)$. Therefore *the global error of the numerical solution constructed with this method is of order one*, or of order $\mathcal{O}(\Delta t)$.

By this argument, in general a method that has a local error of order $\mathcal{O}(\Delta t^n)$ will have a global error of order $\mathcal{O}(\Delta t^{n-1})$.

What is the meaning of the error order? Let us perform a simple analysis for Euler method. Euler method has a global error of order $\mathcal{O}(\Delta t)$, which means that as we increase the resolution of the numerical domain, namely, increase N or equivalently decrease Δt, the numerical solution on D_d should approach the exact solution, the solution in the continuous domain D.

Let us work out an example. Assume we know the exact solution u^e of an IVP like that in (2.1) and calculate two numerical solutions u_i^1 and u_i^2 with Euler method, using resolutions Δt and $\Delta t/2$, respectively. We want to compare the accuracy of both numerical solutions. Knowing the global error is of order $\mathcal{O}(\Delta t)$, the numerical solution at each point of the domain can be written as

$$u_i^1 = u_i^e + E(t_i)\Delta t,$$

$$u_i^2 = u_i^e + E(t_i)\frac{\Delta t}{2},$$

where $E(t_i)$ is the error accumulated from t_0 until t_i. Notice that this comparison requires the solution in the continuous domain D only at points $t_i \in D_d$, where the continuous and discrete domains coincide. From these relations one defines the **convergence factor** of the numerical solutions by

$$CF := \frac{u_i^1 - u_i^e}{u_i^2 - u_i^e} = \frac{\Delta t}{\Delta t/2} = 2. \tag{2.6}$$

The Convergence Factor (CF) provides a criterion to decide when a numerical solution converges toward the solution in the continuum. In order to use this concept in practice, let us calculate numerical solutions of the Newton Cooling Law.

Numerical Solution of Newton Cooling Law. In order to construct the numerical solution, we identify from the theory above the independent variable with time $t \to t$ and the solution function with Temperature $u \to T$. In these terms, the discrete domain D_d will be given by $t_i = i\Delta t$ for $i = 0, \ldots, N$, whereas the temperature will have values $T_i = T(t_i)$.

Then, using formula (2.5) applied to Eq. (2.2) from time t_{i-1} to t_i gives

$$T_i = T_{i-1} + \left.\frac{dT}{dt}\right|_{t_{i-1}} \Delta t$$
$$= T_{i-1} + \kappa(T_a - T_{i-1})\Delta t. \tag{2.7}$$

Notice that $\frac{dT}{dt} = \kappa(T_a - T)$ in (2.2) has an important property, namely, it does not depend explicitly on the independent variable t, which makes easy the calculation of the time derivative of temperature. This property makes the equation **autonomous**.

Let us now use this formula to construct the solution for parameters $T_a = 20C$, $\kappa = 0.3 \, \text{s}^{-1}$ and the initial temperature of the substance $T_0 = 80C$. It is necessary to define the discrete domain, for which we assume $t_f = 25 \, \text{s}$ and $N = 50$, which implies the resolution is $\Delta t = t_f/N = 0.5 \, \text{s}$.

Let us calculate the temperature for the first few values of the discrete time with a calculator

$$T_1 = T_0 + \left.\frac{dT}{dt}\right|_{t_0} \Delta t \tag{2.8}$$
$$= T_0 + k(T_a - T_0)\Delta t$$
$$= 80 + 0.3(20 - 80)0.5$$
$$= 71$$
$$T_2 = T_1 + \left.\frac{dT}{dt}\right|_{t_1} \Delta t$$
$$= T_1 + k(T_a - T_1)\Delta t$$
$$= 71 + 0.3(20 - 71)0.5$$

2.3 Euler Method

$$= 63.35$$
$$T_3 = 56.8475$$
$$T_4 = 51.32037$$
$$\ldots$$
$$T_{20} = 22.32557$$
$$T_{30} = 20.4578$$
$$T_{50} = 20.0177.$$

This solution indicates that at $t_{50} = 50 \, \Delta t = 25$ s, the temperature of the substance is nearly the room temperature $T_a = 20C$. The comparison with the exact solution is shown in Fig. 2.3.

The **global error** of the numerical solution at each point of the discrete domain depends of the resolution and is written as $e(\Delta t) = T_i - T_i^e$, where T_i^e is the exact solution (2.3) evaluated at time $t_i \in D_d$. Now, in order to investigate whether the error of Euler method behaves as predicted by Eq. (2.6), one needs to calculate the numerical solution with higher resolution by increasing N or equivalently decreasing Δt. We then calculate the numerical solution using resolutions $\Delta t_1 = 0.5$ s, $\Delta t_2 = \Delta t_1/2$, $\Delta t_3 = \Delta t_2/2$, and $\Delta t_4 = \Delta t_3/2$, which is equivalent to define the discrete domain D_d with $N = 50, 100, 200,$, and 400 cells, respectively. The error for each of these solutions is shown in Fig. 2.3.

The fact that the error using resolution Δt_1 is twice as big as that using Δt_2 indicates that the criterion of convergence in (2.6) is fulfilled. Notice that in general, the error of the solution calculated with resolution Δt_k is twice as big as the error using resolution Δt_{k+1} for $k = 1, 2, 3$, which proofs the error converges to zero according to the theory in (2.6). When this is the case, it can be said that **the numerical solution converges** to the exact solution with a rate consistent with the error of the numerical method, in this case the global error of Euler method.

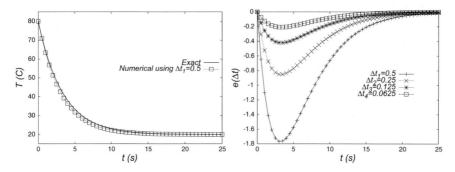

Fig. 2.3 Numerical solution using Euler method. (Left) Comparison of the numerical solution using $\Delta t = 0.5$ and the exact solution. (Right) Error using four different resolutions. The physical parameters in these solutions are $T_a = 20C$, $T_0 = 80C$ and $\kappa = 0.3 \, \text{s}^{-1}$

These results were constructed using the **fortran 90** code shown below, which is the very basic code we start with. The instructions include declaration of variables and arrays, allocation of memory for the arrays, discretization of the numerical domain, setting of initial conditions, and a loop that constructs the numerical solution at each point of the numerical domain and data output. In summary, it contains all the elements needed in the codes for the rest of the book. It is convenient to know that we declare the elements of the numerical domain $t_i \in D_d$ and the numerical solution T_i as entries of arrays of double precision numbers, which helps illustrating the concepts of error an convergence, and later on this approach will be relaxed for ease of programming and memory saving:

```
program NewtCoolLaw_Euler

implicit none

real(kind=8) t0,tmax,dt,Temp0,kappa,Ta
real(kind=8), allocatable, dimension (:) :: t,Temp,Texact,error
integer i,N,resolution_label

! We define Dd and physical parameters
resolution_label = 1
N = 50 ! Number of cells of Dd
t0 = 0.0 ! Initial time
tmax = 25.0 ! Final time
N = 2**(resolution_label-1) * N ! Number of cells for the
discretized domain
kappa = 0.3 ! Cooling rate constant
Ta = 20.0 ! Room temperature
Temp0 = 80.0 ! Initial temperature of the substance

! Allocate memory for the various arrays
! t: time
! Temp: Temperature
! Texact: Exact solution
! error: Temperature - exact solution
allocate(t(0:N),Temp(0:N),Texact(0:N),error(0:N))

! ---> PART A <---
! Definition of the Discrete domain
dt = (tmax - t0) / dble(N)
do i=0,N
   t(i) = t0 + dt * dble(i)
end do
print *, 'dt=',dt ! This will tell you whether dt is correct on
the screen
```

2.3 Euler Method

```
! ---> PART B <---
! Euler method
Temp(0) = Temp0
do i=1,N
  Temp(i) = Temp(i-1) + dt * kappa * (Ta - Temp(i-1))
end do

! ---> Ends Part B <---
! Calculation of the exact solution and error
Texact = Ta + (Temp0-Ta)*exp(-kappa*t)
error = Temp - Texact

! Saving data to a file
open(1,file='NLdata.dat')
  do i=0,N,2**(resolution_label-1)    ! Includes the dawnsampling
of the output
    write(1,*) t(i),Temp(i),Texact(i),error(i)
  end do
close(1)

end program
```

This example shows that Euler method is useful and produces solutions that converge for this particular IVP whose solution is smooth with asymptotic constant value.

How to Check Convergence Using the Code. The program above is written conveniently for checking convergence. In order to understand how it works, consider that one calculates the numerical solution u_i^1 on the numerical domain D_d^1 containing $N+1$ points and resolution Δt_1. Then a second solution u_i^2 is constructed on the domain D_d^2 consisting of $2N+1$ points and resolution $\Delta t_2 = \Delta t_1/2$ and so on by doubling resolution. The solutions $u_i^1, u_i^2, u_i^3, \ldots$ can be compared only at the points of D_d^1 which is completely contained in D_d^2, which in turn is contained in D_d^3 and so on.

The parameter `resolution_label` is an integer that sets the number of points of the numerical domain $D_d^{resolution\_label}$ to $2^{resolution\_label-1}N+1$. Now, the numerical solutions calculated on $D_d^1, D_d^2, D_d^3, \ldots$ can be compared. By construction all the elements of D_d^k are contained in the numerical domain D_d^{k+1}, and therefore D_d^1 is contained in all the subsequent numerical domains.

The parameter `resolution_label` also is used to downsample the output in such a way that the solution is only recorded at the points of D_d^1, where $u_i^1, u_i^2, u_i^3, \ldots$ can be compared. With this method it is easy to plot the numerical solutions calculated on $D_d^1, D_d^2, D_d^3, \ldots$ and their associated errors at exactly the points of D_d^1 as done in Fig. 2.3.

About the Implementation. There is an alternative in the implementation of Euler and the rest of methods described later on, which does not need the use of arrays to allocate the values of the discrete time and temperature; instead it calculates and saves the numerical solution on the fly. This version of the program appears next:

```
program NewtCoolLaw_Euler_NoArrays

implicit none

real(kind=8) t0,tmax,t,dt,Temp0,Temp,Temp_p,Texact,error,kappa,Ta
integer i,N,resolution_label

! We define some parameter values
resolution_label = 4
t0 = 0.0 ! Initial time
tmax = 25.0 ! Final time
N = 50 ! Base number of cells
N = 2**(resolution_label-1) * N ! Effective number of cells for
the dicretized domain
kappa = 0.3 ! Cooling rate constant
Ta = 20.0 ! Room temperature
Temp0 = 80.0 ! Initial temperature of the substance

! Other variables
! t: time
! Temp: Temperature at time n+1
! Temp_p: Temperature at time n
! Texact: Exact solution
! error: Temperature - exact solution

! ---> PART A <---
! Discretization fo domain
dt = (tmax - t0) / dble(N)
print *, 'dt=',dt ! This will tell you whether dt is what you
expect

open(1,file='NLdata.dat')

! Set initial conditions and initial diagnostics
t = t0
Temp = Temp0
Texact = Temp0
error = Temp - Texact

! Save initial data
write(1,*) '# Time T Texact error'
write(1,*) t,Temp,Texact,error

! ---> PART B <---
do i=1,N
```

```
      t = t + dt

      ! Evolution according to Euler method
      Temp_p = Temp
      Temp = Temp_p + dt * kappa * (Ta - Temp_p)

      ! Calculation of the exact solution and error
      Texact = Ta + (Temp0-Ta)*exp(-kappa*t)
      error = Temp - Texact

      ! Saving data to a file
      if (mod(i,2**(resolution_label-1)).eq.0) write(1,*) t,Temp,
   Texact,error

   end do

   close(1)

end program
```

The Style in This Chapter. We will use the first style, the one that stores domain and solutions in arrays, for the illustration of the methods and applications in the reminder of the chapter. The code above shows that it is not difficult to convert one style into the other one, and so it is for all methods and examples below.

2.4 Backward Euler Method

Recall that Euler method constructs the solution u_i at t_i using the equation of a line with slope $f(t_{i-1}, u_{i-1})$, evaluated at the point t_{i-1}. In the Backward Euler method, there is a slight variant. The idea is the same, it uses the equation of a line; however, the slope of this line is $f(t_i, u_i)$, evaluated at point t_i. The expression for the construction of the solution is as follows:

$$u_i \simeq u_{i-1} + f(t_i, u_i)\Delta t. \tag{2.9}$$

In order to implement this method, one needs to first estimate u_i using Euler method and then evaluate $f(t_i, u_i)$, as illustrated with steps 1–3 in Fig. 2.4. The construction of the solution at each t_i can be implemented by substituting the part B of the previous code with the following chunk of code

```
! ---> PART B <---
Temp(0) = Temp0
do i=1,N
   do j=1,2
```

Fig. 2.4 Region near t_i showing the concept of approximate derivative according to (2.9). Number 1 indicates a step using Euler method, number 2 indicates that the derivative has to be evaluated there $f(t_i, u_i)$, and number 3 indicates this slope is used to construct a line passing through (t_{i-1}, u_{i-1}) with slope $f(t_i, u_i)$. The white point indicates the numerical solution

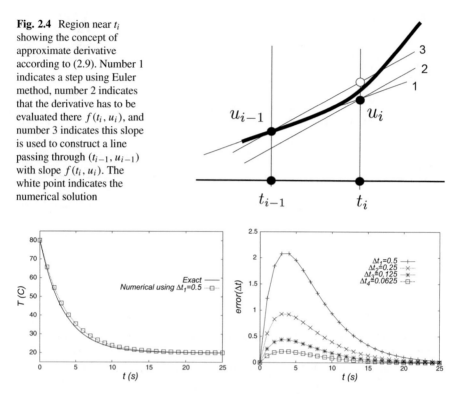

Fig. 2.5 Numerical solution using the Backward Euler method. (Left) Comparison of the numerical solution using $\Delta t = 0.5$ and the exact solution. (Right) Error using four different resolutions

```
      if (j.eq.1) then
         Temp(i) = Temp(i-1) + dt * kappa * (Ta - Temp(i-1))
      else
         Temp(i) = Temp(i-1) + dt * kappa * (Ta - Temp(i))
      end if
   end do
end do
```

Numerical Solution. In order to compare with Euler method, we solve the Newton Cooling Law with the same parameters as before. The results of the numerical solution and the error $e(\Delta t) = T_i - T_i^e$, where T_i^e is the exact solution (2.3) appears in Fig. 2.5 for various resolutions.

Notice that the error using the Backward Euler method has the opposite sign to that when using Euler method. The method also converges with order one, since $e(\Delta t_k) \simeq 2\, e(\Delta t_{k+1}), k = 1, 2, 3$. Finally, observe that the error of both methods is of similar magnitude, which will be exploited by the following method.

2.5 Average Euler Method

Based on these two simple methods, it is possible to construct a more accurate one. From the example above, we know the error using Euler method has the opposite sign of that using the Backward Euler method, as can be seen from Figs. 2.3 and 2.5.

Euler method is based on the idea of using the line equation with the slope evaluated at t_{i-1}, whereas the Backward Euler method uses the line equation with slope evaluated at t_i. One of them underestimates the slope with respect to the exact solution, and the other one overestimates the slope. This can be seen for the monotonically growing functions in Figs. 2.2 and 2.4, where the approximate solution is indicated with the white point. The idea now is to use a line equation with *a slope that is the average of the slopes at t_{i-1} and t_i*.

We denote the slopes at t_{i-1} and t_i by k_1 and k_2, respectively, and the method is written as

$$u_i = u_{i-1} + \frac{1}{2}(k_1 + k_2)\Delta t, \qquad (2.10)$$

$$k_1 = f(t_{i-1}, u_{i-1}),$$

$$k_2 = f(t_i, u_i).$$

In the case of the Newton Cooling Law, $k_1 = \kappa(T_a - T_{i-1})$ and $k_2 = \kappa(T_a - T_i)$. Likewise in the Backward Euler method, this can be programmed using two steps in the part B of the code. It would suffice to replace the code in Part B with the following chunk of code, which will need declaring the new double precision numbers k1 and k2:

```
! ---> PART B <---
Temp(0) = Temp0
do i=1,N
  do j=1,2
    if (j.eq.1) then
       k1 = kappa * ( Ta - Temp(i-1) )
       Temp(i) = Temp(i-1) + dt * k1
    else
       k2 = kappa * ( Ta - Temp(i) )
    end if
    Temp(i) = Temp(i-1) + 0.5d0 * ( k1 + k2 ) * dt
  end do
end do
```

The numerical solution and error are shown in Fig. 2.6. Important information immediately arises; the error is one order of magnitude smaller using resolution Δt_1 compared to that using Euler and Backward Euler methods, and the error decreases

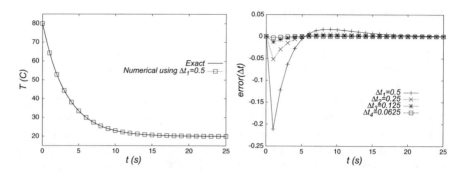

Fig. 2.6 Numerical solution using the Average Euler method. (Left) Comparison of the numerical solution using $\Delta t = 0.5$ and the exact solution. (Right) Error using various resolutions

faster to zero with resolution. Specifically, $e(\Delta t_1) \sim 4\, e(\Delta t_2)$ which means that the numerical error approaches zero twice as faster than the prediction of formula (2.6).

We will later formalize the accuracy of this method, since it is a particular case of the second-order accurate Runge-Kutta class of solvers seen later in Sect. 2.8. By now it is important to notice that this method is *more accurate* than the previous two ones and *converges faster*.

2.6 Workhorse Example 2: Harmonic Oscillator

For a spring obeying Hooke law subject to a damping force, Second Newton Law establishes the equation of motion for a damped harmonic oscillator:

$$m\ddot{x} = -kx - b\dot{x}, \quad \text{or equivalently}$$

$$\ddot{x} + \frac{b}{m}\dot{x} + \frac{k}{m}x = 0, \tag{2.11}$$

where m is the mass of the oscillator and k is the spring's constant, the term $b\dot{x}$ corresponds to a damping term modeled with a linear function of velocity. This equation has unique solution provided initial conditions for $x(0) = x_0$ and $\dot{x}(0) = v_0$.

In the examples below, we assume the initial conditions $x(0) = 1$ and $\dot{x}(0) = 0$, that is, the spring is pulled to the right from the equilibrium position and will be released at initial time.

Case I, the simple harmonic oscillator. In this case $b = 0$, and for the initial conditions, the exact solution and its time derivative are

$$x(t) = \frac{1}{2}e^{i\omega t} + \frac{1}{2}e^{-i\omega t} = \cos(\omega t), \tag{2.12}$$

$$\dot{x}(t) = -\omega \sin(\omega t),$$

2.6 Workhorse Example 2: Harmonic Oscillator

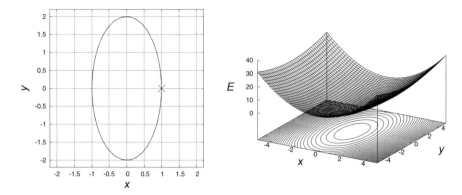

Fig. 2.7 (Left) Phase space diagram of the simple harmonic oscillator ($b = 0$), with $m = 1/2$, $k = 2$ and initial conditions $x(0) = 1$ and $\dot{x}(0) = 0$. The cross indicates the point of the initial conditions. (Right) Energy function (2.13) for the same parameters. Isocontours of E illustrate the trajectories of the system on the phase space for different values of energy, which can be changed at initial time using different initial conditions $x(0)$ and $\dot{x}(0)$. The ellipse on the left corresponds to the initial energy $E = \frac{y^2}{2^2} + \frac{x^2}{1^2}|_{(x=1, y=0)} = 1$

where $\omega = \sqrt{k/m}$. In order to have a numerically well-defined example, we set the mass and spring parameters to $m = 1/2$ and $k = 2$, so that $\omega = 2$. An illustrative portrait of the system's behavior is the phase space diagram. Such diagram is defined by the position and velocity in the $x\dot{x}$−plane. For the particular values of the parameters used and the initial conditions, all the points of the phase space diagram are $(x(t), \dot{x}(t)) = (\cos(2t), -2\sin(2t))$ that correspond to the parametric form of the ellipse in Fig. 2.7, where the cross indicates the position of the system at initial time.

Each point in this diagram corresponds to a state of the system. The system starts in the position $(1, 0)$, and the trajectory moves downward with negative velocity and $x(t)$ starts decreasing. Turning points are those of maximum extension $x = 1$ and maximum compression $x = -1$ where the velocity is zero $\dot{x} = 0$. When the oscillator passes through the equilibrium position at $x = 0$, the velocity is maximum $\dot{x} = 2$ when moving to the right at point $(0, 2)$ and minimum $\dot{x} = -2$ when moving to the left at point $(0, -2)$.

In this particular case of $b = 0$, the system is conservative, with total energy:

$$\begin{aligned} E = T + U &= \frac{1}{2}m\dot{x}^2 + \frac{1}{2}kx^2 \\ &= \frac{1}{4}y^2 + x^2 \\ &= \frac{y^2}{2^2} + \frac{x^2}{1^2}. \end{aligned} \quad (2.13)$$

For a constant value of E, there is a relation between x and $\dot{x}$ that defines an ellipse in the $x\dot{x}$−plane, with horizontal semi-axis $\sqrt{E}$ and vertical semi-axis $2\sqrt{E}$, which is exactly as it appears in Fig. 2.7. Other initial conditions can define a different value of E at initial time and the states of the system would belong to a different ellipse. Isocontours of the energy E in (2.13) define different ellipses as illustrated in the right panel of Fig. 2.7.

Case II, the damped harmonic oscillator. In this case $b \neq 0$, for the initial conditions $x(0) = 1$ and $\dot{x}(0) = 0$, the exact solution reads

$$x(t) = e^{-\alpha t}\left[\frac{1}{\beta}\sin(\alpha\beta t) + \cos(\alpha\beta t)\right] \Rightarrow$$

$$\dot{x}(t) = -\alpha e^{-\alpha t}\left[\frac{1}{\beta}\sin(\alpha\beta t) + \cos(\alpha\beta t)\right]$$

$$+ \alpha e^{-\alpha t}\left[\cos(\alpha\beta t) - \beta\sin(\alpha\beta t)\right], \quad (2.14)$$

where $\alpha = \frac{b}{2m}$ and $\beta = \sqrt{|1 - 4km/b^2|}$. In order to illustrate the solution, we again use the parameters $m = 1/2$, $k = 2$ and $b = 0.1$, and the trajectory given by points $(x(t), \dot{x}(t))$ using (2.14) is shown in Fig. 2.8.

In this case with $b > 0$, the states of the system follow an inward spiral toward the origin of the phase space. This means the energy is being dissipated with rate:

$$\frac{dE}{dt} = \frac{d}{dt}\left(\frac{1}{2}m\dot{x}^2 + \frac{1}{2}kx^2\right)$$

$$= (m\ddot{x} + kx)\dot{x}$$

$$= -b\dot{x}^2,$$

which depends on the value of b.

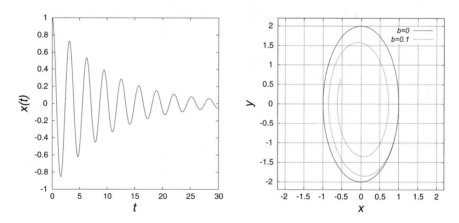

Fig. 2.8 Solution of the damped harmonic oscillator with $b = 0.1$. (Left) Position as function of time. (Right) Trajectory in phase space, which shows an inward espiral toward the origin, starting from the point $(1, 0)$. These plots indicate an attractor behavior toward the equilibrium position of the system. For comparison we draw the trajectory with $b = 0$

2.6 Workhorse Example 2: Harmonic Oscillator

Numerical Solution
The harmonic oscillator equation is second-order, whereas the methods described work for first-order equations. The methods can still be used because it is possible to write a second-order equation of motion as a set of two first-order equations. This is easily done by defining $y = \dot{x}$, then the IVP associated to Eq. (2.11) can be rewritten as

$$\boxed{\begin{aligned} \dot{x} &= y \\ \dot{y} &= -\frac{b}{m}y - \frac{k}{m}x \end{aligned}} \quad (2.15)$$

with initial conditions $x(0) = x_0$ and $y(0) = y_0$, to be solved in the domain $t \in [0, t_f]$. Notice that t is the independent variable and the unknowns are x and y.

In this case one needs to solve for both x_i and y_i in the numerical discrete domain D_d whose elements are t_i, $i = 0, ..., t_N$. One important difference between this problem and the Newton Cooling Law is that now each unknown x and y has its own slope $\frac{dx}{dt} = y$, and $\frac{dy}{dt} = -\frac{b}{m}y - \frac{k}{m}x$.

Euler Method. We first construct the solution using Euler method, which for the system (2.15) reads

$$\begin{aligned} x_i &= x_{i-1} + \left.\frac{dx}{dt}\right|_{t_{i-1}} \Delta t \\ &= x_{i-1} + y_{i-1}\Delta t, \\ y_i &= y_{i-1} + \left.\frac{dy}{dt}\right|_{x_{i-1}} \Delta t \\ &= y_{i-1} + \left(-\frac{b}{m}y_{i-1} - \frac{k}{m}x_{i-1}\right)\Delta t. \end{aligned} \quad (2.16)$$

As an example, we define the following numerical parameters, $t_f = 5$, $N = 200$ or equivalently $\Delta t_1 = 0.025$, and in order to check the convergence, we also define and use resolutions $\Delta t_2 = \Delta t_1/2$, $\Delta t_3 = \Delta t_2/2$, $\Delta t_4 = \Delta t_3/2$. The code oas_E.f90 that calculates the numerical solution is the following:

```
program oas_E
implicit none

real(kind=8) t0,tmax,dt,x0,y0,omega,k,m
real(kind=8), allocatable, dimension (:) :: t,x,y,x_ex,y_ex,
err_x,err_y
integer i,N,j,resolution_label

resolution_label = 1
N = 200
```

```
t0 = 0.0
tmax = 5.0
N = 2**(resolution_label-1) * N ! Number of cells for the
dicretized domain
x0 = 1.0
y0 = 0.0
k = 2.0
m = 0.5
omega = sqrt(k/m)

! Meaning of different variables
! t: time
! x: x
! y: y
! x_ex: Exact x
! y_ex: Exact y
! err_x: x - x_ex
! err_y: y - y_ex

! Allocate memory for the various arrays
allocate(t(0:N),x(0:N),y(0:N),x_ex(0:N),y_ex(0:N),err_x(0:N),
err_y(0:N))

! ---> PART A <---
dt = (tmax - t0) / dble(N)
do i=0,N
  t(i) = t0 + dt * dble(i)
end do
! ---> PART B <---
x(0) = x0
y(0) = y0
do i=1,N
  x(i) = x(i-1) + dt * y(i-1)
  y(i) = y(i-1) + dt * ( -omega**2 * x(i-1) )
end do ! ---> Ends Part B <---

! Exact solution
x_ex = x0 * cos(omega*t)
y_ex = - x0 * omega * sin(omega*t)
err_x = x - x_ex
err_y = y - y_ex

! Saving data to a file
open(1,file='HO_E.dat')
do i=0,N,2**(resolution_label-1)
  write(1,*) t(i),x(i),y(i),x_ex(i),y_ex(i),err_x(i),err_y(i)
end do
close(1)

end program
```

2.6 Workhorse Example 2: Harmonic Oscillator

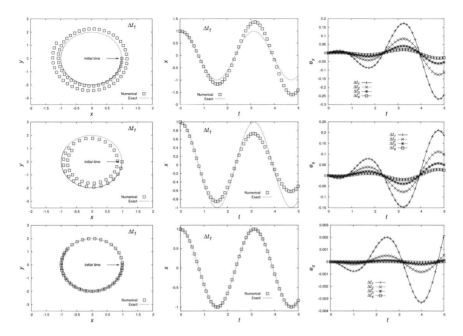

Fig. 2.9 From left to right, we show the phase space diagram of the numerical solution using resolution $\Delta t_1 = 0.025$, the numerical solution for the amplitude of the oscillator also using resolution Δt_1, and the error of the numerical solution using various resolutions. From top to bottom, the results are for Euler, Backward Euler, and Average Euler methods. The time domain uses $t_f = 5$

The results are shown in the first row of Fig. 2.9. At the left we show the phase space diagram of the numerical solution using Euler method. Notice that both the amplitude and velocity of the oscillator grow in time, whereas the exact solution defines an ellipse. The growth of amplitude can be seen in the comparison between the numerical solution and the exact solution for $x(t)$ in the middle panel. At the right we show the error of the numerical solution using various resolutions for the amplitude x_i at time t_i defined as

$$e_x(\Delta t_k) = x_i - x_i^e, \qquad (2.17)$$

which is the difference between the exact solution x^e in (2.12) evaluated at time t_i and the numerical solution x_i also at time t_i. This plot shows that the error converges to zero with fist order, consistently with Eq. (2.6) since $e_x(\Delta t_k) \sim 2e_x(\Delta t_{k+1})$ for $k = 1, 2, 3$.

Backward Euler Method. In this case one needs to calculate the derivatives dx/dt and dy/dt at time t_i. Like in the Newton Cooling Law this is done in two steps:

Advance from t_{i-1} to t_i according to Euler method:

$$x_i = x_{i-1} + y_{i-1}\Delta t,$$

$$y_i = y_{i-1} + \left(-\frac{b}{m}y_{i-1} - \frac{k}{m}x_{i-1}\right)\Delta t.$$

With these values calculate $\frac{dx}{dt}|_{t_i} = y_i$ and $\frac{dy}{dt}|_{t_i} = -\frac{b}{m}y_i - \frac{k}{m}x_i$. Then the solution at time t_i is

$$x_i = x_{i-1} + y_i \Delta t,$$

$$y_i = y_{i-1} + \left(-\frac{b}{m}y_i - \frac{k}{m}x_i\right)\Delta t.$$

In order to code this method, one only needs to substitute part B of the previous code with the following:

```
! ---> PART B <---
x(0) = x0
y(0) = y0
do i=1,N
  do j=1,2
    if (j.eq.1) then
      x(i) = x(i-1) + dt * y(i-1)
      y(i) = y(i-1) + dt * ( -omega**2 * x(i-1) )
    else
      x(i) = x(i-1) + dt * y(i)
      y(i) = y(i-1) + dt * ( -omega**2 * x(i) )
    end if
  end do
end do
```

For comparison we use the same numerical parameters as for the Euler method, $t_f = 5$, $N = 100$ or equivalently $\Delta t_1 = 0.025$, and for convergence checks we also use resolutions $\Delta t_2 = \Delta t_1/2$, $\Delta t_3 = \Delta t_2/2$, $\Delta t_4 = \Delta t_3/2$. The results are shown in the second row of Fig. 2.9. In this case the phase space solution shows an inspiral toward the origin, which is the behavior of the damped oscillator. This can also be observed in the amplitude of $x(t)$, which decreases with respect to the exact amplitude.

Even though the numerical solution is different from that calculated using Euler method, it also converges with first order according to the mid-right plot of Fig. 2.9, which satisfies the theory in (2.6).

Due to the fact that the results are similar to those of a damped harmonic oscillator, the effect produced by the Backward Euler method is called **dissipation**, numerical dissipation.

2.6 Workhorse Example 2: Harmonic Oscillator

Average Euler Method. We now implement the numerical solution using the third of the methods seen so far. In this case of two equations, one needs to average the slopes dx/dt and dy/dt at times t_{i-1} and t_i following formulas (2.10). For the sake of clarity and ease of implementation, we will define the slopes k_{1x}, k_{1y} as the derivatives of x and y at t_{i-1} and k_{2x}, k_{2y} at t_i. The algorithm for the integration is the following:

Execute an Euler step

$$k_{1x} = \left.\frac{dx}{dt}\right|_{t_{i-1}} = y_{i-1},$$

$$k_{1y} = \left.\frac{dy}{dt}\right|_{t_{i-1}} = \left(-\frac{b}{m}y_{i-1} - \frac{k}{m}x_{i-1}\right),$$

$$x_i = x_{i-1} + k_{1x}\Delta t,$$

$$y_i = y_{i-1} + k_{1y}\Delta t.$$

Then calculate the slopes k_{2x} and k_{2y} using these new values, and finally calculate the solution at t_i with the average slopes

$$k_{2x} = \left.\frac{dx}{dt}\right|_{t_i} = y_i,$$

$$k_{2y} = \left.\frac{dy}{dt}\right|_{t_i} = \left(-\frac{b}{m}y_i - \frac{k}{m}x_i\right),$$

$$x_i = x_{i-1} + \frac{1}{2}(k_{1x} + k_{2x})\Delta t,$$

$$y_i = y_{i-1} + \frac{1}{2}(k_{1x} + k_{2x})\Delta t.$$

The implementation is straightforward by defining four double precision numbers k1x, k1y, k2x, k2y, and substituting the following part B in the code:

```
! ---> PART B <---
x(0) = x0
y(0) = y0
do i=1,N
  do j=1,2
    if (j.eq.1) then
      k1x = y(i-1)
      k1y = -omega**2 * x(i-1)
      x(i) = x(i-1) + k1x * dt
      y(i) = y(i-1) + k1y * dt
    else
      k2x = y(i)
      k2y = -omega**2 * x(i)
```

```
         x(i) = x(i-1) + 0.5d0 * ( k1x + k2x ) * dt
         y(i) = y(i-1) + 0.5d0 * ( k1y + k2y ) * dt
      end if
   end do
end do
```

We again construct the numerical solution using exactly the same numerical parameters as for the previous methods. The results are shown in the third row of Fig. 2.9. In this case the phase space of the numerical solutions looks much more similar to the ellipse of the exact solution and $x(t)$ also maintains its amplitude better. Nonetheless, there are errors, which for resolution Δt_1, are one order of magnitude smaller in comparison with those obtained with Euler and Backward Euler methods. There is also the effect of faster convergence rate, since the error of the numerical solution for x is such that $e_x(\Delta t_k) \sim 4e_x(\Delta t_{k+1})$ for $k = 1, 2, 3$.

In summary, using Euler method the amplitude of the numerical solution grows with respect to the exact solution and artificially injects energy to the system.

Using the Backward Euler method, the amplitude of the numerical solution decreases with respect to the exact solution and dissipates energy, producing a damping effect on the amplitude of the oscillations.

Using the Average Euler method the system produces a solution that is more energy conservative compared to the other two methods.

As a final exercise, let us use the three methods to produce a solution in a bigger time domain, measure its amplitude and energy $E = \frac{1}{2}my^2 + \frac{1}{2}kx^2 = \frac{1}{4}y^2 + x^2$.

The results are shown in Fig. 2.10 in the time domain $t \in [0, 50]$, which contains about 16 periods of the motion, using resolution Δt_4. Notice that even with this high resolution, in this big domain, Euler method injects nearly 200% of energy, and the

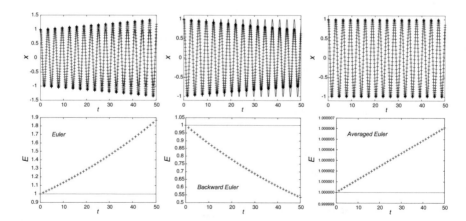

Fig. 2.10 Effects of each method on the amplitude of the numerical solution (top) and the energy (bottom). In all the cases, the continuous line indicates the exact value and the points indicate the numerical solution. From left to right, the method used are Euler, Backward Euler, and Average Euler. The resolution used for this example is $\Delta t_4 = 0.003125$

2.7 Error Theory, Convergence, and Self-Convergence

amplitude of the motion increases nearly by a factor of 1.5. The Backward Euler dissipates energy until it is reduced to 50% of its initial value and the amplitude of the motion is reduced by nearly 20% of its initial value. Finally, the average method injects energy of the order of $\sim 10^{-4}\%$.

The result in this plot is very important, because it tells about the effects produced by the use of various methods, at least for the particular problem of a linear and conservative system. Imagine, for example, that one has to simulate the motion of a spaceship around the Sun or in its travel to Mars; the injection or dissipation of energy due to the method would produce very inaccurate numerical results and possibly wrong predictions.

More dangerously, someone could calculate the numerical solution of the simple harmonic oscillator using the Backward Euler method and supply only the time series in the top-middle of Fig. 2.10 for its interpretation and modeling. One would feel tempted to believe that the time series corresponds to a damped oscillator, instead of the simple oscillator using a considerably dissipative numerical method.

> **Exercise.** It would be interesting to estimate the value of b that the three methods introduce in the solution of the simple harmonic oscillator. This can be done by fitting the data in the first row of plots of Fig. 2.10 using the exact solution (2.14) with b as the fitting parameter.

2.7 Error Theory, Convergence, and Self-Convergence

We have constructed various numerical solutions and used the concept of the convergence of a numerical solution to the solution in the continuum as the resolution is increased. It is time to formalize these ideas.

Convergence. Let us assume we construct the numerical solution of Eq. (2.1) using a method with *global error* of order n. We do this for two resolutions Δt_1 and $\Delta t_2 = \Delta t_1/2$ with numerical solutions u_i^1 and u_i^2 for all $i = 0, \ldots, N$. If the exact solution of the equation is u_i^e, then at every t_i of the numerical domain D_d^1 the following relations hold

$$u_i^1 = u_i^e + E(t_i)\Delta t^n,$$
$$u_i^2 = u_i^e + E(t_i)\left(\frac{\Delta t}{2}\right)^n, \qquad (2.18)$$

that is, the numerical solution is expressed as the exact solution plus an error term, which is proportional to Δt^n. Comparing the two expressions, one finds the convergence factor:

$$CF := \frac{u_i^1 - u_i^e}{u_i^2 - u_i^e} \simeq \frac{\Delta t^n}{\left(\frac{\Delta t}{2}\right)^n} = 2^n \qquad (2.19)$$

If this relation holds, approximately because we dropped terms higher than $\mathcal{O}(\Delta t^n)$ in Eq. (2.18), it is said that the numerical solution **converges with order** n to the exact solution in the continuum, or that **the error converges to zero** with order n. The number 2^n is called **convergence factor**.

In practice, when convergence is confirmed, it is commonly said that resolutions Δx_1 and Δx_2 are appropriate for convergence. This implicitly indicates that the solutions may not converge for arbitrary resolutions. When the numerical solutions converge, it is said that resolutions belong to the **convergence regime**.

With these concepts we can state on solid bases that the solution of the Newton Cooling Law using the methods of Euler and Backward Euler converges to the exact solution with first order. The explanation is that the error for solutions obtained with successive resolutions shown in Figs. 2.3 and 2.5 show a convergence factor $2^1 = 2$, which is consistent with the expression (2.6) and (2.19) for $n = 1$.

In the same manner, Fig. 2.6 shows that the solution of the Newton Cooling Law solved with the Average Euler method converges with second order to the exact solution, since the error between successive resolutions shows a convergence factor 2^2, which corresponds to $n = 2$ in Eq. (2.19).

The solutions for the harmonic oscillator equations solved with Euler and Backward Euler methods, according to Fig. 2.9, show a convergence factor $\simeq 2$ and thus convergence of order one. The solution constructed using the Average Euler method has convergence factor $\simeq 2^2$ and therefore converges with second order to the exact solution.

Self-convergence. The motivation to construct numerical solutions of Differential Equations is that we want to solve equations whose exact solution is unknown. In that case the convergence criterion above would be useless. Nevertheless, convergence tests and the solution of problems whose solution is known are very useful and allow one to understand the strength of the various methods.

Even if the exact solution is unknown, the same idea of the convergence can be generalized as follows. Assume that a third numerical solution u^3 is constructed with resolution $\Delta t_3 = \Delta t_2/2 = \Delta t_1/4$. Such numerical solution can be written as

$$u_i^3 = u_i^e + E(t_i)\left(\frac{\Delta t}{4}\right)^n. \qquad (2.20)$$

2.8 Runge-Kutta Methods

Combining (2.18) and (2.20), one obtains the **self-convergence factor**

$$\boxed{SCF := \frac{u_i^1 - u_i^2}{u_i^2 - u_i^3} \simeq \frac{\Delta t^n - \left(\frac{\Delta t}{2}\right)^n}{\left(\frac{\Delta t}{2}\right)^n - \left(\frac{\Delta t}{4}\right)^n} = \frac{1 - \frac{1}{2^n}}{\frac{1}{2^n} - \frac{1}{4^n}} = 2^n} \quad (2.21)$$

When the solutions constructed using three successive resolutions related by a factor 2 obey the relation (2.21), it is said the **the numerical solution self-converges with order** n to the solution in the continuum.

This is a criterion used to verify that a numerical solution converges **without the need of the exact solution**.

> **Exercise.** Determine whether the numerical solution for x in the top row of Fig. 2.10 self-converges using one of the three methods. For this, calculate the solutions using resolutions Δt_2, Δt_3, Δt_4, and verify whether the approximate relation (2.21) is satisfied.

2.8 Runge-Kutta Methods

So far, for the generic IVP (2.1) on the discrete numerical domain D_d, we have constructed useful methods, specifically Euler, Backward Euler, and an Average Euler, that provide an expression for u_i in terms of known information, namely, the slope of the solution at t_{i-1} in the case of the Euler method, at t_i in the case of the Backward Euler method and the average of these two slopes in the Average Euler method. The slope is used to predict the solution from t_{i-1} to t_i using the equation of a line.

The Runge-Kutta (RK) method is a generalization of this strategy and consists in proposing a more general expression for the slope of a line such that the prediction of the numerical solution u_i is of the type

$$u_i = u_{i-1} + g(t_{i-1}, u_{i-1}, \Delta t)\Delta t, \quad (2.22)$$

where g is the slope of a line that we use to predict the value u_i. The RK method consists in assuming g to be a linear combination of slopes

$$g = a_1 k_1 + a_2 k_2 + \ldots + a_m k_m, \quad (2.23)$$

where k_j represents the slope of the numerical solution evaluated at a given point within the interval $[t_{i-1}, t_i]$ and specifically defined as

$$k_1 = f(t_{i-1}, u_{i-1}), \tag{2.24}$$
$$k_2 = f(t_{i-1} + b_1 \Delta t, u_{i-1} + c_{11} k_1 \Delta t),$$
$$k_3 = f(t_{i-1} + b_2 \Delta t, u_{i-1} + c_{21} k_1 \Delta t + c_{22} k_2 \Delta t),$$

$$\vdots$$

$$k_m = f(t_{i-1} + b_{m-1} \Delta t, u_{i-1} + c_{m-1,1} k_1 \Delta t + c_{m-1,2} k_2 \Delta t + \ldots$$
$$+ c_{m-1,m-1} k_{m-1} \Delta t).$$

Notice that the slope at t_{i-1}, namely, k_1, is that of Euler method, whereas $k_2, k_3, \ldots$ depend on the previous slopes. Coefficients b_j determine the location where the slope is to be evaluated, and c_{jk} are coefficients of a linear combination of all the slopes calculated previously. These constants are to be determined in terms of the accuracy of the method.

Second-Order Runge-Kutta (RK2)

Following [1], let us now illustrate the construction of the method in Eqs. (2.22), (2.23), and (2.24) for a second-order accurate solver. In this case Eq. (2.22) reduces to an expression with two slopes:

$$u_i = u_{i-1} + (a_1 k_1 + a_2 k_2) \Delta t, \tag{2.25}$$
$$k_1 = f(t_{i-1}, u_{i-1}),$$
$$k_2 = f(t_{i-1} + b_1 \Delta t, u_{i-1} + c_{11} k_1 \Delta t).$$

In order to determine the constants a_1, a_2, b_1, c_{11}, the expression for u_i in (2.25) is compared with a Taylor expansion to second order of u around t_{i-1} as follows:

$$k_1 = f(t_{i-1}, u_{i-1}),$$
$$k_2 = f(t_{i-1} + b_1 \Delta t, u_{i-1} + c_{11} k_1 \Delta t)$$
$$= f(t_{i-1}, u_{i-1}) + b_1 \frac{\partial f}{\partial t}\bigg|_{t_{i-1}} \Delta t + c_{11} k_1 \frac{\partial f}{\partial u}\bigg|_{t_{i-1}} \Delta t + \mathcal{O}(\Delta t^2),$$

2.8 Runge-Kutta Methods

which implies that u_i in (2.25) is

$$u_i = u_{i-1} + \left[a_1 f(t_{i-1}, u_{i-1}) + a_2 \left(f(t_{i-1}, u_{i-1}) + b_1 \frac{\partial f}{\partial t} \bigg|_{t_{i-1}} \Delta t \right. \right.$$

$$\left. \left. + c_{11} k_1 \frac{\partial f}{\partial u} \bigg|_{t_{i-1}} \Delta t \right) \right] \Delta t$$

$$= u_{i-1} + [a_1 f(t_{i-1}, u_{i-1}) + a_2 f(t_{i-1}, u_{i-1})] \Delta t$$

$$+ \left(a_2 b_1 \frac{\partial f}{\partial t} + a_2 c_{11} k_1 \frac{\partial f}{\partial u} \right) \bigg|_{t_{i-1}} \Delta t^2 + \mathcal{O}(\Delta t^3)$$

$$= u_{i-1} + (a_1 + a_2) f(t_{i-1}, u_{i-1}) \Delta t + \left(a_2 b_1 \frac{\partial f}{\partial t} + a_2 c_{11} \frac{du}{dt} \frac{\partial f}{\partial u} \right) \bigg|_{t_{i-1}}$$

$$\times \Delta t^2 + \mathcal{O}(\Delta t^3). \tag{2.26}$$

On the other hand, the Taylor series expansion of u_i around t_{i-1} reads

$$u_i = u_{i-1} + \frac{du}{dt} \bigg|_{t_{i-1}} \Delta t + \frac{1}{2!} \frac{d^2 u}{dt^2} \bigg|_{t_{i-1}} \Delta t^2 + \mathcal{O}(\Delta t^3), \tag{2.27}$$

where

$$\frac{du}{dt} \bigg|_{t_{i-1}} = f(t_{i-1}, u(t_{i-1})) = f(t_{i-1}, u_{i-1}),$$

$$\frac{d^2 u}{dt^2} \bigg|_{t_{i-1}} = \frac{df(t_{i-1}, u_{i-1})}{dt} = \left(\frac{\partial f(t, u)}{\partial t} + \frac{\partial f(t, u)}{\partial u} \frac{du}{dt} \right) \bigg|_{t_{i-1}},$$

which substituted into Eq. (2.27) gives

$$u_i = u_{i-1} + f(t_{i-1}, u_{i-1}) \Delta t + \left(\frac{\partial f(t, u)}{\partial t} + \frac{\partial f(t, u)}{\partial u} \frac{du}{dt} \right) \bigg|_{t_{i-1}} \frac{\Delta t^2}{2} + \mathcal{O}(\Delta t^3). \tag{2.28}$$

Comparison of (2.26) and (2.28) implies that

$$a_1 + a_2 = 1,$$
$$a_2 b_1 = \frac{1}{2},$$
$$a_2 c_{11} = \frac{1}{2},$$

which is an underdetermined system of three equations with four unknowns $a_1, a_2, b_1,$ and c_{11}, which has an infinity of solutions. Following [1], this system can be written in terms of a_2:

$$a_1 = 1 - a_2,$$
$$b_1 = c_{11} = \frac{1}{2a_2}, \qquad (2.29)$$

so that for a given value of a_2, different flavors of the method can be constructed, all of them within the family of Runge-Kutta methods with second-order accuracy. A few specific examples are described next.

Midpoint method. It is obtained using $a_2 = 1$, which implies that $a_1 = 0$ and $b_1 = c_{11} = 1/2$. The method in (2.25) becomes

$$u_i = u_{i-1} + (a_1 k_1 + a_2 k_2)\Delta t + \mathcal{O}(\Delta t^3)$$
$$= u_{i-1} + k_2 \Delta t + \mathcal{O}(\Delta t^3),$$
$$k_1 = f(t_{i-1}, u_{i-1}),$$
$$k_2 = f\left(t_{i-1} + \frac{1}{2}\Delta t, u_{i-1} + \frac{1}{2}k_1 \Delta t\right), \qquad (2.30)$$

where k_2 is the slope calculated at the midpoint between t_{i-1} and t_i.

Ralston method. It is obtained assuming $a_2 = 2/3$, which implies $a_1 = 1/3$ and $b_1 = c_{11} = 3/4$. The method in (2.25) becomes fully determined through the following expressions:

$$u_i = u_{i-1} + (a_1 k_1 + a_2 k_2)\Delta t + \mathcal{O}(\Delta t^3)$$
$$= u_{i-1} + \left(\frac{1}{3}k_1 + \frac{2}{3}k_2\right)\Delta t + \mathcal{O}(\Delta t^3),$$
$$k_1 = f(t_{i-1}, u_{i-1}),$$
$$k_2 = f\left(t_{i-1} + \frac{3}{4}\Delta t, u_{i-1} + \frac{3}{4}k_1 \Delta t\right). \qquad (2.31)$$

Heun method. In this case $a_2 = 1/2$, which implies that $a_1 = 1/2, b_1 = c_{11} = 1$ and the method (2.25) reduces to

$$u_i = u_{i-1} + (a_1 k_1 + a_2 k_2)\Delta t + \mathcal{O}(\Delta t^3)$$
$$= u_{i-1} + \left(\frac{1}{2}k_1 + \frac{1}{2}k_2\right)\Delta t + \mathcal{O}(\Delta t^3),$$
$$k_1 = f(t_{i-1}, u_{i-1}),$$
$$k_2 = f(t_{i-1} + \Delta t, u_{i-1} + k_1 \Delta t). \qquad (2.32)$$

2.8 Runge-Kutta Methods

In all these expressions, the error $\mathcal{O}(\Delta t^3)$ is local from t_{i-1} to t_i, which, according to the analysis in Sect. 2.3, implies that the global error is of second order $\mathcal{O}(\Delta t^2)$.

Observation. Notice that the expansions above, including the slopes k_2, involve the evaluation of f in the two arguments (t, u) shifted by an amount proportional to Δt. In all the examples so far, f has been function only of the unknown u. Specifically, in the Newton Cooling Law, the right-hand side in (2.2) depends only on the temperature T not on the independent variable t. For the Harmonic oscillator, notice that the right-hand sides in (2.15) depend only on x and y, but not on t. It means that the examples so far are associated to *autonomous* systems of equations.

Notice, for example, that Heun method (2.32), for an autonomous system, that is when $f = f(u)$, is written as

$$u_i = u_{i-1} + \frac{1}{2}(k_1 + k_2)\Delta t + \mathcal{O}(\Delta t^3),$$
$$k_1 = f(u_{i-1}),$$
$$k_2 = f(u_{i-1} + k_1 \Delta t), \qquad (2.33)$$

is the Average Euler method used to solve the Newton Cooling Law and the harmonic oscillator before.

The description of the three particular cases of second-order RK2 methods above illustrates how to implement the numerical solution of nonautonomous systems, whose slopes depend on the independent variable as well as on the unknown u. We will work out examples of this type later in this chapter, but as for now, it is important to illustrate how the three flavors of RK2 methods work for the damped harmonic oscillator. These examples will also indicate how to easily program the evaluation of the slopes k_1 and k_2 that will become more elaborate when dealing with higher-order methods.

We will use this moment to improve the programming. For this we will implement the three flavors of second-order Runge-Kutta explained before. This time, instead of defining k_{1x}, k_{2x}, and k_{1y}, k_{2y}, we will define two arrays k1(:) and k2(:) of sizes the number of equations NE. For the case of the harmonic oscillator, one needs only NE=2, since there are two Eqs. (2.15), one for x and one for y.

Moreover, we will not use arrays x and y; instead we define a two-dimensional array u(:,:) of size NE and N.

We also define an array of slopes rhs(:) of size NE that will provide the right-hand sides of the two Eqs. (2.15), one entry of the array for each equation.

A more important sophistication is the use of a module where global variables are declared, which makes easy for any subroutine to make use of such variables.

The following code rk2_HO.f90 contains the implementation of the three RK2 flavors described above, midpoint (2.30), Ralston (2.31), and Heun (2.32) for the solution of the damped harmonic oscillator (2.15):

```fortran
module numbers

! arrays
real(kind=8), allocatable, dimension (:) :: t
real(kind=8), allocatable, dimension (:,:) :: u
real(kind=8), allocatable, dimension (:) :: k1,k2,rhs
real(kind=8), allocatable, dimension (:) :: x_ex,y_ex,err_x,err_y

! parameters
real(kind=8) x0,y0,omega,m,b,k,alpha,beta

! Domain
real(kind=8) t0,tmax,dt
integer N,resolution_label,NE

end module

! --> Program starts <--
program rk2_HO

use numbers
implicit none

! Counters defined locally
integer i,j

resolution_label = 1
NE = 2
N = 4000
t0 = 0.0d0
tmax = 200.0d0
N = 2**(resolution_label-1) * N ! Number of cells for the
discretized domain
x0 = 1.0d0
y0 = 0.0d0
m = 0.5d0
k = 2.0d0
b = 0.05d0
alpha = 0.5d0 * b / m ! Parameter for the exact solution
beta = sqrt(abs(1.0d0 - 4.0d0*k*m/b**2)) ! Parameter for the
exact solution

! Allocate memory for the various arrays
! t: time
! u(1,:): x
! u(2,:): y

allocate(t(0:N),u(1:NE,0:N),rhs(1:NE),k1(1:NE),k2(1:NE))
allocate(x_ex(0:N),y_ex(0:N),err_x(0:N),err_y(0:N))

! ---> PART A <---
```

2.8 Runge-Kutta Methods

```
dt = (tmax - t0) / dble(N)
do i=0,N
  t(i) = t0 + dt * dble(i)
end do

! ---> PART B <---
! Set initial conditions
u(1,0) = x0
u(2,0) = y0
do i=1,N
  do j=1,4
    if (j.eq.1) then
      call calcrhs( t(i-1) ,u(:,i-1))
      k1 = rhs
    else
      ! if MIDPOINT uncomment the following three lines
      call calcrhs( t(i-1) + 0.5d0 * dt, u(:,i-1) + 0.5d0 * k1(:)
 * dt)
      k2 = rhs
      u(:,i) = u(:,i-1) + k2 * dt
      ! if RALSTON uncomment the following three lines
      ! call calcrhs( t(i-1) + 0.75d0 * dt, u(:,i-1) + 0.75d0 *
k1(:) * dt)
      ! k2 = rhs
      ! u(:,i) = u(:,i-1) + ( 1.0d0/3.0d0 * k1(:) + 2.0d0/3.0d0 *
k2(:) ) * dt
      ! if HEUN uncomment teh following three lines
      ! call calcrhs( t(i-1) + dt, u(:,i-1) + k1(:) * dt)
      ! k2 = rhs
      ! u(:,i) = u(:,i-1) + 0.5d0 * ( k1(:) + k2(:) ) * dt
    end if
  end do
end do

! ---> Ends Part B <---
! Exact solution
x_ex = exp(-alpha*t)*(sin(alpha*beta*t)/beta + cos(alpha*beta*t))
y_ex = -alpha*exp(-alpha*t) * (sin(alpha*beta*t)/beta
   + cos(alpha*beta*t)) &
   + alpha*exp(-alpha*t) * (cos(alpha*beta*t) -
   beta*sin(alpha*beta*t))
err_x = u(1,:) - x_ex
err_y = u(2,:) - y_ex

! Saving data to a file
open(1,file='HO_rk2.dat')
do i=0,N,2**(resolution_label-1)
  write(1,*) t(i),u(1,i),u(2,i),x_ex(i),y_ex(i),err_x(i),err_y(i)
end do
close(1)

end program
```

```
! --> Ends program <--

! --> Subroutine that calculates the right hnd sides <--

subroutine calcrhs(my_t,my_u)

use numbers
implicit none

real(kind=8), intent(in) :: my_t
real(kind=8), dimension(NE), intent(in) :: my_u

  rhs(1) = my_u(2)
  rhs(2) = - b / m * my_u(2) - k / m * my_u(1)

end subroutine calcrhs
```

Notice that the subroutine `calcrhs` consists in the right-hand sides of Eqs. (2.15), whose arguments are evaluated in the independent variable t and the unknowns x, y at specific arguments, with names `my_t`, `my_u(1)` and `my_u(2)` in the code, respectively. These arguments are given by formulas (2.30) for Midpoint, (2.31) for Ralston and (2.32) for Heun methods.

The resulting solutions are concentrated in Fig. 2.11 where we show the error of the numerical solution for the amplitude x, using different resolutions Δt_1, Δt_2, and Δt_3. We rescale the errors in order to show the second-order convergence of the three methods according to the convergence factor in (2.19) for $n = 2$. Notice also that the amplitude of the error is pretty similar for the three flavors of RK2 described here.

Exercise. Construct your own flavor of RK2. Define an original value for a_2 in expression (2.29), and construct the formulas equivalent to (2.30), (2.31), and (2.32) for your own method. Program the resulting formulas and produce convergence tests similar to those in Fig. 2.11 using the same resolutions.

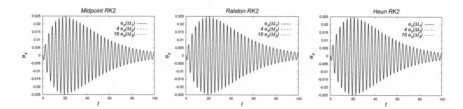

Fig. 2.11 Convergence tests of the solution calculated using the three flavors of RK2 with resolutions $\Delta t_1 = 0.05$, $\Delta t_2 = 0.025$, and $\Delta t_3 = 0.0125$. In the plot we show $e_x(\Delta t_1)$, $2^2 e_x(\Delta t_2)$, and $4^2 e_x(\Delta t_3)$ as defined in Eq. (2.17). The three curves superpose for the three methods, midpoint, Ralston, and Heun

Fourth-Order Runge-Kutta (RK4)

We have described in detail the construction of second-order RK methods. Higher-order methods can be constructed in a similar way, something we will not develop here; see, e.g., [2] for a derivation. Instead we will only use the final result for the most popular flavor of fourth-order RK methods. For the generic equation $\frac{du}{dt} = f(t, u)$ the method follows the rule:

$$u_i = u_{i-1} + \frac{1}{6}(k_1 + 2k_2 + 2k_3 + k_4)\Delta t + \mathcal{O}(\Delta t^5), \tag{2.34}$$

$$k_1 = f(t_{i-1}, u_{i-1}),$$

$$k_2 = f\left(t_{i-1} + \frac{1}{2}\Delta t, u_{i-1} + \frac{1}{2}k_1 \Delta t\right),$$

$$k_3 = f\left(t_{i-1} + \frac{1}{2}\Delta t, u_{i-1} + \frac{1}{2}k_2 \Delta t\right),$$

$$k_4 = f(t_{i-1} + \Delta t, u_{i-1} + k_3 \Delta t),$$

whose local error is of order $\mathcal{O}(\Delta t^5)$ and global error of order $\mathcal{O}(\Delta t^4)$.

Based on the previous code, the implementation of this formula is straightforward as illustrated with the following problem.

2.9 Workhorse Example 3: Damped and Forced Harmonic Oscillator

It is time to solve an IVP associated to a nonautonomous system of equations. For this consider the damped and forced harmonic oscillator whose equation of motion reads

$$\ddot{x} + \frac{b}{m}\dot{x} + \frac{k}{m}x = \frac{F_0}{m}\sin(\Omega t). \tag{2.35}$$

This problem can be transformed into the IVP with the following first-order system of equations

$$\boxed{\begin{aligned} \dot{x} &= y \\ \dot{y} &= -\frac{b}{m}y - \frac{k}{m}x + \frac{F_0}{m}\sin(\Omega t) \end{aligned}} \tag{2.36}$$

to be solved in the domain $D = [0, t_f]$, with initial conditions $x(0) = x_0$ and $y(0) = y_0$.

As a particular example, we again fix the physical parameters $m = 1/2$, $k = 2$, initial conditions $x(0) = 1$, $y(0) = 0$, and the domain $t_f = 200$. For the numerical solution, we define the numerical domain D_d with $N = 4000$ or equivalently with resolution $\Delta t_1 = 0.05$ and higher resolutions $\Delta t_2 = \Delta t_1/2$, $\Delta t_3 = \Delta t_2/2$, for convergence tests. The values of b, F_0, and Ω will be defined later for various scenarios.

For the solution we implement, the fourth-order Runge-Kutta described in Eq. (2.34). In this case one needs four slopes k_j for each of the variables, which needs the definition of arrays k1(:), k2(:), k3(:) and k4(:) of size NE=2.

We also define an array of slopes rhs(:) of size NE=2 that will contain the right-hand sides of the two Eqs. (2.36), one entry of the array for each equation.

We write down the whole program rk4_HO.f90 because it will be the basis of the forthcoming particular examples for which only parts of the code will be modified:

```
module numbers

! arrays
real(kind=8), allocatable, dimension (:) :: t
real(kind=8), allocatable, dimension (:,:) :: u
real(kind=8), allocatable, dimension (:) :: k1,k2,k3,k4,rhs

! parameters
real(kind=8) x0,y0,omega,m,b,k,F0,CapOmega

! Method and code vars
real(kind=8) t0,tmax,dt
integer N,resolution_label,NE

end module

! --> Program starts <--
program rk4_HO

use numbers
implicit none

! Counters defined locally
integer i,j

resolution_label = 2
NE = 2
N = 4000
t0 = 0.0d0
tmax = 200.0d0
N = 2**(resolution_label-1) * N ! Number of cells for the dicretized domain
x0 = 1.0d0
y0 = 0.0d0
m = 0.5d0
```

2.9 Workhorse Example 3: Damped and Forced Harmonic Oscillator

```fortran
k = 2.0d0
b = 0.05d0 ! <- 0.0 for the undamped HO
F0 = 0.2d0 ! <- 0.0 for the unforced HO
CapOmega = 0.5d0 ! <- case A 0.5 case B 2.0

! Allocate memory for the various arrays
! t: time
! u(1,:): x
! u(2,:): y

allocate(t(0:N),u(1:NE,0:N),rhs(1:NE),k1(1:NE),k2(1:NE),k3(1:NE),
k4(1:NE))

! ---> PART A <---
dt = (tmax - t0) / dble(N)
do i=0,N
  t(i) = t0 + dt * dble(i)
end do

! ---> PART B <---
! Set initial conditions
u(1,0) = x0
u(2,0) = y0
do i=1,N
  do j=1,4
    if (j.eq.1) then
      call calcrhs( t(i-1) ,u(:,i-1))
      k1 = rhs
    else if (j.eq.2) then
      call calcrhs( t(i-1) + 0.5d0 * dt, u(:,i-1) + 0.5d0 * k1(:)
* dt)
      k2 = rhs
    else if (j.eq.3) then
      call calcrhs( t(i-1) + 0.5d0 * dt, u(:,i-1) + 0.5d0 * k2 *
dt)
      k3 = rhs
    else
      call calcrhs( t(i-1) + dt, u(:,i-1) + k3 * dt)
      k4 = rhs
      u(:,i) = u(:,i-1) + (1.0d0/6.0d0)*( k1 + 2.0d0 * k2 + 2.0d0
* k3 + k4 ) * dt
    end if
  end do
end do
! ---> Ends Part B <---

! Saving data to a file
open(1,file='HO_rk4.dat')
do i=0,N,2**resolution_label
  write(1,*) t(i),u(1,i),u(2,i)
end do
close(1)
```

```
end program

! --> End program <--

! --> Subroutine that calculates the right hand sides <--

subroutine calcrhs(my_t,my_u)

use numbers
implicit none
real(kind=8), intent(in) :: my_t
real(kind=8), dimension(NE), intent(in) :: my_u

rhs(1) = my_u(2)
rhs(2) = - b / m * my_u(2) - k / m * my_u(1) + F0 / m *
    sin( CapOmega * my_t )

end subroutine calcrhs
```

Notice that the subroutine calcrhs calculates the slope at every point required by the RK4. The arguments of this subroutine are the values in the independent variable and the unknowns for each intermediate iteration of (2.34). This is the crucial subroutine because it defines the equations that are being solved, in our case equations (2.36), and this simple program will be generalized to solve a variety of IVPs later on.

Before we solve the damped-forced oscillator, it would be interesting to double check that fourth-order convergence is achieved for the simple case $b = F_0 = 0$ case, whose exact solution is given by Eq. (2.12).

In Fig. 2.12 we show the error e_x of the amplitude x and the fourth-order convergence toward the exact solution. According to theory, since the RK4 method has a global error of fourth order, the convergence factor in (2.19) is expected to

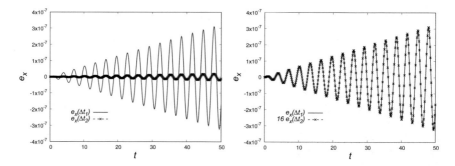

Fig. 2.12 On the left we show the error of the numerical solution for $b = F_0 = 0$, using resolutions Δt_1 and Δt_2. On the right we show that $e_x(\Delta t_1)$ and $2^4 e_x(\Delta t_2)$ lie approximately on top of each other, which shows that the numerical solution fulfills (2.19)

2.9 Workhorse Example 3: Damped and Forced Harmonic Oscillator

be 2^4 when doubling resolution. The error using resolution Δt_2 is actually 16 times smaller than the error using Δt_1 as shown in the second plot of Fig. 2.12, where we have multiplied $e_x(\Delta t_2)$ by 2^4.

We now solve the damped-forced oscillator for two cases.

Case A corresponds to a friction dominated scenario. For this we set $b = 0.05$, $F_0 = 0.2$, and $\Omega = 0.5$. These parameters allow one to see in the time series of $x(t)$, the two important phases of this oscillator, a first one in which the amplitude is damped by the term $b\dot{x}$, where the natural frequency of the oscillator $\omega_0 = \sqrt{k/m} = 2$ dominates, and a second phase where the motion is driven by the harmonic source $F_0 \sin(\Omega t)$, which has a slower frequency $\Omega = 0.5$. The results for the time series and phase space appear in Fig. 2.13. Notice how the phase space picture of the system differs significantly from those of the simple or the damped oscillator.

Case B associated to a growing amplitude regime, for which we use $\Omega = 2$ and keep the values of all other parameters. The results are shown in Fig. 2.13. The phase space picture begins erratical and asymptotically in time it approaches an ellipse twice as big as that of the simple case $b = F_0 = 0$.

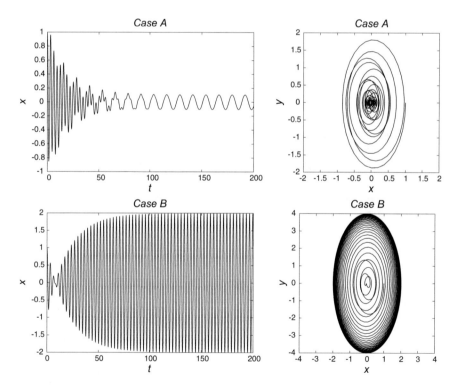

Fig. 2.13 Numerical solution x for Cases A and B and the phase space pictures. For these plots we use resolution $\Delta t_1 = 0.05$

Exercise. It would be interesting to program a different source, for instance, $F_0 \exp(-(t - t_I)/\sigma^2)$, which is a Gaussian pulse of amplitude F_0, centered at time t_I of approximate duration σ. The expected result is that the oscillator will respond to the Gaussian pulse, and then the damping term will rule the further evolution of the system.

Improve. Construct the exact solution of the problem (2.35) by adding the solution to the homogenous equation and the particular solution.

Self-Convergence of the Solution

This is a sufficiently complicated problem whose exact solution is not straightforward. Moreover the IVP is associated to a *nonautonomous* system of equations because the right-hand sides of Eqs. (2.36) contain the independent variable t, and the evaluation of the slopes used by RK methods in the independent variable, as seen in Eqs. (2.30), (2.31), (2.32), and (2.34) becomes useful.

Knowing that the RK4 method has a global error of order 4, we expect Eq. (2.21) to be satisfied for $n = 4$.

For the self-convergence test, we calculate the numerical solutions for Cases A and B with the code written above. We name x_1 the numerical solution for x calculated with resolution $\Delta t_1 = 0.05$, x_2 the solution using resolution $\Delta t_2 = \Delta t_1/2$, and x_3 the solution calculated with $\Delta t_3 = \Delta t_2/2$.

The self-convergence test is shown in Fig. 2.14. First we show the difference $x_1 - x_2$ between the numerical solutions x_1 and x_2, together with the difference $x_2 - x_3$. We also show a zoomed region with the comparison between $x_1 - x_2$ and $2^4(x_2 - x_3)$, showing that Eq. (2.21) is fulfilled. This type of plots illustrate a **typical self-convergence test of a numerical solution**.

2.10 Extrapolation

The previous example illustrates the self-convergence of numerical solutions, and it is possible to go further. Once we have a self-convergent solution, it is possible to approximately estimate the solution in the continuum out of the numerical solutions with various resolutions. The theory is as follows.

Assume we have calculated numerical solutions of the IVP problem (2.1), u_i^1 using resolution Δt_1, u_i^2 using resolution $\Delta t_1/2$ and u_i^3 using resolution $\Delta t_2/2$, and that self-convergence has been verified. Then an approximate solution in the continuum u_i^0 can be constructed with two of the three numerical solutions. Notice

2.10 Extrapolation

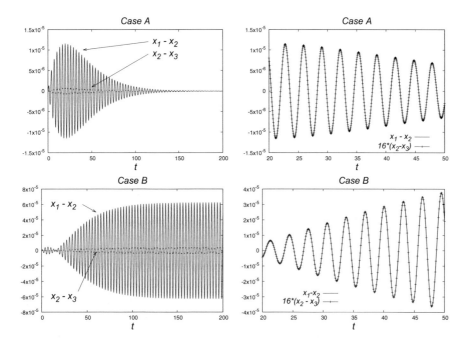

Fig. 2.14 Self-convergence test for Cases A and B. On the left the differences between numerical solutions in the whole time domain without rescaling. On the right we show an inset with the difference between $x_1 - x_2$ compared with $2^4(x_2 - x_3)$. The fact of being on top of each other indicates that the relation (2.21) is fulfilled

that we distinguish between the exact solution u_i^e and u_i^0, which we use to denote an approximation in the continuum, only.

If the global error of the method used to construct the numerical solutions is n, then we have seen that the numerical solutions can be written as

$$u_i^1 = u_i^0 + E(\Delta t)^n,$$
$$u_i^2 = u_i^0 + E\left(\frac{\Delta t}{2}\right)^n.$$

Multiplying the second relation by 2^n and subtracting

$$u_i^1 - 2^n u_i^2 = u_i^0(1 - 2^n) \quad \Rightarrow$$

$$\boxed{u_i^0 = \frac{u_i^1 - 2^n u_i^2}{1 - 2^n}} \tag{2.37}$$

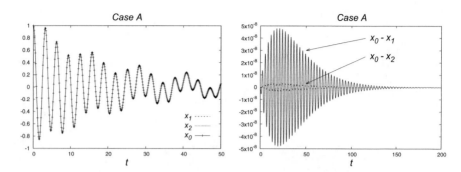

Fig. 2.15 (Left) Numerical solutions x_1 and x_2 produced with resolutions $\Delta t_1 = 0.025$ and $\Delta t_2 = \Delta t_1/2$, respectively, and the extrapolated solution calculated with formula (2.37) for $n = 4$. Difference between the extrapolated solution and the numerical solutions for Case A

This is an estimate of the solution in the continuum calculated out of u_i^1 and u_i^2. This expression is called the **Richardson extrapolation formula** and u_i^0 is the **extrapolated solution**. As an exercise, we use this formula to calculate the extrapolated solution for the amplitude x of the damped-forced harmonic oscillator from the previous section for Case A, whose numerical solution is illustrated in Fig. 2.13.

The extrapolated solution for Case A is shown in the left plot of Fig. 2.15, calculated using the numerical solutions x_1 and x_2, with resolutions $\Delta t_1 = 0.05$ and $\Delta t_2 = \Delta t_1/2$. In the same plot, we show the extrapolated solution that uses (2.37) with $n = 4$. The two numerical solutions appear on top of each other in the left plot, since the numerical resolution used is good. In the right plot, we can see that the differences $x_0 - x_1$ and $x_0 - x_2$ are nonzero.

> **Exercise.** Calculate the extrapolated solution for Case B.

2.11 Applications

In this section we present the numerical solution of various Initial Value Problems involving ODEs.

2.11 Applications

2.11.1 Predator-Prey Model

This model assumes there are two species, species x representing the population of prey and species y representing the population of predators. The traditional equations describing the dynamics of the two populations are the following:

$$\boxed{\begin{aligned} \frac{dx}{dt} &= ax - bxy \\ \frac{dy}{dt} &= -cy + dxy \end{aligned}} \tag{2.38}$$

where a, b, c, d are positive constant coefficients and $x, y \geq 0$. The term $ax > 0$ represents a Malthusian model for the reproduction of the prey; notice that this term alone would imply an exponential growth of population x. The term $-bxy < 0$ indicates that the interaction between the two species will decrease the prey population x. The term $-cx < 0$ indicates that the population of predators, living alone, would decrease due to the lack of food; should this term be the only one would imply an exponential decrease of population y toward extinction. The term $dxy > 0$ indicates that the interaction between the two species will represent energy and therefore reproduction for predators y.

It is straightforward to program a code that solves this system of equations, based on the code used for the damped-forced oscillator. One only needs to replace the parameters of the system and declare values for a, b, c, d, along with the initial conditions $x_0 = x(t = 0)$ and $y_0 = y(t = 0)$. The major and most important modification takes place in the subroutine calcrhs that should read as follows:

```
subroutine calcrhs(my_t,my_u)
use numbers
implicit none
real(kind=8), intent(in) :: my_t
real(kind=8), dimension(NE), intent(in) :: my_u
   rhs(1) =   a * my_u(1) - b * my_u(1) * my_u(2)
   rhs(2) = - c * my_u(2) + d * my_u(1) * my_u(2)
end subroutine calcrhs
```

Let us use the code to solve the predator-prey system with the following parameters and initial conditions $a = 1.25$, $b = 0.7$, $c = 0.4$, $d = 0.25$, $x_0 = 3$, and $y_0 = 2$, in the domain $t \in [0, 100]$ using resolution $\Delta t = 0.1$. The results are in Fig. 2.16. The interpretation of the solution is that both populations oscillate in time with a phase such that when the number of predators y grows, the population of prey decreases. When food is not sufficient, the population of predators starts decreasing, whereas when the number of predators decreases, the prey have the chance to reproduce and increase in number. A common picture of the behavior of the system is the phase space portrait, shown in Fig. 2.16, which describes a closed cycle.

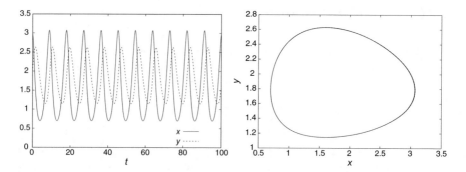

Fig. 2.16 (Left) Solution for the predator y and prey x populations as functions of time. (Right) Phase space portrait of the solution

Fig. 2.17 Phase space picture of the solution for three different values of $d = 0.25,\ 0.15,\ 0.1$, keeping a, b, c as before

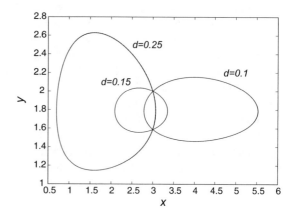

The period of the cycle, so as the maximums and minimums of x and y, depends on the value of the parameters a, b, c, d. For instance, the effect of changing the influence of d, that is, the interaction between the two species from the predator's point of view is shown in Fig. 2.17, producing various solutions. Likewise, the other parameters in one way or another also affect the interaction between the two species and the amplitude of oscillations of the two populations.

The Fixed Point

In the study of dynamical systems, it is said that there is a fixed point when the derivatives with respect to t in Eqs. (2.38) are zero. The importance of fixed points is whether they are stable, unstable, or saddle points, which can be studied with a stability analysis (see, e.g., [3, 4]). System (2.38) has a fixed point in a situation where both x and y are time-independent. These conditions are obtained by setting

$$\frac{dx}{dt} = 0 = ax - bxy,$$

$$\frac{dy}{dt} = 0 = -cy + dxy, \qquad (2.39)$$

2.11 Applications

which has the solution $x_f = c/d$ and $y_f = a/b$, where the subindex stands for fixed. The theory shows that this point is stable and one can prove this result numerically.

> **Exercise.** Using the initial conditions $x(0) = x_f$ and $y(0) = y_f$, solve the system (2.38) with the code, and demonstrate that $x(t)$ and $y(t)$ will remain nearly constant in time, or equivalently in the phase space the solution remains at one point. Modify the initial conditions slightly with $x(0) = x_f + \epsilon$ and $y(0) = y_f + \epsilon$, with $\epsilon = 0.01$, and show the solution describes a closed trajectory around the point x_f, y_f. This would be a numerical evidence of (x_f, y_f) being a stable fixed point. The formal proof of linear stability for this system and others in [3, 4] is recommended.

An **interesting variant** of this model include the collaborative systems, where $b, c < 0$, in which the growth of one population implies the growth of the other as well. An example of this in reality is the bees-flowers system.

The results discussed here can be reproduced with the code `rk4_PP.f90`, available as Electronic Supplementary Material described in Appendix B.

2.11.2 SIR Model for the Spread of Diseases

This is a very basic model for the propagation of an epidemic outbreak. This model assumes the existence of three interacting populations: individuals susceptible to be infected S, the population of infected individuals I, and a population of removed individuals R. The three populations interact according to the following system of equations:

$$\frac{dS}{dt} = \mu N - \mu S - \beta \frac{SI}{N},$$

$$\frac{dI}{dt} = \beta \frac{SI}{N} - \gamma I - \mu I,$$

$$\frac{dR}{dt} = \gamma I - \mu R, \tag{2.40}$$

subject to the conditions of $N = S + I + R$ constant and that all the coefficients μ, β, γ are positive and constant.

Coefficient μ indicates in the first equation that the number of susceptibles increases with the number of infected and removed; notice that $\mu(N - S) = \mu(I + R)$. In the third equation, it indicates that population R decreases and is transferred to population S, which might mean that the infected individuals recover and become susceptible again. Coefficient β in the first equation indicates the susceptible people, in contact with infected people produces a decrease of susceptible population. Notice that the same term appears in the second equation with a positive sign, which means that the individuals in the set S are transferred to the set I by interaction between infected and susceptibles. Finally γ indicates how quickly individuals are transferred from infected to removed.

A normalization that helps analyzing the system consists in defining normalized variables $s = S/N$, $i = I/N$ and $r = R/N$. The system above written in these variables reads

$$\begin{aligned} \frac{ds}{dt} &= \mu - \mu s - \beta s i \\ \frac{di}{dt} &= \beta s i - \gamma i - \mu i \\ \frac{dr}{dt} &= \gamma i - \mu r \end{aligned} \qquad (2.41)$$

In this normalization $\sigma = \frac{\beta}{\mu+\gamma}$ is interpreted as the amount of contacts among the normalized populations i and s. The number $1/\mu$ is interpreted as the mean lapse of time needed for a susceptible individual to get infected.

Likewise in the previous example, different values of coefficients β, γ, μ and different initial conditions $s_0 = s(t = 0)$, $i_0 = i(t = 0)$ and $r_0 = r(t = 0)$ will imply different results.

For the numerical solution of this system, one can again construct a code based on the previous one that would only need a few adjustments. Firstly the parameter defining the number of equations has to be set to NE=3. Secondly one needs to declare variables for the parameters μ, β, γ in the system (2.41). And finally one needs the routine that calculates the right-hand sides of system (2.41) as follows:

```
subroutine calcrhs(my_t,my_u)

use numbers
implicit none
real(kind=8), intent(in) :: my_t
real(kind=8), dimension(NE), intent(in) :: my_u

rhs(1) = mu - mu * my_u(1) - beta * my_u(1) * my_u(2)
rhs(2) = -mu * my_u(2) - gamma * my_u(2) + beta
    * my_u(1) * my_u(2)
rhs(3) = gamma * my_u(2) - mu * my_u(3)

end subroutine calcrhs
```

For illustration let us consider two scenarios.

2.11 Applications

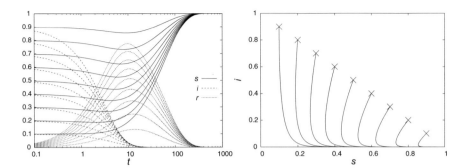

Fig. 2.18 Solution of the SIR model, Eqs. (2.41), for Scenario 1. First we show the behavior of the three populations in the time domain. Also shown is the dynamics in the si-plane. The infected population i approaches zero, whereas the s approaches one, which means the whole population becomes healthy asymptotically in time

Scenario 1. We assume $1/\mu = 60$, $\gamma = 1/3$, and $\sigma = \frac{\beta}{\mu+\gamma} = 1/2$. As initial conditions we start with $r_0 = 0$ and use various initial combinations of susceptible and infected populations such that $s_0 + i_0 = 1$, with $s_0 = 0.1, 0.2, ..., 0.9$. The results for these initial conditions are shown in Fig. 2.18, both the evolution in time of the three populations and a projection of the phase space diagram on the si-plane. Initial conditions are indicated in the phase space portrait with crosses.

With these values of $\sigma, \mu, \gamma, \beta$, the infected and removed populations approach zero and all the population becomes healthy with s approaching one. This can be seen both, in the time domain and in the phase space. Notice that the asymptotic value of the populations is independent of the nine different initial conditions used.

Scenario 2. In this case we set the parameters to $1/\mu = 60$, $\gamma = 1/3$ and $\sigma = 3$, which means that the amount of contacts, parametrized by σ is six times bigger than in Scenario 1. Again we use initial conditions such that $s_0 + i_0 = 1$ with $s_0 = 0.1, ..., 0.9$. The results for different initial conditions lead each population to a constant nonzero asymptotic value. This behavior is said to be endemic because the three populations coexist with nonzero populations. The results are shown in Fig. 2.19. In the time domain $t \in [0, 1000]$, the solution shows that the three populations tend asymptotically to a constant. In the phase si-plane, the attractor behavior is illustrated by the inward spiral.

What also happens in this case is that the asymptotic values of the populations are *independent* of the initial conditions. This indicates that parameters μ, β, σ, which are parameters of the disease or of the population behavior, determine the asymptotic values.

This model allows one to learn that the combination of infection's characteristics and population's behavior would have different coefficients. For instance, an aggressive infection in an undisciplined population would imply a big β, whereas a disciplined health system that recommends individuals to remain at home to avoid contact in order to avoid the infection of others will imply a small β.

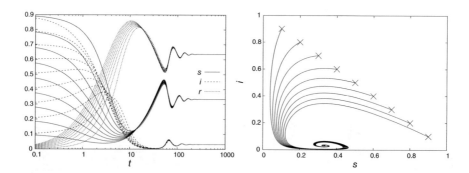

Fig. 2.19 On the left we show the evolution of populations for Scenario 2 and how they converge asymptotically in time to a finite value. On the right the trajectory in the phase si–plane for all combinations of initial conditions

All the results discussed here can be reproduced with the code `rk4_SIR.f90` listed in Appendix B, where we define initial conditions subject to the restriction $i_0 + s_0 = 1$. Then given different values of s_0, one obtains the associated initial value of the infected population $i_0 = 1 - s_0$. The code for the implementation of these initial conditions can be as follows:

```
do k=0,8
  ! initial conditions u(1,0), u(2,0), u(3,0)
  u(1,0) = s0 + dble(k) * 0.1
  u(2,0) = 1.0 - u(1,0)
  u(3,0) = r0
  do i=1,N
    ! Here the Runge-Kutta integration
  end do
end do
```

As simple as this, the dynamics of populations can be modeled with ODEs as long as diffusion effects are not considerable, which needs the solution of PDEs and will be seen later in this book.

Exercise. Search for coefficients μ, γ, σ that result in a highly infected population.

Challenge. Adapt the code to reproduce the results for the COVID-19 spread model in [5], which includes vaccines and mutation.

2.11 Applications

2.11.3 Lorenz System and Chaos

While studying atmospheric models, in 1962 Edward Lorenz arrived to the following system of equations:

$$\begin{aligned}\frac{dx}{dt} &= a(y - x) \\ \frac{dy}{dt} &= bx - y - xz \\ \frac{dz}{dt} &= xy - cz\end{aligned} \quad (2.42)$$

where x is associated to the convective intensity, y to temperature, and z measures the deviation from linearity of a vertical profile flow. The parameters are the Prandtl number a; the Rayleigh number b, associated to the addition of heat to the system; and the aspect ratio of fluid's shear c, concepts that are worth learning from fluid dynamics, but not needed to understand the mathematical behavior of the solutions.

This system of equations became interesting and famous because it shows Chaos and because it has solutions with strange attractor behavior. Chaotic behavior is related to sensibility to initial conditions and dimension of the system, whereas the strange attractor property is related to a bounded phase space-diagram with no closed trajectories. See [3, 4] for a complete discussion of this system.

In order to illustrate the numerical solution of this IVP, we define two scenarios, one in which the solution shows an attractor in time solution and a second one in which the system shows strange attractor behavior. These two scenarios can be achieved by assuming the initial conditions $(x_0, y_0, z_0) = (7, 6, 5)$ with fixed parameters $a = 15$, $c = 3$ and two values of the parameter b that determine the two different behaviors.

Case A, the fixed point attractor regime. In this case we set parameter $b = 20$ and solve the system in the time domain $t \in [0, 100]$. The resolution is an important issue, and in order to achieve convergence, one needs to use a rather high one; specifically we set $\Delta t = 0.001$ or equivalently $N = 100,000$.

The numerical solution with these parameter values is shown in Fig. 2.20, where the three variables x, y, z approach a fixed point value in the time domain. Also in the phase space portrait, the trajectory oscillates between two loops and finally spirals down toward the center of one of them.

This is a nonlinear sensitive system and we want to practice a self-convergence test. For this we construct solutions x_1, x_2, x_3 that use resolutions $\Delta t_1 = 0.001$, $\Delta t_2 = \Delta t_1/2$, $\Delta t_3 = \Delta t_2/2$, respectively. We show the results in Fig. 2.20, in the time window before the asymptotic behavior $t \in [0, 35]$. Notice that $x_1 - x_2$ is nearly 2^4 times bigger than $x_2 - x_3$, which confirms the theory in (2.21) and validates self-convergence.

Case B, the strange attractor regime. This case can be illustrated with the parameter value $b = 30$ and all other parameters and initial conditions as before

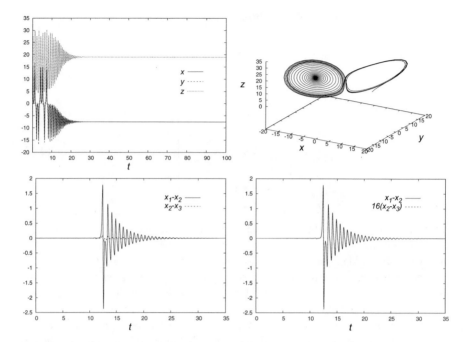

Fig. 2.20 (Top) Solution of Lorenz system (2.42) for Case A. First we show the dynamics in the time domain and then the phase space diagram. The behavior in both pictures shows a late-time attractor behavior toward a fixed point. (Bottom) Self-convergence test for the variable x showing $(x_1 - x_2)$ and $(x_2 - x_3)$. On the right we show $(x_1 - x_2)$ and $2^4(x_2 - x_3)$ that approximately superpose

$a = 15$, $c = 3$, $x_0 = 7$, $y_0 = 6$, $z_0 = 5$, in the time domain $t \in [0, 100]$ using resolution $\Delta t = 0.001$. In this case the three time series do not approach a fixed attractor value; instead, the time-series of x, y, z bounce between very different values as shown in Fig. 2.21. In the same Figure, we also show the famous Lorenz attractor phase space diagram.

This is one of the canonical examples used to illustrate Chaos, partly because the solution does not seem to follow a pattern and partly because the system is sensitive to initial conditions. In order to show the later property, we solve the system with parameters $a = 15$, $b = 30$, $c = 3$ and initial conditions slightly different from $(x_0, y_0, z_0) = (7, 6, 5)$:

$$(x_0, y_0, z_0) = (7.001, 6, 5),$$
$$(x_0, y_0, z_0) = (7, 6.001, 5).$$

The results for the variable x using these two initial conditions are shown in Fig. 2.22, where by $t \sim 10$ it can be seen that the two solutions depart from each other, which illustrates the sensibility to initial conditions.

2.11 Applications

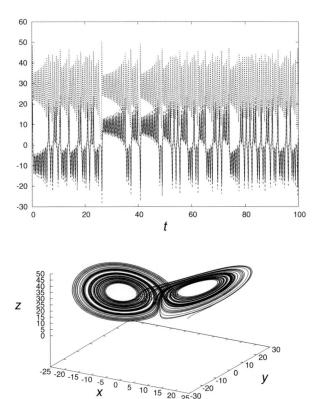

Fig. 2.21 Solution of Lorenz system (2.42) for Case B using the initial conditions $(x_0, y_0, z_0) = (7, 6, 5)$. We show the dynamics in time domain of x, y, z and the phase space diagram. The behavior in both pictures does not show a late-time attractor toward a fixed point, instead it shows the strange attractor behavior

These results can be reproduced with the code `rk4_Lorenz.f90`, available as Electronic Supplementary Material described in Appendix B. The difference of this code with the previous ones is the part that calculates the right-hand sides of system (2.42) and the part providing parameter values and initial conditions. In this case we identify `u(1,:)`, `u(2,:)`, `u(3,:)` with x, y and z respectively, and the routine that calculates the right-hand sides of the system reads:

```
subroutine calcrhs(my_t,my_u)

use numbers
implicit none
real(kind=8), intent(in) :: my_t
real(kind=8), dimension(NE), intent(in) :: my_u

rhs(1) = a * ( my_u(2) - my_u(1) )
rhs(2) = b * my_u(1) - my_u(2) - my_u(1) * my_u(3)
```

```
rhs(3) = my_u(1) * my_u(2) - c * my_u(3)

end subroutine calcrhs
```

In this very important and famous case, it is not enough to show the phase space picture; instead one has to make sure the numerical solution self-converges in the regime where one sees departure of the solutions with very similar initial conditions. We thus prepared a self-convergence test with the numerical solutions for the variable x labeled x_1, x_2, x_3 calculated using resolutions $\Delta t_1 = 0.001$, $\Delta t_2 = 0.0005$, $\Delta t_3 = 0.00025$, respectively. The results are shown in the bottom of Fig. 2.22, where we show that $x_1 - x_2$ is nearly $16 = 2^4$ times bigger than $x_2 - x_3$, which, according to theory (2.21), means the numerical solutions for x self-converges in this time-window.

Then one *can* claim that we have constructed a self-convergent numerical solution of the Lorenz system with the parameters and initial conditions of Case B corresponding to the strange attractor regime in the time window $t \in [0, 20]$.

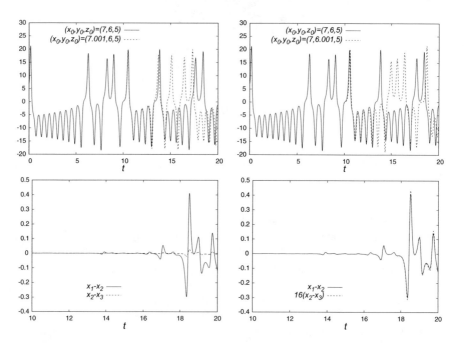

Fig. 2.22 (Top) Two solutions for the variable x using slightly different initial conditions. Notice that they start being different after $t \sim 10$. (Bottom) Self-convergence test for x showing $(x_1 - x_2)$ and $(x_2 - x_3)$ for the initial conditions $(x_0, y_0, z_0) = (7, 6, 5)$. On the right we show $(x_1 - x_2)$ and $2^4(x_2 - x_3)$ that approximately superpose. Similar plots can be constructed for the variables y and z

2.11 Applications

> **Exercise.** Investigating what happens after $t = 20$ is interesting. Verify that the numerical solution does not self-converge after $t \sim 30$ even using further refined resolutions $\Delta t_4 = \Delta t_1/2^3$, $\Delta t_5 = \Delta t_1/2^4$, $\Delta t_6 = \Delta t_1/2^5$, $\Delta t_7 = \Delta t_1/2^6$, $\Delta t_8 = \Delta t_1/2^7$.

With the experience gained in this section, we would like to invite the reader to solve the Lorenz system using higher-order methods that should be easy to implement now, analyze the solution in bigger time windows, and check self-convergence. An example of a sixth-order accurate Runge-Kutta method that can be easily implemented can be found in [6].

2.11.4 Simple Pendulum

This is a case similar to the harmonic oscillator. A clear important difference is that the spatial coordinate is now periodic, which leads to interesting physical results we are to explore numerically.

The degree of freedom of the system is associated to the angle of the pendulum. Following the convention in Fig. 2.23, where $\phi = 0$ corresponds to the vertical position, the equation of motion for the angle of the simple pendulum is

$$\frac{d^2\phi}{d\tilde{t}^2} + \frac{g}{l}\sin\phi = 0,$$

that can be simplified by rescaling time $t = \tilde{t}\sqrt{g/l}$, such that the equation reduces to

$$\frac{d^2\phi}{dt^2} + \sin\phi = 0. \tag{2.43}$$

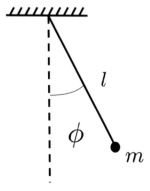

Fig. 2.23 Illustration of the pendulum and the convention we use to measure the angle

The system is conservative and it is possible to define an energy function as follows:

$$\frac{d\phi}{dt}\frac{d^2\phi}{dt^2} + \frac{d\phi}{dt}\sin\phi = 0 \implies \frac{d}{dt}\left(\frac{1}{2}\left(\frac{d\phi}{dt}\right)^2 - \cos\phi\right) := \frac{dE}{dt} = 0,$$

where E is the energy. Now we write Eq. (2.43) as a system of two first-order equations:

$$\boxed{\begin{aligned} \frac{d\phi}{dt} &= \eta \\ \frac{d\eta}{dt} &= -\sin\phi \end{aligned}} \qquad (2.44)$$

where η is the angular velocity. The energy in terms of these variables reads

$$E(\phi, \eta) = \frac{1}{2}\left(\frac{d\phi}{dt}\right)^2 - \cos\phi = \frac{1}{2}\eta^2 - \cos\phi,$$

and the theory indicates that constant values of this functional determine the trajectories of the system in the phase space $\phi\eta$. We illustrate this in Fig. 2.24, where we show the isocontours of $E(\phi, \eta)$ in the domain $(\phi, \eta) \in [-10, 10] \times [-4, 4]$.

Fig. 2.24 Energy of the pendulum as function of ϕ and η. The isocontours of E correspond to accessible trajectories of the system in the phase space

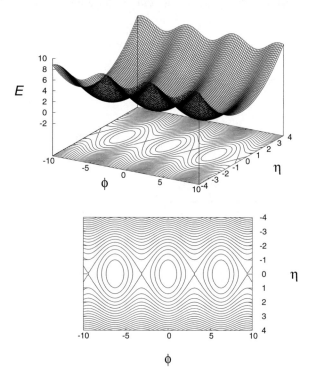

2.11 Applications

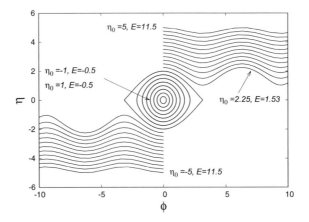

Fig. 2.25 Phase space diagram of the frictionless pendulum obtained by solving Eqs. (2.44) for $\phi_0 = 0$ and various initial values of η_0

There are two types of interesting points. In the Figure, points $(\phi, \eta) = (-2\pi, 0)$, $(\phi, \eta) = (0, 0)$, $(\phi, \eta) = (2\pi, 0)$ correspond to local minimums of E and the phase space trajectories near them are ellipses, like those for the simple harmonic oscillator.

Points like $(\phi, \eta) = (-3\pi, 0)$, $(\phi, \eta) = (-\pi, 0)$, $(\phi, \eta) = (\pi, 0)$, $(\phi, \eta) = (3\pi, 0)$ are saddle points in the phase space that can be seen in the left plot of Fig. 2.24. They are local maximums along the line $\eta = 0$ and minimums along the lines $\phi = -3\pi, -\pi, \pi, 3\pi$.

We proceed to solve the system (2.44) using the RK4 solver for given initial conditions $\phi(0) = \phi_0$ and $\eta(0) = \eta_0$. Let us assume initial conditions $\phi_0 = 0$ and η_0 taking various values. This means that the initial position of the pendulum is vertical and has a given initial angular velocity. For this we set the initial velocity to $\eta_0 = -5 + k\delta\eta_0$, with $\delta\eta_0 = 0.25$ and $k = 0, \ldots, 40$. The resulting phase space diagram of the numerical solutions appears in Fig. 2.25.

The initial velocity η_0 ranges from negative to positive values, which results in trajectories moving initially to the right in the $\phi\eta$-plane and trajectories moving to the left, respectively. Some trajectories remain bounded, and others extend toward the outskirts of the domain in the Figure. The cases of trapped trajectories have a negative energy, whereas those extending toward infinity have positive energy. We indicate the values of the Energy in the Figure for some of the trajectories.

Notice finally that these phase space trajectories resemble those pictured in Fig. 2.24 although partially. The phase space can be completed by solving the IVP with a bigger set of initial conditions. Also observe that for negative energies we only draw the trajectories around the minimum of E at $(\phi, \eta) = (0, 0)$. In order to construct trajectories around $(\phi, \eta) = (-2\pi, 0)$ and $(\phi, \eta) = (2\pi, 0)$, one has to start with values of ϕ_0 at a different branch. Using the same initial conditions for η_0 but this time $\phi_0 = -2\pi$, one will produce the trajectories at the top-left zone of Fig. 2.25, whereas setting $\phi_0 = 2\pi$ will fill the right-bottom zone of the diagram.

We do not discuss the code used for solving this system here, because we next solve the damped pendulum, where we describe the code that includes the ideal pendulum as a particular case.

Exercise. Within the code provided below, program the appropriate initial conditions required to fill the blank spots in the phase space diagram of Fig. 2.25.

2.11.5 Damped Pendulum

We start by generalizing the simple pendulum Eq. (2.43) by adding a friction term as follows:

$$\frac{d^2\phi}{dt^2} + b\frac{d\phi}{dt} + \sin\phi = 0, \tag{2.45}$$

where the term in the middle plays the role of friction for $b > 0$ and will damp the motion out. Recalling that the energy of the system is $E = \frac{1}{2}(\frac{d\phi}{dt})^2 - \cos\phi$, the loss of energy goes like

$$\frac{dE}{dt} = \frac{d}{dt}\left(\frac{1}{2}\left(\frac{d\phi}{dt}\right)^2 - \cos\phi\right) = -b\dot\phi^2,$$

which eventually provokes the pendulum to settle at rest independently of the initial conditions.

Unlike the simple pendulum, isocontours of E do not define the phase space trajectories since E is not conserved in this case, and the phase space diagram will be constructed by solving the evolution equation for a bundle of initial conditions.

Let us explore the phase space diagram to see the effects of the damping for $b = 0.1$. Numerical solutions are calculated using $\Delta t = 0.1$ in the time domain $t \in [0, 100]$. The results are shown in the left plot of Fig. 2.26. We show the diagram for the same initial conditions used to generate Fig. 2.25. The loss of energy brings the system toward attractor points, which happen to be the minimums of the energy functional of the ideal pendulum in Fig. 2.24. The trajectories that were ellipses in the frictionless case are now inspirals toward the central attractor point $(0, 0)$. Some of the trajectories that led the system toward the far left or far right of the diagram in Fig. 2.25 in the frictionless case are now trapped at the attractor points $(\phi, \eta) = (\pm 2k\pi, 0)$.

A more appealing phase space diagram can be generated easily by choosing a wider range of initial conditions. At the right of Fig. 2.26, we show the diagram for

2.11 Applications

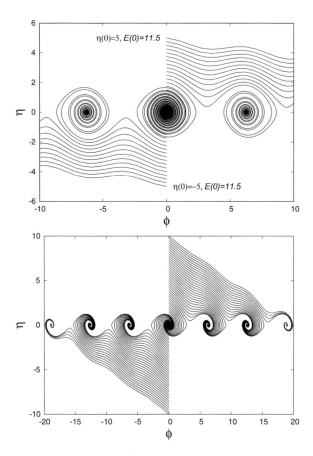

Fig. 2.26 (Top) Phase space diagram for the damped pendulum. The case for $b = 0.1$ and the same initial conditions as for the friction-less case in Fig. 2.25. (Bottom) The case with $b = 0.5$ and a wider range of initial conditions that illustrate how the system evolves toward the attractor points of the system

$b = 0.5$, with initial conditions for $\phi_0 = 0$ and η_0 ranging from -10 to 10. The result shows a good-looking sequence of attractor points.

The code rk4_Pendulum.f90, available as Electronic Supplementary Material described in Appendix B, generates these solutions and phase space diagrams uses a right-hand side routine that reads as follows:

```
subroutine calcrhs(my_t,my_u)
use numbers
implicit none
real(kind=8), intent(in) :: my_t
real(kind=8), dimension(NE), intent(in) :: my_u

  rhs(1) = my_u(2)
```

```
rhs(2) = - b * my_u(2) - sin( my_u(1) )

end subroutine calcrhs



DO ics = 0,80 ! 40 for simple, 80 for nice pic
! ---> PART B <---
! Set initial conditions
u(1,0) = phi0
u(2,0) = dotphi0 + dble(ics) * 0.25 ! Changes the value of
   dotphi0
! Integration loop along t
  do i=1,N
    ! RK4 loop
    ! Save data
  end do
END DO
```

Basin of Attraction. We use the case $b = 0.5$, whose phase space trajectories are on the right of Fig. 2.26, to illustrate how a given domain of initial conditions approaches asymptotically in time toward one of the attractor points. It can be seen that the initial conditions with $\eta_0 = -10, -9.75, -9.5, -9.25$ in the phase space end up in the attractor point $(-6\pi, 0)$. It is then said that these initial conditions belong to the *basin of attraction of this particular attractor point*.

One **challenge** is to investigate the basin of attraction of each attractor point. By slightly generalizing the code, one can explore the initial conditions for which the system ends at the attractors points.

A simple strategy is to define a bundle of uniformly distributed initial conditions within the domain $(\phi_0, \eta_0) \in [-10, 10] \times [-10, 10]$. For example, by setting initial conditions $\phi_{0,j} = -10 + 0.1j$ and $\eta_{0,j} = -10 + 0.1j$, with $j = 0, ..., 200$. This implies solving 200×200 IVPs, from which one can select those that asymptotically in time approach a particular attractor point. Keeping in mind that the time domain is $t \in [0, 100]$, and labeling an attractor point as (ϕ_A, η_A), we check whether the condition

$$(\phi(100) - \phi_A)^2 + (\eta(100) - \eta_A)^2 < \epsilon, \qquad (2.46)$$

where $\epsilon > 0$ is a tolerance, is fulfilled or not for each of the initial conditions around each attractor point. If the condition is fulfilled, we save the position of initial conditions in a file we can plot; otherwise we do not save it. With this criterion we show in Fig. 2.27 the basin of attraction for the three attractor points in the chunk of phase space $[-10, 10] \times [-10, 10]$ and for initial conditions also restricted to this domain.

There are still other interesting subjects to analyze with the code, for example, it would be interesting to study the behavior of the system at bifurcation points $((2k + 1)\pi, 0)$.

2.11 Applications

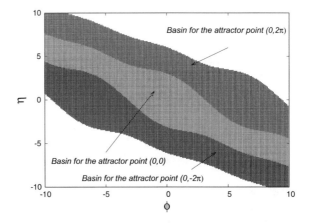

Fig. 2.27 Basin of attraction of the three attractor points in the domain $[-10, 10] \times [-10, 10]$, for the damped pendulum with $b = 0.5$

Exercise. Program the appropriate initial conditions required to fill the blank spots in the phase space diagram of Fig. 2.26.

Exercise. Reproduce Fig. 2.27.

2.11.6 Polytropic Star Models with Spherical Symmetry

The goal of this example is to show that three-dimensional problems with certain symmetries can be rewritten as ODEs and also how to handle subtleties emerging at the coordinate origin sometimes when using spherical coordinates.

For example, if one wants to solve Poisson equation $\nabla^2 \phi = 4\pi\rho$ for a spherically symmetric system, it is helpful to use spherical coordinates. The spherical symmetry means that $\phi = \phi(r)$ and $\rho = \rho(r)$; then writing the Laplace operator in spherical coordinates, Poisson equation reduces to

$$\frac{1}{r^2}\frac{d}{dr}\left(r^2\frac{d\phi}{dr}\right) = 4\pi\rho(r), \tag{2.47}$$

where all the terms depending on the angular coordinates vanish. Then the 3D problem is reduced to an ordinary differential equation that can be solved as an IVP.

The idea of this application is to solve the equations of a Tolman-Oppenheimer-Volkov (TOV) star model [7]. The TOV star model is a stationary solution to the Euler-Poisson system. The system consists of a fluid that generates its own gravitational field, in a situation where the gravitational field and the pressure of the fluid compensate each other, so that the fluid remains in hydrostatic equilibrium.

The star is made of a fluid that obeys Euler equations of hydrodynamics for an ideal fluid:

$$\partial_t \rho + \nabla \cdot (\rho \mathbf{v}) = 0,$$
$$\partial_t (\rho \mathbf{v}) + \nabla \cdot (\mathbf{v} \otimes \rho \mathbf{v}) + \nabla p = -\rho \nabla \phi,$$
$$\partial_t E + \nabla \cdot [(E + p)\mathbf{v}] = -\rho \mathbf{v} \cdot \nabla \phi, \quad (2.48)$$

where ρ, $\mathbf{v}$, and p are the density, velocity field, and pressure of a fluid volume element, with total energy $E = \rho(\frac{1}{2}v^2 + e)$, where e is the internal energy related to ρ and p through an equation of state (EoS), and finally ϕ is a potential responsible for a force acting on the fluid, which in our case will be the gravitational potential sourced by the fluid itself, which is solution of Poisson equation $\nabla^2 \phi = 4\pi \rho$. In general, all variables involved depend on time and space.

We will come to these general equations in a later chapter where we deal with PDEs; here we only look for spherically symmetric and stationary solutions of these equations. The spherically symmetric version of Eqs. (2.48), in which all the variables depend on time and the radial coordinate reads

$$\frac{\partial \rho}{\partial t} + \frac{1}{r^2} \frac{\partial}{\partial r}(r^2 \rho v) = 0, \quad (2.49)$$

$$\frac{\partial}{\partial t}(\rho v) + \frac{1}{r^2} \frac{\partial}{\partial r}(r^2 \rho v^2) + \frac{\partial p}{\partial r} = -\rho \frac{\partial \phi}{\partial r},$$

$$\frac{\partial E}{\partial t} + \frac{1}{r^2} \frac{\partial}{\partial r}(r^2 v(E + p)) = -\rho v \frac{\partial \phi}{\partial r},$$

$$\frac{\partial}{\partial r}\left(r^2 \frac{\partial \phi}{\partial r}\right) = 4\pi G r^2 \rho,$$

where v is the radial component of the velocity field. A key issue in problems related to hydrodynamics is that this system of equations is underdetermined. There are four equations for five variables ρ, v, E, p, ϕ. An EoS is used to close the system that relates ρ, E, p. The polytropic EoS $p = K\rho^\gamma$ is very popular, suitable for isentropic processes, where $K = p_0/\rho_0^\gamma$ is a constant defining a reference thermodynamical state of the system and $\gamma = c_p/c_v$ is the adiabatic index resulting from the ratio between the specific heat capacities at constant pressure and volume. This is the EoS used to construct the TOV star model. For the adiabatic index, we will use $\gamma = 5/3$, which is a suitable value for a simple gas made of particles moving at nonrelativistic speeds.

Assume additionally in (2.49) that the system is time-independent if one expects the pressure and gravitational field to compensate each other in equilibrium. This implies that the variables do not depend on time and that $v = 0$, which reduces the

2.11 Applications

system to two nontrivial ODEs:

$$\frac{dp}{dr} = -\rho \frac{d\phi}{dr},$$

$$\frac{d}{dr}\left(r^2 \frac{d\phi}{dr}\right) = 4\pi G r^2 \rho, \qquad (2.50)$$

which are usually rewritten considering $\frac{d\phi}{dr} = G\frac{m(r)}{r^2} = -\frac{\phi}{r}$, because the force produced by a sphere of radius r containing the mass $m(r)$ acting on a test particle of mass m_{test} at position r is $F = -\frac{d\phi}{dr} m_{test} = -G\frac{m(r)m_{test}}{r^2}$. Notice that the relation between $\frac{d\phi}{dr}$ and $m(r)$ is actually the definition of the first derivative of ϕ as a new variable, as done in all the previous examples involving second-order equations. This implies that

$$\frac{d}{dr}\left(r^2 G \frac{m}{r^2}\right) = 4\pi G r^2 \rho, \qquad (2.51)$$

and therefore one ends with a system of three first-order ODEs:

$$\boxed{\begin{aligned} \frac{dm}{dr} &= 4\pi r^2 \rho \\ \frac{dp}{dr} &= -G\frac{m}{r^2}\rho \\ \frac{d\phi}{dr} &= G\frac{m}{r^2} = -\frac{\phi}{r} \end{aligned}} \qquad (2.52)$$

which is a system of equations for $m(r)$, $p(r)$ and $\phi(r)$. We set the domain of integration from the origin until a maximum radius $r \in [0, r_{max}]$ that we discretize like we have done so far, $D_d = \{r_i = i\Delta r\}, i = 0, ..., N$, with $\Delta r = r_{max}/N$ for a given N. The IVP is now formulated with the following initial conditions corresponding to values of the unknown functions at the origin, and the numerical solution will be m_i, p_i, ϕ_i. Given the central value of density ρ_c, the initial/central conditions are:

1. For the mass $m(0) = 0$, since the integrated mass at the origin is zero.
2. For the pressure we use the polytropic EoS $p(0) = K\rho_c^\gamma$, where ρ_c is the central density of the star.
3. Finally, the potential $\phi(0) = \phi_c$ is an arbitrary constant. Its value at the origin is irrelevant because according to (2.52), the potential can be rescaled multiplying by a constant, for example, using any value of ϕ_c one can always rescale the potential:

$$\phi_{rescaled} = \frac{m(r_{max})}{r_{max}\phi(r_{max})}\phi(r) \qquad (2.53)$$

so that the known value of the potential at the outermost point of the domain is $\phi(r_{max}) = -\frac{m(r_{max})}{r_{max}}$. This scaling works as long as ϕ_c is sufficiently negative so that $\phi(r_{max})$ is negative too and the previous expression does not change sign.

Given γ and K, properties of the fluid, the central value of mass density ρ_c is the only free parameter of the solution that distinguishes one star from another. This means that each ρ_c is the parameter that labels the family of solutions for the pair (γ, K).

The **origin** is another important concern because Eqs. (2.52) seem to be singular at $r = 0$. Fortunately, since $m(r)$ is also zero at the origin, one can investigate whether the equations are regular at the origin, for which we develop the Taylor expansion of the mass function

$$m'(0) = 4\pi r^2 \rho|_0 = 0,$$

$$m''(0) = 4\pi \left(2r\rho + r^2 \frac{d\rho}{dr}\right)|_0 = 0,$$

$$m'''(0) = 4\pi(2\rho + 2r\rho' + r^2\rho'' + 2r\rho')|_0 = 8\pi\rho \quad \Rightarrow$$

$$m \simeq m(0) + m'(0) + \frac{1}{2}m''(0)r^2 + \frac{1}{6}m'''(0)r^3 + O(r^4)$$

$$= \frac{4\pi\rho}{3}r^3 + O(r^4).$$

This expansion indicates that the singularity can actually be compensated by the smallness of the mass near the origin, because the mass approaches zero faster than the denominators in the singular terms of the equations. Therefore the equations for p and ϕ when $r \ll 1$ can be approximated as follows:

$$\frac{dp}{dr} = -\frac{4\pi G}{r^2}\rho \frac{\rho r^3}{3} = -\frac{4}{3}\pi G\rho^2 r,$$

$$\frac{d\phi}{dr} = \frac{G}{r^2} \cdot \frac{4\pi \rho r^3}{3} = \frac{4}{3}\pi G\rho r, \tag{2.54}$$

which are regular and can be used instead of those in (2.52) for the first grid points of D_d where $r \ll 1$, when integrating the ODEs. In practice the right-hand sides in (2.54) are used only to integrate the equation from the first point of the discrete domain $r_0 = 0$ to the second one $r_1 = \Delta r$. From this point of the discrete domain on, the right-hand sides used for the integration are those in (2.52).

The code that solves these equations uses the same technology as in the previous examples. One only needs to define initial conditions, which in this case are values of the functions at the origin, and use the appropriate right-hand sides there. We consider the domain $r_{max} = 20$ that we discretize with $N = 1000$ or equivalently

2.11 Applications

the resolution $\Delta r = 0.02$. The initial conditions in the code are

```
resolution_label = 1
NE = 3
N = 1000
t0 = 0.0d0
tmax = 20.0d0
m0 = 0.0d0
rho0 = 0.01d0
phi0 = -5.0d0
K = 10.0d0
gamma = 1.666666666
N = 2**(resolution_label-1) * N ! Number of cells for the
    dicretized domain
pi = acos(-1.0d0)
GG = 1.0d0 ! We use the gravitational constant set to one
```

For the right-hand sides, there are two subtleties. First, notice that ρ appears in the equations for m and p. This density can be calculated from the equation of state $\rho = (p/K)^{1/\gamma}$ before calculating the right-hand sides (2.52) and (2.54). Second, since the density is positive, it can only be calculated when $p > 0$. During the numerical integration, it might happen that the pressure could become negative because it is a number resulting from a numerical calculation. In that situation we have to prevent the density from becoming negative. Then we need to lock its value with the condition that $\rho = \max(0, (p/K)^{1/\gamma})$. In fact, the radius of the star r_{star} can be defined as the radius at which the density is zero.

Considering these details, the code for the right-hand sides of this problem reads

```
subroutine calcrhs(my_t,my_u)

use numbers
implicit none
real(kind=8), intent(in) :: my_t
real(kind=8), dimension(NE), intent(in) :: my_u
real(kind=8) rho_here

if (my_u(2).gt.0.0d0) then
  rho_here = ( my_u(2) / K )**(1.0d0/gamma)
else
  rho_here = 0.0d0
end if

rhs(1) = 4.0d0 * pi * my_t**2 * rho_here
if (my_t.le.dt) then
  rhs(2) = - 4.0d0 * pi * GG * rho_here**2 * my_t / 3.0d0
  rhs(3) = 4.0d0 * pi * GG * rho_here * my_t / 3.0d0
else
```

```
        rhs(2) = - GG * my_u(1) * rho_here / my_t**2
        rhs(3) = GG * my_u(1) / my_t**2
     end if

end subroutine calcrhs
```

where we keep our notation my_t for the independent variable, in this case the coordinate r. Notice that in order to avoid overflows, we first ask whether the pressure is positive or not and make the choice to set the density to the value from the EoS or to zero. Also notice that we use (2.54) for pressure and gravitational potential when $r \leq \Delta r$ and (2.52) otherwise.

It is now straightforward to construct equilibrium configurations of TOV stars. Results for the case with $\rho_c = 0.01$, $K = 10$, and $\gamma = 5/3$, using $\phi_c = -5$ are shown in Fig. 2.28.

The central density starts with the central value ρ_c and decays with radius, whereas the pressure follows a similar profile, since these two quantities are related through the EoS $p = K\rho^\gamma$. These functions decay until they acquire a small enough value. The mass function $m(r)$ is the integral of ρ and grows until the density is very small. From there on, the mass is nearly a constant function of r. The mass of the star can be defined as $M = m(r_{max})$.

For the integration of the gravitational potential, we use the central value $\phi(0) = -5$, and after rescaling with (2.53), the central value is nearly -0.7. Remember that this rescaling was set for the potential satisfying the condition $\phi(r_{max}) = -M/r_{max}$ at r_{max}.

Going further, it is usual to construct a whole family of equilibrium solutions for different values of ρ_c for each pair of K and γ. This family is presented in M vs ρ_c and M vs r_{star} diagrams. In Fig. 2.29 these two diagrams are shown for $K = 10, \gamma = 5/3$. Notice that the mass grows with the central density, whereas the radius of the star decreases. This behavior is contained in the third plot in this Figure, where the compactness of the star M/r_{star} vs ρ_c indicates that the star's compactness grows with the central density.

The diagrams can be implemented by adding a loop that changes the value of ρ_c within a range and a discrete set of values separated by $\delta\rho_c$.

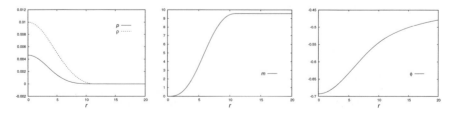

Fig. 2.28 Density ρ and pressure p, mass m and gravitational potential ϕ, for a TOV star with parameters $\rho_c = 0.01$, $K = 10$, and $\gamma = 5/3$

2.11 Applications

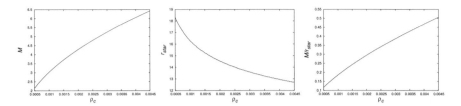

Fig. 2.29 Diagrams M vs ρ_c, r_{star} vs ρ_c and compactness for the family of solutions with $K = 10$, $\gamma = 5/3$, and for a range of values of the central density

The code used to construct these solutions is rk4_NewTOV.f90, available as Electronic Supplementary Material described in Appendix B.

> **Exercise.** Construct M vs ρ_c curves like the first one in Fig. 2.29 using values of γ between 1 which corresponds to a pressure-less fluid and 4/3, appropriate for relativistic gases.

2.11.7 Polytropic Relativistic TOV Star Models with Spherical Symmetry

A relativistic Tolman-Oppenheimer-Volkoff (TOV) relativistic star is a solution of Einstein's equations $G_\mu = 8\pi T_{\mu\nu}$, sourced by a perfect fluid defined by the stress energy tensor:

$$T^{\mu\nu} = \rho_0 h u^\mu u^\nu + p g^{\mu\nu}, \quad (2.55)$$

usually written also as

$$T^{\mu\nu} = (\rho + p) u^\mu u^\nu + p g^{\mu\nu}.$$

Each fluid element is described by the rest mass density ρ_0, total energy density $\rho = \rho_0(1 + e)$, specific enthalpy $h = 1 + e + p/\rho_0$, internal energy e, pressure p, 4-velocity u^μ, and $g^{\mu\nu}$ is the 4-metric of the space-time, being all these quantities local at the position of a fluid element [7].

Like in the Newtonian case of Sect. 2.11.6, the polytropic name arises from the isentropic **equation of state** (EoS) imposed on the fluid that relates pressure and rest mass density $p = K\rho_0^\gamma$ where γ is the adiabatic index and K a polytropic constant.

Because it is useful, we remind the reader that the line element of the space-time $ds^2 = g_{\mu\nu} dx^\mu dx^\nu$, for x^μ labeling coordinates of the space-time, for the

Schwarzschild black hole in Schwarzschild coordinates reads

$$ds^2 = -\left(1 - \frac{2M}{r}\right)dt^2 + \frac{dr^2}{\left(1 - \frac{2M}{r}\right)} + r^2(d\theta^2 + \sin^2\theta d\phi^2) \tag{2.56}$$

where (t, r, θ, ϕ) are the coordinates of the space-time. The assumptions in the construction of this solution are that (i) the space-time is assumed to be vacuum, that is $T_{\mu\nu} = 0$, (ii) the space-time is spherically symmetric, and (iii) the space-time is static. Birkoff's theorem ensures that Schwarzschild solution is the only solution under these conditions [8].

TOV stars will be constructed by relaxing condition (i) at the price of adding new conditions, namely, (a) that the solution will be regular at the origin and (b) that asymptotically the solution will approach Schwarzschild solution. Condition (a) is consistent with the idea that stars have no singularities, whereas condition (b) is consistent with the fact that far away from the star's surface the solution will correspond to vacuum and consequently should match the solution in vacuum.

The relaxation of condition (i) will be introduced as follows, in the line element (2.56), where M is a constant of integration (see, e.g., [8]). We relax the condition by allowing M to be a function of r, $M \to m(r)$. In this way the line element becomes

$$ds^2 = -\alpha(r)^2 dt^2 + \frac{dr^2}{1 - \frac{2m(r)}{r}} + r^2 d\theta^2 + r^2 \sin^2\theta d\phi^2, \tag{2.57}$$

where $g_{tt} = \alpha(r)^2$ is a positive function associated to the gravitational potential that not necessarily will equal $1 - \frac{2m(r)}{r}$ like in the vacuum case (2.56).

Assuming spherical symmetry means that all the state variables of the fluid depend only on r. Notice that none of the metric functions depends on time t, which is consistent with stationarity, whereas the fact that ds^2 is invariant under the transformation $t \to -t$ ensures staticity of the space-time.

Now we are in position of writing Einstein's equations $G^{\mu\nu} = 8\pi T^{\mu\nu}$. The resulting equations can be cast in a form very similar to that of equations for Newtonian TOV models of the previous Section. The $\mu = \nu = 0$ and $\mu = \nu = r$ components of Einstein's equations, and the $\mu = r$ component of the divergence free condition on the stress energy tensor $T^{\mu\nu}{}_{;\nu} = 0$ become the following set of three independent equations (see, e.g., [8])

$$\frac{dm}{dr} = 4\pi r^2 \rho, \tag{2.58}$$

$$\frac{dp}{dr} = -\frac{m}{r^2}(\rho + p)\left(1 + \frac{4\pi r^3 p}{m}\right)\left(1 - \frac{2m}{r}\right)^{-1} = -(\rho + p)\frac{m + 4\pi r^3 p}{r(r - 2m)}, \tag{2.59}$$

2.11 Applications

$$\frac{1}{\alpha}\frac{d\alpha}{dr} = -\frac{1}{\rho+p}\frac{dp}{dr} = \frac{m+4\pi r^3 p}{r(r-2m)}. \tag{2.60}$$

This is a system of ordinary equations whose solutions will represent relativistic stars. Like in the Newtonian case we integrate the equations as an IVP in the domain $r \in [0, r_{max}]$, outwards from $r = 0$ up to $r = r_{max}$, on the numerical discrete domain $D_d = \{r_i = i\Delta r\}$ for $i = 0, \ldots, N$, and resolution $\Delta r = r_{max}/N$, using the fourth-order Runge-Kutta method. Assuming we provide a central value of the rest mass density ρ_{0c}, the conditions at the origin are:

1. For the mass $m(0) = 0$, since the mass at the origin is zero. The consequence is that $g_{rr}(0) = 1$, which means the space is flat there.
2. For the pressure $p(0) = K\rho_{0c}^\gamma$, where ρ_{0c} is the central value of the rest mass density of the star.
3. Finally, $\alpha(0) = \alpha_0$ is arbitrary. Its value at the origin is irrelevant because according to (2.60), the solution can be rescaled multiplying by a constant.

Given γ and K, properties of the fluid, the central value of rest mass density ρ_{0c} is the only free parameter of the solution. This means that each TOV star will be characterized by ρ_{0c} and one can construct families of solutions in terms of this parameter for the pair (γ, K).

Notice that the right-hand side in the equation for m is the total energy density ρ, which *must be distinguished from the rest mass density* ρ_0. The total energy density is the rest mass density plus the internal energy $\rho = \rho_0(1 + e)$. The calculation of internal energy needs an additional condition, for example, the ideal gas equation of state $p = \rho_0 e(\gamma - 1)$, from which the internal energy is $e = \frac{p}{\rho_0(\gamma-1)}$. We use this expression for the internal energy to write the total energy density ρ in terms of the variables involved in the system:

$$\rho = \rho_0 + \rho_0 e = \left(\frac{p}{K}\right)^{1/\gamma} + \frac{p}{\gamma - 1}, \tag{2.61}$$

which is the expression used in the right-hand side of Eqs. (2.58)–(2.60).

Notice also that Eqs. (2.59) and (2.60) seem to be singular at $r = 0$. Nevertheless a Taylor series expansion reveals that the factor $(m+4\pi r^3 p)/(r(r-2m))$ is regular. For this we write the Taylor series expansion of this factor around $r = 0$:

$$\frac{m + 4\pi r^3 p}{r^2 - 2mr} \simeq \frac{m(0) + m'(0)r + \frac{1}{2}m''(0)r^2 + \frac{1}{6}m'''(0)r^3 + O(r^4) + 4\pi r^3 p}{r^2 - 2r(m(0) + m'(0)r + \frac{1}{2}m''(0)r^2 + \frac{1}{6}m'''(0)r^3 + O(r^4))}$$

$$= \frac{4\pi\rho r/3 + 4\pi r p}{1 - 8\pi\rho r^2/3}, \tag{2.62}$$

where the derivatives of m use Eq. (2.58), that is, $dm/dr|_{r=0} = 4\pi r^2 \rho|_{r=0} = 0$, $d^2m/dr^2|_{r=0} = 4\pi(2r\rho + r^2\rho)|_{r=0} = 0$, and $d^3m/dr^3|_{r=0} = 4\pi(2\rho + 2r\rho + r^2\rho)|_{r=0} = 8\pi\rho$. With these results, Eqs. (2.59) for pressure and (2.60) for α near the origin are

$$\frac{dp}{dr} = -(\rho + p)\frac{4\pi\rho r/3 + 4\pi r p}{1 - 8\pi\rho r^2/3},$$

$$\frac{1}{\alpha}\frac{d\alpha}{dr} = \frac{4\pi\rho r/3 + 4\pi r p}{1 - 8\pi\rho r^2/3}, \qquad (2.63)$$

regular at the $r = 0$. These equations are used to solve the system from the origin at $r_0 = r(0)$, until $r_1 = \Delta r$, which is the second point of the discrete domain D_d.

Like in the Newtonian case, a subtlety is that when integrating the pressure, it may become negative, and therefore the rest mass density would be negative as well, which would be unphysical. In order to avoid this problem, one imposes a condition for the density to be non-negative, known as the implementation of an atmosphere. In order to guarantee the positiveness of rest mass density, we set

$$\rho_0 = \max\left(\left(\frac{p}{K}\right)^{1/\gamma}, 0\right). \qquad (2.64)$$

With this, we now have **all** the ingredients to integrate the equations and construct TOV star configurations.

Diagnostics. With the numerical solution, it is possible to calculate physical properties of the resulting configurations. The **radius** of the star is calculated as the smallest value of $r = R$ for which $\rho_0 = 0$. The **total mass** of the star can be defined as $M = m(R)$ and the **rest mass** of the star is given by the integral of the rest mass density in the numerical domain $M_0 = 4\pi \int_0^R \rho_0 r^2 \sqrt{g_{rr}} dr$. The difference between the total mass M, and the rest mass M_0 of the star is the **binding energy** of the star $E_B = M - M_0$.

Finally an additional scaling of α. In order for the space-time line element (2.57) to mimic the property $g_{tt} = 1/g_{rr}$ of Schwarzschild metric (2.56), one can scale the solution α such that $\alpha(r_{max}) = 1/\sqrt{1 - 2m(r_{max})/r_{max}}$ as follows:

$$\alpha(r) = \frac{\alpha(r)}{\alpha(r_{max})}\sqrt{1 - \frac{2m(r_{max})}{r_{max}}}.$$

Let us finally construct TOV stars for two invented cases with $\gamma = 2$ and $K = 1$ and two central densities $\rho_{0c} = 0.3, 0.6$, in the domain with $r_{max} = 5$ using $N = 1000$ or equivalently resolution $\Delta r = 0.005$. The results are shown in Fig. 2.30.

Notice that the metric functions $g_{tt} = \alpha^2$ and $g_{rr} = \frac{1}{1 - \frac{2m(r)}{r}}$ tend to one as r grows, approaching, in these coordinates to the Minkowski flat space-time metric. Notice also that the configuration with higher central density ρ_{0c} has space-time metric components g_{tt} and g_{rr} that depart more from 1.

2.11 Applications

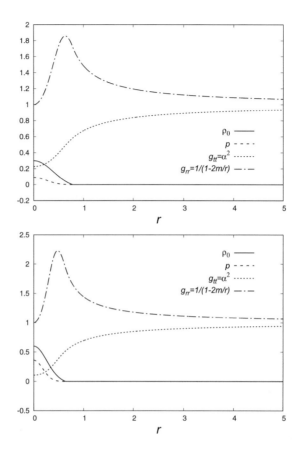

Fig. 2.30 Solution of the TOV equations for $\gamma = 2$, $\alpha_c = 0.5$, $K = 1$. On the left the case $\rho_{0c} = 0.3$, with $M = 0.1636$, $M_0 = 0.1798$, and $R = 0.78$. On the right the case $\rho_{0c} = 0.6$, with $M = 0.1538$, $M_0 = 0.1663$, and $R = 0.64$

Another interesting property is that in the two examples $M_0 > M$, which implies that the binding energy $E_B = M - M_0 < 0$, and consequently the system is gravitationally bound.

Beyond these two examples, it is possible to construct families of solutions in terms of the central density ρ_{0c} for given pairs of K and γ. This is usually done by solving the TOV equations for various values of ρ_{0c} and compiling the results into a M vs ρ_{0c} diagram.

The results for two families of solutions are shown in Fig. 2.31. The maximum of the curves, marked with squares, indicate the stability threshold, those configurations to the left from the maximum define the stable branch of configurations, whereas those to the right from the maximum define the unstable branch of the family of solutions.

Other important point indicated with a star is that where the binding energy $E_B = M - M_0$ changes sign from negative to positive. This means that solutions to the right from the star have a positive binding energy, and theoretically they should explode when perturbed. Those configurations between the square and the star are

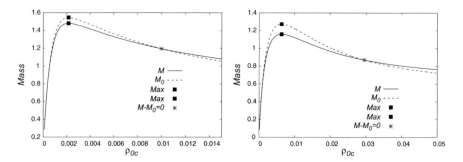

Fig. 2.31 Diagrams M vs ρ_{0c} for the families of solutions with $K = 10$, $\gamma = 5/3$, and $K = 50$, $\gamma = 2$

gravitationally bound and unstable, and their fate is theoretically the collapse into a black hole as seen in [9].

The results discussed here can be reproduced with the code `rk4_RelTOV.f90` available as Electronic Supplementary Material described in Appendix B.

> **Exercise.** Likewise in the Newtonian case, construct M vs ρ_{0c} plots for different values of γ. This time monitor the change of the location of the maximum mass and the star point that indicates where the binding energy changes sign.

2.11.8 The String: An Eigenvalue Problem

An eigenvalue problem is defined by an ODE of the form $Lu = lu$ that satisfies boundary conditions in the domain $x \in [x_0, x_f]$, with L a linear differential operator and l an eigenvalue. A traditional example of this type of problem is Helmholtz equation:

$$\frac{d^2u}{dx^2} + k^2 u = 0, \tag{2.65}$$

defined on the numerical domain $x \in [0, 1]$ with **Dirichlet** boundary conditions $u(0) = u(1) = 0$. This problem has a general solution of the form $u = A\cos(kx) + B\sin(kx)$, and the boundary conditions fix the constants to

$$u(0) = A = 0,$$

$$u(1) = B\sin(kx)|_{x=1} = B\sin(k) = 0, \tag{2.66}$$

2.11 Applications

which restricts the values of k to $k = n\pi$ with n an integer. Notice that this problem can actually be cast as an eigenvalue problem with $L = \frac{d^2}{dx^2}$ and $l = -k^2$:

$$\frac{d^2 u}{dx^2} = -k^2 u, \tag{2.67}$$

for which we know that the solutions are the eigenfunctions $u_k = B\sin(kx)$, each of them associated with the eigenvalue $-k^2 = -n^2\pi^2$.

This is a problem that will help illustrate the **shooting method**. So far we have illustrated the integration of IVPs associated ODEs, but eigenvalue problems are boundary value problems in which the unique solution of an ODE is determined with Dirichlet conditions for the solution function at two points, instead of conditions for the solution and its derivative at the starting point of integration.

The strategy to solve this problem is based on the solution of the IVP for Eq. (2.65) starting at $x = 0$, using the initial condition $u(0) = 0$. However, in order for the solution to match the boundary condition $u(1) = 0$ at $x = 1$, one searches for values of k that fulfill such condition.

Observe that we have solved IVPs associated to second-order equations before; however, we have always required initial conditions for the unknown function and its derivative. In this case though, the value of $u'(0)$ is not playing any role. The name "shooting" suggests that one tries to hit a target with a parabolic shooting by choosing the appropriate derivative or inclination of a throw.

This is not the case. What is being searched for is the value of k that allows the solution to fulfill the boundary conditions at $x = 0$ and $x = 1$. The derivative will end up playing no essential role and will control only the amplitude of the solution. The reason is that (2.67) is linear, and therefore given the solution u, the function au is also a solution for an arbitrary amplitude a.

In order to solve the problem, we cast the equation as a system of two first-order ODEs by defining $v = du/dx$. In this case Eq. (2.67) can be written as

$$\boxed{\begin{aligned}\frac{du}{dx} &= v \\ \frac{dv}{dx} &= -k^2 u\end{aligned}} \tag{2.68}$$

for $u_0 = 0$, v_0 free and k the important number to search for.

In order to show how this process works, let us solve system (2.68) with $v_0 = 0.1$ and two values of $k = 2, 4$ on purpose, for illustration, because we already know for this problem that k is multiple integer of π, and we choose one value smaller and one bigger than π.

We construct the solutions for the domain $D = [0, 1]$ on the discrete domain $D_d = \{x_i = i\Delta x\}$ for $i = 0, \ldots, N$ with resolution $\Delta x = 1/N$. For the examples below, we use $N = 100$ or resolution $\Delta x = 0.01$ with the RK4 method to solve the IVP, and the results are shown in Fig. 2.32. In these two cases, one can see that

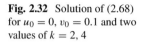

Fig. 2.32 Solution of (2.68) for $u_0 = 0$, $v_0 = 0.1$ and two values of $k = 2, 4$

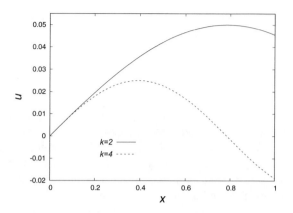

the slope of the solution is the same for the two values of k at $x = 0$, but none of the solutions hits the target at $x = 1$ such that $u(1) = 0$. For $k = 2$, $u(1) > 0$, whereas for $k = 4$, $u(1) < 0$, and this result is useful because if the transition between positive and negative is continuous, there would be a value of k that fulfills the boundary condition within a small tolerance $|u(1)| < \epsilon$, because it should be expected that the boundary condition at $x = 1$ can only be fulfilled with a certain finite accuracy. The tolerance we use for this exercise will be set to $\epsilon = 10^{-8}$ in what follows, unless specified otherwise.

The problem now is the search of the appropriate value of k, which can be found using a bisection search. The value of k, using bisection works as follows (see, e.g., [1]):

1. Solve the system for an initial $k = k_0$. Let us assume $k_0 > 0$ is small so that the solution at $x = 1$ is positive.
2. Evaluate the solution at the right boundary $u(1)$.
3. If $|u(1)| < \epsilon$, k is a good eigenvalue and the process can finish.
4. If $u(1) > \epsilon$, increase $k = k + dk$
5. Else, if $u(1) < -\epsilon$, there was a change of sign at the right boundary.

    - Return to the previous value of k, that is $k = k - dk$
    - Redefine the step dk, for example $dk = dk/2$
    - Restart from step 4

6. If $|u(1)| < \epsilon$, k is a good eigenvalue and the process can finish.

An example of the solution to this problem using this strategy is shown in Fig. 2.33 assuming $k_0 = 0.1$ and initial $dk = 0.1$, using $v_0 = 0.1$ and $v_0 = 0.5$. The number of iterations was less than 80, and $k = 3.1415924540399374$ is the first value of k that fulfills $|u(1)| < \epsilon$. This plot illustrates the aforementioned role of the derivative at $x = 0$, namely, it regulates the amplitude of the solution which is bigger for $v_0 = 0.5$ than for $v_0 = 0.1$.

The effect of reducing the tolerance, for example, if $\epsilon = 10^{-12}$, has an improvement on the value of $k = 3.1415926790940412$ and a slight increase in

2.11 Applications

Fig. 2.33 Solution for $v_0 = 0.1$ and $v_0 = 0.5$. The effect of v_0 on the solution is reflected in the amplitude. In both cases the boundary condition is fulfilled with $\epsilon = 10^{-8}$ after less than 80 iterations

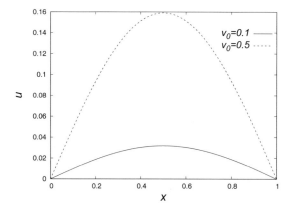

the number of iterations. For a fair comparison in the accuracy of the calculation of π, we use the common double precision value of π in fortran 90 that we calculate as $\pi = \arccos(-1.0d0) = 3.1415926\mathbf{535897931}$. The digits in bold are those in which the value of the eigenvalue and the value of π differ. Notice that the improvement is in only one digit when reducing the tolerance four orders of magnitude.

Let us go further. So far we have used $v_0 > 0$, which launches initially positive solutions of u. If $v_0 < 0$ it is important to take care of the bisection process because the change of sign at $x = 1$ will happen from negative to positive.

Nodes. Now, as indicated by the exact solution, the eigenvalue of the problem is $k = n\pi$, and so far we have only solved a problem for $n = 1$, corresponding to zero nodes of u. For the construction of u with nodes two elements are needed:

(i) Count the number of zeros of the solution for each k used for the integration.
(ii) Keep increasing the value of k until the desired number of nodes is achieved and then execute steps 4 to 6 in the algorithm above.

Locating the zeroes of the solution u for each value of k can be done by counting the changes of sign of u within the RK4 integration process, by asking whether $u_i u_{i-1} \leq 0$ at each point of the domain, taking care at the borders $x_0 = 0$ and $x_N = 1$, where we expect the solution to be near to zero. This is the reason why in the code below, a buffer of two points from the borders is used to make sure the zeroes are in the interior of the domain.

The code that solves this problem `rk4_Helmholtz.f90`, available as Electronic Supplementary Material described in Appendix B, has the loop that controls the values of k and counts the number of nodes. The Part B of the code is the following:

```
! Set initial conditions
open(1,file='Helmholtz.dat')
u(1,0)  = x0     ! u0
u(2,0)  = y0     ! v0
```

```
k = k0
! ---> PART B <---
DO l=1,k_it_max   ! k_it_max is a maximum number of throws
  zeroes = 0
  do i=1,N
    do j=1,4
      if (j.eq.1) then
        call calcrhs( t(i-1) ,u(:,i-1))
        k1 = rhs
      else if (j.eq.2) then
        call calcrhs( t(i-1) + 0.5d0 * dt, u(:,i-1) + 0.5d0 * k1(:)
* dt)
        k2 = rhs
      else if (j.eq.3) then
        call calcrhs( t(i-1) + 0.5d0 * dt, u(:,i-1) + 0.5d0 * k2 *
dt)
        k3 = rhs
      else
        call calcrhs( t(i-1) + dt, u(:,i-1) + k3 * dt)
        k4 = rhs
        u(:,i) = u(:,i-1) + (1.0d0/6.0d0)*( k1 + 2.0d0 * k2 +
2.0d0 * k3 + k4 ) * dt
      end if
    end do
    ! The following counts zeros
    if ( (u(1,i)*u(1,i-1).le.0.0d0).and.(i.ge.2).and.(i.le.N-2) )
then
      zeroes = zeroes + 1
    end if
  end do
  ! ---> Ends Part B <---

  ! Changing k
  x1_p = u(1,N)
  print *,l-1,k,u(1,N)

  if ( (zeroes.eq.num_of_nodes) .and. ((-1)**(num_of_nodes)
* x1_p.lt.0.0) ) then
    k = k - dk
    dk = 0.5d0 * dk
  else
    k = k + dk
  end if

  do i=0,N,2**resolution_label
    write(1,*) t(i),u(1,i),u(2,i)
  end do
  write(1,*)
  write(1,*)
  ! Tolerance
  if (abs(u(1,N)).lt.epsilon) then
    exit
  end if
```

2.11 Applications

```
END DO

close(1)
print *, zeroes * acos(-1.0d0). ! Only for comparison with the
exact solution n*pi
end program
```

Notice the subtle control on the change of sign at $x = 1$. When the solution is passing from positive to negative and the number of nodes is even, the result is correct. However, when the number of nodes is odd, the solution should be passing from negative to positive. Of course all this is valid when $v_0 > 0$ and would be the opposite otherwise.

Finally, the right-hand sides of the system (2.68) are coded as follows:

```
subroutine calcrhs(my_t,my_u)

use numbers
implicit none
real(kind=8), intent(in)  :: my_t
real(kind=8), dimension(NE), intent(in) :: my_u

rhs(1) = my_u(2)
rhs(2) = -k**2 * my_u(1)

end subroutine calcrhs
```

Solutions corresponding to various iterations during the solution of the problem for cases with 1, 2, 3, and 4 nodes are shown in Fig. 2.34. The values $v_0 = 0.1$, initial $k_0 = 0.1$, and initial $dk = 0.1$ were used in all these solutions, in order to show that during the first iterations, when the value of k is growing, the solutions are the same from 20 to 40 iterations, and they only start differing when the solution acquires the appropriate number of nodes. The final values of k where $k = 6.2831864339588037 \simeq 2\pi$, $k = 9.4247834656355280 \simeq 3\pi$, $k = 12.566397281980471 \simeq 4\pi$, and $k = 15.708041615903994 \simeq 5\pi$, correspond to solutions with 1, 2, 3, and 4 nodes, respectively.

> **Exercise.** Later on in the book we show some exact solutions of the stationary Schrödinger equation. Using the shooting method, solve Eq. (3.64) for the particle in a one-dimensional box, and show that the allowed energies are discrete according to Eq. (3.65).

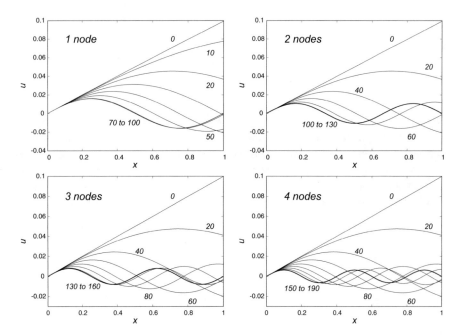

Fig. 2.34 Solutions for 1, 2, 3, and 4 nodes that fulfill the boundary condition $|u(1)| < \epsilon = 10^{-8}$ for different values of k. Numeric labels near certain curves indicate the iteration number in the bisection loop

2.11.9 Test Particle Orbiting a Central Object

In this case we consider the motion of a particle of mass m subject to a central Newtonian gravitational field. Let us first review the elements of the central field problem.

The Lagrangian of a test particle of mass m subject to a central field is

$$\mathcal{L} = \frac{1}{2}m\dot{\mathbf{x}} \cdot \dot{\mathbf{x}} - V(|\mathbf{x}|), \tag{2.69}$$

where $\dot{\mathbf{x}}$ is the velocity of the particle and the potential V depends only on the distance to the origin of coordinates, where the source of the field is assumed to lie. If this Lagrangian is considered to be written in spherical coordinates (r, θ, ϕ), then $V = V(r)$. Since $\mathcal{L}$ does not depend on ϕ, the angular momentum $\ell = \partial \mathcal{L}/\partial \dot{\phi} = mr^2\dot{\phi}$ is a constant of motion, and because it is conserved, the trajectory of the particle is a curve on a plane and also leads to the second Kepler's law. That the motion happens on a plane is the property we exploit below.

Even though the description of motion is usually written in polar coordinates, we use Cartesian coordinates to program a code. We assume the gravitational field is due to a point-like source of mass M located at the origin of coordinates. According

2.11 Applications

to second and gravitation Newton's laws, the position **x** of a test particle with mass m evolves according to the equation:

$$m\frac{d^2\mathbf{x}}{dt^2} = -G\frac{Mm}{|\mathbf{x}|^3}\mathbf{x} \quad \Rightarrow \quad \frac{d^2\mathbf{x}}{dt^2} = -G\frac{M}{|\mathbf{x}|^3}\mathbf{x}. \tag{2.70}$$

Here we assume the test particle has mass m with the condition $m \ll M$, which implies that the position of the particle of mass M sourcing the dominant gravitational field can be considered fixed in space at the origin of coordinates. Given initial conditions for the position $\mathbf{x}(0) = \mathbf{x}_0$ and velocity $\dot{\mathbf{x}}(0) = \mathbf{v}_0$ of the test particle, one has a well-posed IVP consisting of three second-order equations that can be decomposed into six first-order equations by defining $\mathbf{x} = (x, y, z)$ and $\dot{\mathbf{x}} = (v_x, v_y, v_z)$:

$$\boxed{\begin{aligned}\frac{dx}{dt} &= v_x \\ \frac{dy}{dt} &= v_y \\ \frac{dz}{dt} &= v_z \\ \frac{dv_x}{dt} &= -\frac{GMx}{(x^2+y^2+z^2)^{3/2}} \\ \frac{dv_y}{dt} &= -\frac{GMy}{(x^2+y^2+z^2)^{3/2}} \\ \frac{dv_z}{dt} &= -\frac{GMz}{(x^2+y^2+z^2)^{3/2}}\end{aligned}} \tag{2.71}$$

with initial conditions $x(0) = x_0$, $y(0) = y_0$, $z(0) = z_0$, $v_x(0) = v_{x0}$, $v_y(0) = v_{y0}$, $v_z(0) = v_{z0}$.

Let us study some cases of interest and review some theoretical results that will help to verify the numerical solution and set initial conditions. In polar coordinates, the trajectory of the test particle is

$$\mathbf{x} = r\cos\phi\,\hat{x} + r\sin\phi\,\hat{y}, \tag{2.72}$$

and can be parameterized with the formula

$$\frac{p}{r} = 1 + e\cos\phi, \tag{2.73}$$

where p is the parameter and e the eccentricity of the trajectory, which are given by

$$p = \frac{\ell^2}{GMm^2}, \quad e = \sqrt{1 + \frac{2E\ell^2}{G^2M^2m^3}}, \quad (2.74)$$

and $E = \frac{1}{2}m\dot{r}^2 + \frac{\ell^2}{2mr^2} - G\frac{Mm}{r}$ is the total energy of the test particle. For elliptical trajectories $0 \leq e \leq 1$ and in such case the parameter is given by $p = a(1 - e^2)$ where a is the semimajor axis of the ellipse. In this case

$$\frac{p}{r} = \frac{a(1-e^2)}{r} = 1 + e\cos\phi \quad \Rightarrow \quad r = \frac{a(1-e^2)}{1+e\cos\phi} \quad (2.75)$$

whereas ϕ evolves according to the constancy of the angular momentum ℓ:

$$\dot\phi = \frac{\ell}{mr^2} = \frac{\sqrt{GMm^2 p}}{mr^2} = \frac{\sqrt{GMa(1-e^2)}}{r^2} = \sqrt{\frac{GM}{a^3(1-e^2)^3}}(1+e\cos\phi)^2. \quad (2.76)$$

Initial conditions $\mathbf{x}_0$ for the position of a test particle in an elliptical trajectory can be defined in terms of a and e. For example, by initializing the position of the particle at the angle $\phi = 0$ and using Eq. (2.75) to determine r, one has that $r = \frac{a(1-e^2)}{1+e}$. On the other hand, the initial velocity $\mathbf{v}_0$ at $\phi = 0$ is tangent to the trajectory and has only y–component with magnitude $r\dot\phi$. These two conditions and the use of Eqs. (2.75) and (2.76) together give

$$\mathbf{x}_0 = \frac{a(1-e^2)}{1+e}\hat{x},$$

$$\mathbf{v}_0 = \frac{a(1-e^2)}{1+e}\sqrt{\frac{GM}{a^3(1-e^2)^3}}(1+e)^2\hat{y} = \sqrt{\frac{GM}{a(1-e^2)}}(1+e)\hat{y}. \quad (2.77)$$

Some particular examples we can use for illustration are the trajectories of some important objects orbiting around the Sun, whose orbital parameters needed to set initial conditions appear in Table 2.1. We use MKS units, consider $G = 6.6743 \times 10^{-11} \frac{\text{m}^3}{\text{kg s}^2}$ and $M = M_\odot = 1.98847 \times 10^{30}$ kg.

Table 2.1 Orbital parameters of the trajectories of some important objects around the Sun

Object	a(m)	e	Θ	Ω
Earth	$1.49598023 \times 10^{11}$	0.0167086	0	-11.26064^0
Jupiter	7.78479×10^{11}	0.0489	1.303^0	100.464^0
Pluto	5.906423×10^{12}	0.2488	17.16^0	110.299^0
Comet Halley	2.66792×10^{12}	0.96714	162.26^0	58.42^0

2.11 Applications

We construct the numerical solution using the RK4 method like in the previous examples on the domain $t \in [0, t_{max}]$ with units in seconds, discretized as $D_d = \{t_i = i\Delta t\}$, with resolution $\Delta t = 100\,\text{s}$ and t_{max} appropriate for each object's orbit.

Integration of Eqs. (2.71) using initial conditions (2.77) leads to the results in Fig. 2.35. We show the coordinates of the objects as function of time on the equatorial plane, as well as the trajectory of each one on its own orbital plane. For comparison among the orbits, we show the four trajectories together in Fig. 2.36.

Now considering that the orbital plane of each of these objects is not the same adds extra complexity to the problem, specifically to establish initial conditions. Assuming the reference plane is Earth's orbit plane, the inclination of the orbital planes of Jupiter, Pluto, and Halley's comet with respect to Earth's plane is Θ and appears in the fourth column of the Table.

Another parameter is the orientation of the plane of each object, including Earth, with respect to a reference direction, because not all ellipses have the same orientation as assumed in the construction of Fig. 2.35. Assuming that the reference direction is the First Point of Aries, Ω is the orientation angle of the orbit and is specified for each object in the Table.

With these two new parameters, the initial conditions need to be modified. First we still assume that the initial angle within the orbit is $\phi = 0$. Initial position and velocity of the test particle (2.77) need two successive rotations, one rotation R_y, Θ around the y-axis by the angle Θ and then a rotation $R_{z,\Omega}$ around the z-axis by the angle Ω as follows:

$$\bar{\mathbf{x}}_0 = R_{z,\Omega} R_{y,\Theta} \mathbf{x}_0 = R_{z,\Omega} \begin{bmatrix} \cos\Theta & 0 & -\sin\Theta \\ 0 & 1 & 0 \\ \sin\Theta & 0 & \cos\Theta \end{bmatrix} \begin{bmatrix} \frac{a(1-e^2)}{1+e} \\ 0 \\ 0 \end{bmatrix}$$

$$= \begin{bmatrix} \cos\Omega & -\sin\Omega & 0 \\ \sin\Omega & \cos\Omega & 0 \\ 0 & 0 & 1 \end{bmatrix} \begin{bmatrix} \frac{a(1-e^2)}{1+e}\cos\Theta \\ 0 \\ \frac{a(1-e^2)}{1+e}\sin\Theta \end{bmatrix} = \begin{bmatrix} \cos\Omega \frac{a(1-e^2)}{1+e}\cos\Theta \\ \sin\Omega \frac{a(1-e^2)}{1+e}\cos\Theta \\ \frac{a(1-e^2)}{1+e}\sin\Theta \end{bmatrix},$$

$$\bar{\mathbf{v}}_0 = R_{z,\Omega} R_{y,\Theta} \mathbf{v}_0 = R_{z,\Omega} \begin{bmatrix} \cos\Theta & 0 & -\sin\Theta \\ 0 & 1 & 0 \\ \sin\Theta & 0 & \cos\Theta \end{bmatrix} \begin{bmatrix} 0 \\ \sqrt{\frac{GM}{a(1-e^2)}}(1+e) \\ 0 \end{bmatrix}$$

$$= \begin{bmatrix} \cos\Omega & -\sin\Omega & 0 \\ \sin\Omega & \cos\Omega & 0 \\ 0 & 0 & 1 \end{bmatrix} \begin{bmatrix} 0 \\ \sqrt{\frac{GM}{a(1-e^2)}}(1+e) \\ 0 \end{bmatrix} = \begin{bmatrix} -\sin\Omega \sqrt{\frac{GM}{a(1-e^2)}}(1+e) \\ \cos\Omega \sqrt{\frac{GM}{a(1-e^2)}}(1+e) \\ 0 \end{bmatrix},$$

(2.78)

where the bar indicates the new initial conditions of the test particle used to integrate Eqs. (2.71). The resulting trajectories in three dimensions appear in Fig. 2.37.

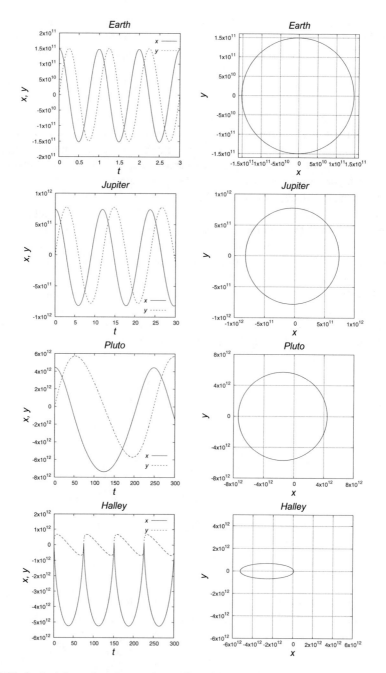

Fig. 2.35 On the left we show the x and y position of the three planets and the comet as function of time and on the right the trajectory on the $z = 0$ plane. Time is in years and position in meters. The Sun is fixed at the origin of the xy-plane. The orbit is constructed assuming that each object travels on its own xy-plane, whereas the initial conditions according to (2.77) for each object are set separately, and then what is seen are the coordinates and orbit of each object in its own orbital plane and all of them with the same orientation

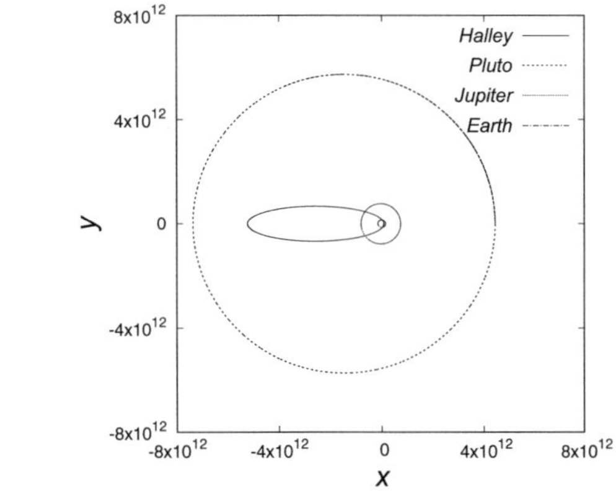

2.11 Applications

Fig. 2.36 Orbits of the four objects together in order to compare among spatial scales. Units are in meters

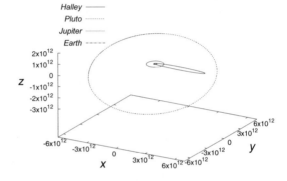

Fig. 2.37 The orbits of the four objects together in a traditional Solar System perspective. Units are in meters

For programming the solution of this problem, we use the alternative approach described for the Newton Cooling Law, because the time domain can be so large that it might be unpractical to define arrays with an arbitrarily big number of entries. Instead we integrate the equations using the RK4 method by defining only two time levels of the variables u for u_i^{n+1} and u_p for u_i^n. Before t was defined as an array to store the numerical domain, it is no longer an array but a double precision number that is being advanced in the time domain $t \rightarrow t + \Delta t$. The code rk4_SunPlanets.f90 listed in Appendix B, illustrates this approach.

Exercise. Add the orbits of the planets not considered here and complete the trajectories of all the planets in the Solar System. The values of Θ and Ω can be found in online data bases.

Exercise. Assuming that an electron and a proton obey Coulomb law, integrate the trajectory of the electron around the Proton of the Hydrogen atom that travels on a circular orbit with radius equal to the Bohr radius.

2.11.10 Two Body Problem

The system consists of two particles with masses m_1, m_2 and positions $\mathbf{x}_1(t)$ and $\mathbf{x}_2(t)$. The interaction between the two particles depends only on the distance between them $|\mathbf{x}_1 - \mathbf{x}_2|$ and on the law of interaction. A scheme with these conventions appears in Fig. 2.38. Assuming these particles interact gravitationally according to the universal gravitational law, second's Newton law establishes that the equations of motion of the particles are

$$m_1 \frac{d\mathbf{x}_1}{dt^2} = G \frac{m_1 m_2}{|\mathbf{x}_1 - \mathbf{x}_2|^3} (\mathbf{x}_2 - \mathbf{x}_1),$$

$$m_2 \frac{d\mathbf{x}_2}{dt^2} = G \frac{m_1 m_2}{|\mathbf{x}_1 - \mathbf{x}_2|^3} (\mathbf{x}_1 - \mathbf{x}_2), \qquad (2.79)$$

where the direction of the acceleration is included in the sign of $(\mathbf{x}_1 - \mathbf{x}_2)$. This system defines an IVP in the time domain $t \in [0, t_f]$, associated to a six second-order ODEs that can be translated into a system of twelve first-order equations. Using the definitions $\mathbf{x}_1 = (x_1, y_1, z_1)$, $\mathbf{x}_2 = (x_2, y_2, z_2)$, $\mathbf{v}_1 = \dot{\mathbf{x}}_1 = (v_{1x}, v_{1y}, v_{1z})$

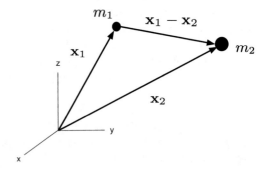

Fig. 2.38 Scheme of the two body problem

2.11 Applications

and $\mathbf{v}_2 = \dot{\mathbf{x}}_2 = (v_{2x}, v_{2y}, v_{2z})$, the system of Eqs. (2.79) is explicitly

$$\begin{aligned}
\frac{dx_1}{dt} &= v_{x1}, & \frac{dv_{x1}}{dt} &= Gm_2 \frac{x_2 - x_1}{r^3}, \\
\frac{dy_1}{dt} &= v_{y1}, & \frac{dv_{y1}}{dt} &= Gm_2 \frac{y_2 - y_1}{r^3}, \\
\frac{dz_1}{dt} &= v_{z1}, & \frac{dv_{z1}}{dt} &= Gm_2 \frac{z_2 - z_1}{r^3}, \\
\frac{dx_2}{dt} &= v_{x2}, & \frac{dv_{x2}}{dt} &= Gm_1 \frac{x_1 - x_2}{r^3}, \\
\frac{dy_2}{dt} &= v_{y2}, & \frac{dv_{y2}}{dt} &= Gm_1 \frac{y_1 - y_2}{r^3}, \\
\frac{dz_2}{dt} &= v_{z2}, & \frac{dv_{z2}}{dt} &= Gm_1 \frac{z_1 - z_2}{r^3},
\end{aligned} \quad (2.80)$$

where $r = \sqrt{(x_1 - x_2)^2 + (y_1 - y_2)^2 + (z_1 - z_2)^2}$. Initial conditions are needed for all six coordinates $x_{10}, y_{10}, z_{10}, x_{20}, y_{20}, z_{20}$ and six velocities $v_{x10}, v_{y10}, v_{z10}, v_{x20}, v_{y20}, v_{z20}$.

Before we solve the problem numerically we review the theory of the problem.

The two body problem in Fig. 2.38 has total mass $M = m_1 + m_2$, reduced mass $\mu = \frac{m_1 m_2}{m_1 + m_2}$ and $\mathbf{x} := \mathbf{x}_1 - \mathbf{x}_2$ is the vector between the position of the two particles. Being $\mathbf{x}_{cm}$ the center of mass, in the reference system located at the center of mass, we have

$$\begin{aligned}
\mathbf{x}_{cm} &= m_1 \mathbf{x}_1 + m_2 \mathbf{x}_2 = \mathbf{0} & \Rightarrow \\
m_1(\mathbf{x} + \mathbf{x}_2) + m_2 \mathbf{x}_2 &= \mathbf{0} & \Rightarrow \\
(m_1 + m_2)\mathbf{x}_2 + m_1 \mathbf{x} &= \mathbf{0} & \Rightarrow \\
\mathbf{x}_2 &= -\frac{m_1}{M}\mathbf{x} & \Rightarrow \\
\mathbf{x}_1 = \mathbf{x} + \mathbf{x}_2 &= \mathbf{x} - \frac{m_1}{M}\mathbf{x} & \Rightarrow \\
\mathbf{x}_1 &= \frac{m_2}{M}\mathbf{x}, & (2.81)
\end{aligned}$$

that is, the position $\mathbf{x}_1$ and $\mathbf{x}_2$ of the two particles appears in terms of $\mathbf{x}$. Since gravity is a central field, the Lagrangian of the system becomes

$$\begin{aligned}
\mathcal{L} &= \frac{1}{2}m_1 \dot{\mathbf{x}}_1 \cdot \dot{\mathbf{x}}_1 + \frac{1}{2}m_2 \dot{\mathbf{x}}_2 \cdot \dot{\mathbf{x}}_2 - V(|\mathbf{x}|) \\
&= \frac{1}{2}\mu \dot{\mathbf{x}} \cdot \dot{\mathbf{x}} - V(|\mathbf{x}|).
\end{aligned} \quad (2.82)$$

This is the Lagrangian of a virtual particle of mass μ, whose position is $\mathbf{x}$, subject to a central potential V. If the space is described in cylindrical coordinates (r, ϕ, z) with basis vectors $\{\hat{r}, \hat{\phi}, \hat{z}\}$ then

$$\mathbf{x} = r\hat{r} + z\hat{z} \quad \Rightarrow$$
$$\dot{\mathbf{x}} = \dot{r}\hat{r} + r\dot{\phi}\hat{\phi} + \dot{z}\hat{z} \quad \Rightarrow$$
$$\dot{\mathbf{x}} \cdot \dot{\mathbf{x}} = \dot{r}^2 + r^2\dot{\phi}^2 + \dot{z}^2. \tag{2.83}$$

The expression (2.83) and the fact that $|\mathbf{x}| = \sqrt{r^2 + z^2}$ imply that the Lagrangian (2.82) is independent of ϕ, which in turn implies that the angular momentum along $\hat{z}$

$$p_\phi = \frac{\partial \mathcal{L}}{\partial \dot{\phi}} = \mu r^2 \dot{\phi} := \ell \tag{2.84}$$

is a constant of motion. Therefore, like in the Lagrangian of a test particle around a central field (2.69), the motion of this virtual particle of mass μ subject to the potential $V(|\mathbf{x}|)$ occurs on a plane that we choose to be the equatorial plane at $z = 0$ and consequently $\dot{z} = 0$. With this conventions, the energy of the system is written as

$$E = \frac{1}{2}\mu \dot{\mathbf{x}} \cdot \dot{\mathbf{x}} + V(r)$$
$$= \frac{1}{2}\mu \dot{r}^2 + \frac{\ell^2}{2\mu r^2} - G\frac{m_1 m_2}{r}. \tag{2.85}$$

The trajectory of this particle can be described with the position vector in the Cartesian basis but cylindrical coordinates

$$\mathbf{x} = r\cos\phi \hat{x} + r\sin\phi \hat{y}, \tag{2.86}$$

and once $\mathbf{x}$ is known, the position in time of each of the two particles can be calculated using (2.81). The trajectory of the virtual particle can be written in parametric form according to Eqs. (2.74)–(2.75):

$$\frac{p}{r} = 1 + e\cos\phi, \tag{2.87}$$

where p is the parameter and e is the eccentricity of the orbit, which in terms of the angular momentum (2.84) and energy (2.85) of the system read

$$p = \frac{\ell^2}{G\mu m_1 m_2}, \tag{2.88}$$

$$e = \sqrt{1 + \frac{2E\ell^2}{\mu G^2 m_1^2 m_2^2}}. \tag{2.89}$$

2.11 Applications

From now on we consider only the case of elliptical trajectories, that is, $0 \leq e \leq 1$. In this case $p = a(1 - e^2)$ where a is the semimajor axis of the elliptical trajectory. From the parametric Eq. (2.87), one has $r = p/(1 + e \cos \phi)$, which implies that

$$r = \frac{a(1 - e^2)}{1 + e \cos \phi}. \tag{2.90}$$

With this result one can construct the equation of motion for ϕ as follows:

$$\dot{\phi} = \frac{\ell}{\mu r^2} = \frac{\sqrt{\mu G m_1 m_2 p}}{\mu r^2}$$

$$= \frac{\sqrt{G m_1 m_2 a(1 - e^2)}}{\sqrt{\frac{m_1 m_2}{M}} r^2} = \frac{\sqrt{G M a(1 - e^2)}}{r^2}$$

$$= \sqrt{\frac{G M}{a^3 (1 - e^2)^3}} (1 + e \cos \phi)^2. \tag{2.91}$$

Given a and e one obtains ϕ integrating (2.91) in time, and then calculate r using (2.90), and it is possible to calculate the position $\mathbf{x}$ using (2.86). Finally the position of each particle using (2.81):

$$\mathbf{x}_1 = \frac{m_2}{M} r (\cos \phi \hat{x} + \sin \phi \hat{y}) = \frac{m_2 a (1 - e^2)}{M (1 + e \cos \phi)} [\cos \phi \hat{x} + \sin \phi \hat{y}], \tag{2.92}$$

$$\mathbf{x}_2 = -\frac{m_1}{M} r (\cos \phi \hat{x} + \sin \phi \hat{y}) = -\frac{m_1 a (1 - e^2)}{M (1 + e \cos \phi)} [\cos \phi \hat{x} + \sin \phi \hat{y}]. \tag{2.93}$$

We use these results to set initial conditions for the evolution of two gravitationally interacting point particles. We choose the value of the angle $\phi = 0$ at $t = 0$, and then using (2.92) and (2.93), we calculate the initial positions of the particles:

$$\mathbf{x}_{1,0} = \frac{m_2 a (1 - e)}{M} \hat{x},$$

$$\mathbf{x}_{2,0} = -\frac{m_1 a (1 - e)}{M} \hat{x}. \tag{2.94}$$

The initial velocity is constructed using the tangent velocity, which has only y−component, and we use (2.91) to define the initial angular velocity:

$$\dot{\phi}_0 = \sqrt{\frac{G(m_1 + m_2)}{a^3 (1 - e^2)^3}} (1 + e)^2, \tag{2.95}$$

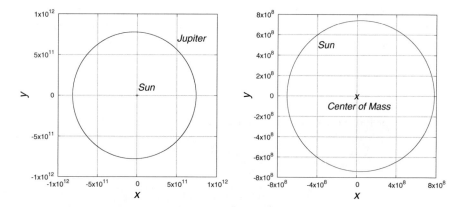

Fig. 2.39 Trajectories of the Sun-Jupiter two-body system. On the left the trajectories of the two objects. On the right a zoom of the Sun's trajectory

thus, the initial velocity of each particles is

$$\mathbf{v}_{1,0} = \frac{m_2 a(1-e)}{M}\dot{\phi}_0 \hat{y},$$

$$\mathbf{v}_{2,0} = -\frac{m_1 a(1-e)}{M}\dot{\phi}_0 \hat{y}. \tag{2.96}$$

These initial conditions are used to explore the numerical solution in a couple of scenarios.

Sun-Jupiter case. We solve the problem considering the solar mass $M_\odot = 1.98847 \times 10^{30}$ kg and Jupiter's mass $M_J = 1.8982 \times 10^{27}$ kg. Using these values, together with a and e from Table 2.1 for Jupiter, we integrate equations (2.80) and the results are shown in Fig. 2.39.

First, notice that the center of mass is at the origin of coordinates as defined in (2.81). The trajectory of Jupiter is pretty much that of Fig. 2.35, however differences in the trajectory appear with a zoom because now the planet orbits around the center of mass, not around the Sun.

The zoom of the Sun's trajectory in the Figure reveals that the Sun orbits around the center of mass on a finite size trajectory. In the previous section, the Sun was assumed to be so dominant that it would occupy the position of the center of mass; however, when the two bodies interact, this is no longer the case. Consider that the solar radius is $R_\odot = 6.957 \times 10^8$ m and thus the orbit of the Sun around the center of mass has size of the order of its radius.

The case of GW150914. We can simulate the Newtonian version of the trajectory of the two black holes first detected as source of Gravitational Waves. The mean parameters inferred from the observations are the black hole masses $m_1 = 36 M_\odot$ and $m_2 = 29 M_\odot$, estimated assuming eccentricity $e = 0$. The trajectory is in reality not elliptical, and initial conditions need to be set in order to fit the Gravitational

2.12 How to Improve the Methods in This Chapter

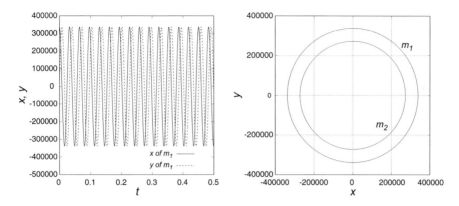

Fig. 2.40 On the left the coordinates of the hole with mass m_1 as function of time. On the right the trajectory of the two holes on the xy-plane. Notice that since $e = 0$, the trajectories are circular. Units are in meters and seconds

Wave strain. We here assume that initially the holes are in orbit with $a = 6.1 \times 10^5$ m.

Using these parameters we integrate equations (2.80), and the results appear in Fig. 2.40. These would be the trajectories of the black holes if the process would not emit gravitational radiation and both, energy and angular momentum, are conserved, unlike in the case with emission of gravitational waves, which carry energy and angular momentum outwards, leading to the eventual merger of the two holes. Modeling the loss of angular momentum and energy can only be done in the general relativistic regime by solving Einstein equations for two black holes (see, e.g., [10]).

The results discussed here can be reproduced with the code rk4_TwoBodies.f90 available as Electronic Supplementary Material described in Appendix B.

> **Exercise.** Compare the time series of $x(t)$ of Jupiter's orbit with that obtained in the previous section and determine the shift of the orbit along the x-axis in meters.

2.12 How to Improve the Methods in This Chapter

The methods developed and used in the applications are very simple and yet very useful in a wide number of scenarios. Other problems might need specific methods due to the stiffness of the equations or can depend on the conservation of a specific functional, like energy, where dissipation becomes important.

The few methods described and used here are limited by their stability and dissipation. Stability limits the resolution and can be improved using implicit methods. Dissipation is extremely important in problems that need a very accurate conservation of energy, and the methods here can be replaced by symplectic methods.

Yet, the methods here can improve importantly. One possibility involves the adaptive step size that can be easily implemented within the codes learned here and are well described in [11]. Another straightforward possibility is the implementation of higher-order Runge-Kutta methods.

2.13 Projects

With the methods described in this chapter, it is possible to carry out the following projects.

2.13.1 Synchronization of Chaos and Message Encryption

An exciting project involves the synchronization of Chaos and the transmission of messages hidden within signals that are solution of the Lorenz system. Needless to mention that the project is inspired in the set 9.6 of exercises in [3]. The idea was implemented by Kevin Cuomo and Alan Oppenheim and is based on synchronization of Chaos. Consider the Lorenz system with parameters in the chaotic regime, as done for the construction of Fig. 2.21. Despite the irregularity of the signal, it is possible to construct a second system similar to the Lorenz system, whose solution approaches the solution of the original one. The process is called synchronization in the chaotic regime. Assume we solve the Lorenz system that we call *emitter*, rewritten with a different set of tag variables u, v, w:

$$\dot{u} = a(v - u),$$
$$\dot{v} = bu - v - uw,$$
$$\dot{w} = uv - cw. \tag{2.97}$$

Once the solution is calculated, one constructs a similar system called *receiver*:

$$\dot{u}_r = a(v_r - u(t)),$$
$$\dot{v}_r = bu(t) - v_r - u(t)w,$$
$$\dot{w}_r = u(t)v_r - cw_r, \tag{2.98}$$

2.13 Projects

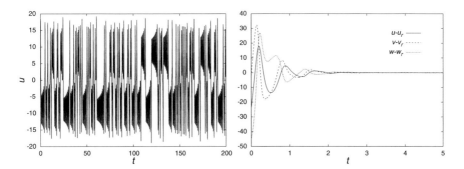

Fig. 2.41 On the left we show the time series of u showing the system is in the chaotic regime. On the right $u - u_r$, $v - v_r$, $w - w_r$ as functions of time that illustrate the process of synchronization, because the differences between the variables of emitter and receiver become indistinguishable after a finite time. The parameters used are $a = 10, b = 28, c = 3$, with initial conditions $u_0 = 7$, $v_0 = 6$, $w_0 = 5$ for the emitter, whereas initial conditions for the receiver are arbitrarily $u_{r0} = 3u_0 + 10$, $v_{r0} = 2v_0 - 4$, $w_{r0} = 7w_0 + 15$

whose essential property is that $u(t)$ is the solution of the emitter system (2.97). It can happen that with the pass of time, the solution u_r, v_r, w_r tends toward the solution u, v, w. When this happens it is said that the *emitter* and *receiver* have **synchronized**.

Synchronization occurs even when initial conditions are different for the two systems. In Fig. 2.41 we show an example of synchronization of the Lorenz system and the differences between *emitter* and *receiver* variables, that is, $u - u_r$, $v - v_r$, $w - w_r$, as functions of time. Notice that the differences approach zero rapidly. The synchronization happens for a set of rather different initial conditions for the emitter and the receiver.

Sending a Message. The emitter can superpose a **message** $m(t)$ to the chaotic signal $u(t)$ that we choose to be a harmonic function modulated by a Gaussian that we arbitrarily center at time $t \sim 100$:

$$u(t) \rightarrow u(t) + m(t), \quad m(t) = A \sin(\omega t) e^{-(t-100)^2}.$$

Thanks to synchronization, the *receiver* will calculate u_r, v_r, w_r that we know tend toward u, v, w, and therefore the **message** can be subtracted from the receiver solution u_r:

$$m_r(t) = u(t) - u_r. \tag{2.99}$$

In Fig. 2.42 we show the emitted and received signals for messages using two different frequencies. It is unfortunate that the error in the quality of the recovered signal depends on the frequency of the signal used to send a message. But it has an explanation. It happens that the Fourier Transform of the solution in Fig. 2.41

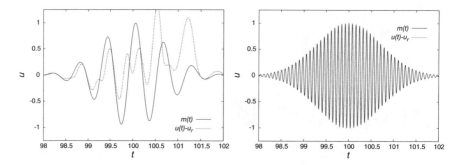

Fig. 2.42 Examples of message, both original and recovered for low $\omega = 10$ and high $\omega = 80$ frequencies. The resolution used for the solution of the equations is $\Delta t = 4 \times 10^{-5}$

Fig. 2.43 Fourier Transform of the solution of Fig. 2.41. Notice that the power is smaller for higher frequencies. Then a message can be recovered more accurately if it is transmitted using high frequencies, bigger than 60 for example

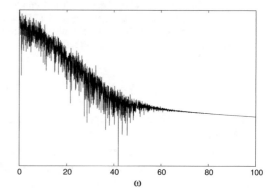

has strong power for small frequencies as we show in Fig. 2.43. It happens that the message with small frequencies can be confused more easily when transmitted in small frequencies than when using high frequencies as found in [12].

One project can include the measurement of fidelity message recovery as function of frequency and amplitude of the message. Plots of L_1 and L_2 norms of the error $e = m(t) - m_r(t)$ would reveal the desired dependency.

The project can include the transmission of more elaborate messages, not only those modulated with a Gaussian centered at time $t = 100$. For example, the superposition of modulated harmonic functions $m(t) = \sum_k = A_k \sin(\omega_k t) \exp(-(t - t_k)^2/\sigma_k^2)$, with different amplitudes, frequencies, and modulation pulses.

2.13.2 Gravitational Waves Emitted By a Binary System Using the Quadrupolar Formula

Detected Gravitational Waves are sourced by binary systems of compact objects, either black holes, neutron stars or a combination of the two. The process of

2.13 Projects

merger, horizon growth, and production of Gravitational Waves is calculated by solving Einstein's field equations, an elaborate system of nonlinear coupled PDEs. Nevertheless, in the orbital phase prior to merger, post-Newtonian and quadrupolar formulas explain well the strain of signals.

In the Newtonian regime, the problem can be defined using the notation of the two body problem illustrated in Fig. 2.38. When the system emits Gravitational Waves, energy and angular momentum are radiated away, and consequently the trajectory of the particles is strongly modified, the semimajor axis a, eccentricity e, angular momentum ℓ, and period P of the orbit change in time. The system of equations that model the evolution of the time average on a period P of these quantities can be found in [8] and [13]:

$$\langle \dot{a} \rangle = -\frac{64}{5} \frac{G^3}{c^5} \frac{\mu M^2}{a^3 (1-e^2)^{7/2}} \left(1 + \frac{73}{24} e^2 + \frac{37}{96} e^4\right),$$

$$\langle \dot{e} \rangle = -\frac{304}{15} \frac{G^3}{c^5} \frac{\mu M^2}{a^4 (1-e^2)^{5/2}} \left(1 + \frac{121}{304} e^2\right) e,$$

$$\langle \dot{E} \rangle = -\frac{32}{5} \frac{G^4}{c^5} \frac{\mu^2 M^3}{a^5 (1-e^2)^{7/2}} \left(1 + \frac{73}{24} e^2 + \frac{37}{96} e^4\right),$$

$$\langle \dot{P} \rangle = -\frac{192\pi}{5} \frac{G^{5/2}}{c^5} \frac{\mu M^{3/2}}{a^{5/2} (1-e^2)^{7/2}} \left(1 + \frac{73}{24} e^2 + \frac{37}{96} e^4\right),$$

$$\langle \dot{\ell} \rangle = -\frac{32}{5} \frac{G^{7/2}}{c^5} \frac{\mu^2 M^{5/2}}{a^{7/2} (1-e^2)^2} \left(1 + \frac{7}{8} e^2\right), \quad (2.100)$$

where μ and M are the reduced and total masses of the binary system and G, c are the fundamental constants. Suitable initial conditions are that the semimajor axis and eccentricity have some values a_0 and e_0 and that the energy is that of the virtual particle of mass μ on a circular trajectory at initial time, and consistent initial value of the period and angular momentum for a_0 and e_0 are as follows:

$$a(0) = a_0,$$

$$e(0) = e_0,$$

$$E(0) = -\frac{1}{2} \frac{G\mu M}{a_0},$$

$$P(0) = \frac{2\pi}{\sqrt{GM}} a_0^{3/2},$$

$$\ell(0) = \sqrt{Ga_0(1-e_0^2)\,\mu^2 M}. \quad (2.101)$$

Notice that for the case of two point particles, the parameters a_0 and e_0 determine the initial conditions for all other variables. In order to know the trajectory of the virtual particle, one has to couple the equation of the trajectory (2.91) assuming the parameters of the orbit are those integrated from the system (2.100) as averaged over one orbital period P:

$$\dot{\phi} = \sqrt{\frac{GM}{\langle a \rangle^3 (1 - \langle e \rangle^2)^3}} (1 + \langle e \rangle \cos(\phi))^2. \tag{2.102}$$

From this, using (2.92) and (2.93) one can track the location of each particle. On the other hand, the strain of the Gravitational Waves radiated has two polarizations with the time dependent parameters of the orbit, following [8]:

$$h_+ = -(2.2)^2 \frac{4\mu M}{\langle a \rangle R} \cos(2t\omega(t)), \tag{2.103}$$

$$h_\times = (2.2)^2 \frac{4\mu M}{\langle a \rangle R} \sin(2t\omega(t)), \tag{2.104}$$

where $\omega = \frac{2\pi}{P}$ and R is the distance from the binary source to the Earth. The time domain is restricted by the time it takes for the two particles to collide, the so-called collision time that reads [13]:

$$t_{coll} = \frac{5}{256} \frac{c^5}{G^3} \frac{a_0^4}{\mu M^2}, \tag{2.105}$$

so that $t \in [0, t_{coll}]$. The solution of Eqs. (2.100) in this time domain can lead to gravitational wave signals similar to those in observations.

The case of GW150914. In this case the parameters of the binary source inferred from the first observation of Gravitational Waves [14] are the masses of the black holes $m_1 = 36^{+5}_{-4} M_\odot$, $m_2 = 29^{+4}_{-4} M_\odot$, and the distance to the source $R = 410^{+160}_{-180} \times 10^6$pc. Assuming an initial separation of $a_0 = 610$km, one can obtain a minimum number of cycles of the strain as in the observation. The time for collision in this case is $t_{coll} = 0.13283^{+0.0004}_{-0.0015}$s. One can explore the case with or without eccentricity.

For example, the case $e_0 = 0$ leads to the strain and trajectory of the two black holes as illustrated in Figures 4.5 to 4.7 of [15]. The strains h_+ and $h_\times$ are integrated for combinations of m_1 and m_2 and various values of the distance to the source R. Similar signals can be obtained for the various other combinations of m_1 and m_2 within the uncertainties reported in [14].

The project consists in producing a catalog of strain signals that might serve for comparison with observations in the interferometers of Gravitational Waves.

References

1. S.C. Chapra, R.P. Canale, *Numerical Methods for Engineers*, 7th edn. (Mac Graw Hill, New York, 2010)
2. L. Ling-Hsiao, in *Numerical Simulation of Space Plasmas (I)*. Lecture Notes. Institute of Space Science, National Central University. http://www.ss.ncu.edu.tw/~lyu/lecture_files_en/lyu_NSSP_Notes/Lyu_NSSP_AppendixC.pdf
3. S.H. Strogatz, *Nonlinear Dynamics and Chaos*, 2nd edn. (Westview, Boulder, 2015)
4. E. Ott, *Chaos in Dynamical Systems*, 2nd edn. (Cambridge University, Cambridge, 2002)
5. Y.R. Kim, Y.-J. Choi, Y. Min, A model of COVID-19 pandemic with vaccines and mutant viruses. PLoS ONE **17**(10), e0275851 (2022). https://doi.org/10.1371/journal.pone.0275851
6. H.A. Luther, An explicit sixth-order Runge-Kutta formula. Math. Comp. **22**(102), 434–436 (1968)
7. S.L. Shapiro, S.A. Teukolsky, *Black Holes, White Dwarfs and Neutron Stars: The Physics of Compact Objects* (Wiley-VCH, New York, 2004)
8. B.F. Schutz, *A First Course in General Relativity*, 2nd edn. (Cambridge University, Cambridge, 2009)
9. F.S. Guzmán, F.D. Lora-Clavijo, M.D. Morales, Revisiting spherically symmetric relativistic hydrodynamics. Rev. Mex. Fis. **E 58**, 84–98 (2012). arXiv:1212.1421 [gr-qc] https://doi.org/10.48550/arXiv.1212.1421
10. F. Pretorius, Evolution of Binary Black-Hole Spacetimes. Phys. Rev. Lett. **95**, 121101 (2005). arXiv:gr-qc/0507014, https://doi.org/10.48550/arXiv.gr-qc/0507014
11. W.H. Press, S.A Teukolsky, W.T. Vettering, B.P. Flannery, *Numerical Recipes in Fortran 77: The Art of Scientific Computing*, 2nd edn. (Cambridge University, Cambridge, 2003)
12. J.-P. Yeh, K.-L. Wu, A simple method to synchronize chaotic systems and its application to secure communications. Math. Comp. Model. **47**, 894–902 (2008). https://doi.org/10.1016/j.mcm.2007.06.021
13. P. Peters, Gravitational radiation and the motion of two point cases. Phys. Rev. **136**, B1224–B1232 (1964). https://doi.org/10.1103/PhysRev.136.B1224
14. B.P. Abbott et al. (LIGO Scientific Collaboration and Virgo Collaboration), Observation of gravitational waves from a binary black hole merger. Phys. Rev. Lett. **116**, 061102 (2016). https://doi.org/10.1103/PhysRevLett.116.061102
15. Karla Sofía Zavala Alvarez. Undergraduate Thesis, *Study of gravitational waves produced by a binary system in the weak field regime*. Universidad Michoacana de San Nicolás de Hidalgo (2017). https://sites.google.com/umich.mx/fsguzman/group

Chapter 3
Simple Methods for Initial Value Problems Involving PDEs

Abstract In this chapter we implement the simplest methods used to solve Initial Value Problems associated to Partial Differential Equations, restricted to the case of one spatial and one temporal dimensions. We start by defining the problem on a discrete domain of finite size for a general problem and show how to construct a discrete version of the problem. To our view, the wave equation is the simplest and most illustrative equation, the reason why we use it to describe the implementation of the Finite Differences method in the construction of a global solution on a finite domain, of initial conditions, as well as various types of boundary conditions. We also use this problem to illustrate the stability of evolution schemes and the construction of the Crank-Nicolson implicit method, whose most important property is stability. We take advantage of this discussion to exemplify the stability of simple and implicit schemes in the solution of the Diffusion equation. We finish the chapter with the solution of Schrödinger equation. In this chapter, the specific IVPs studied are the paradigms of hyperbolic, parabolic, and complex equations.

Keywords Partial differential equations · Basic methods · Wave equation · Schrödinger equation · Diffusion equation · Error theory · Convergence

In this chapter we implement the simplest methods used to solve Initial Value Problems associated to Partial Differential Equations, restricted to the case of one spatial and one temporal dimensions. We start by defining the problem on a discrete domain of finite size for a general problem and show how to construct a discrete version of the problem. To our view, the wave equation is the simplest and most illustrative equation, the reason why we use it to describe the implementation of the Finite Differences method in the construction of a global solution on a finite domain, of initial conditions, as well as various types of boundary conditions. We also use this problem to illustrate the stability of evolution schemes and the construction of the Crank-Nicolson implicit method, whose most important property is stability. We take advantage of this discussion to exemplify the stability of simple and implicit schemes in the solution of the Diffusion equation. We finish the chapter with the solution of Schrödinger equation. In this chapter, the specific IVPs studied are the paradigms of hyperbolic, parabolic, and complex equations.

3.1 Discretization of an IVP

Consider the IVP associated to a sufficiently general second-order PDE defined on a two-dimensional domain, being t and x the independent variables that label time and space:

$$\begin{aligned}
&\text{combination of } \left(f, \frac{\partial f}{\partial t}, \frac{\partial f}{\partial x}, \frac{\partial^2 f}{\partial t^2}, \frac{\partial^2 f}{\partial x^2}\right) = g(x,t) \quad f = f(x,t) \\
&D = x \in [x_{min}, x_{max}] \times t \in [0, t_f] \quad &\text{Domain} \\
&f(x,0) = f_0(x) \quad &\text{Initial Conditions} \\
&\dot{f}(x,0) = \dot{f}_0(x) \quad & \\
&f(x_{min}, t), \; f(x_{max}, t) \quad &\text{Boundary Conditions}
\end{aligned}$$

(3.1)

It is possible to construct an approximate solution of this problem on a discrete domain using a Finite Differences approach. Two ingredients are needed:

1. Define a discrete version of the domain D_d.
2. Construct a discrete version of the PDE, together with an evolution scheme across time labels, as well as of initial and boundary conditions.

1. The discrete domain D_d. A simple discrete domain is the set of points $D_d = \{(x_i, t^k)\}$, where $x_i = x_{min} + i\Delta x$ with $i = 0, \ldots, N_x$ and $t^n = n\Delta t$, $\Delta x = (x_{max} - x_{min})/N_x$, $\Delta t = t_f/N_t$, are spatial and time resolutions. Notice that the domain is defined in a very similar way as done for ODEs. Again N_x is the number of cells of size Δx that cover the domain along the spatial direction, or equivalently there are $2N_x + 1$ points at all times that define D_d. An illustration of D_d is shown in Fig. 3.1, where particular points of the domain are indicated.

The key importance of this domain is that the functions involved on the IVP are defined at points of D_d only. This means that f and g are functions evaluated at points of D_d, explicitly $f = f(x_i, t^n)$ and $g = g(x_i, t^n)$. For ease, without confusion, for a generic function f we denote $f(x_i, t^n) = f_i^n$.

2. Discrete version of the PDE. One needs to construct a discrete version of the equation on the discrete domain D_d, which includes the spatial and time derivatives of f in the PDE of problem (3.1). We construct now the discrete version of first- and second-order derivatives of f that will be needed in the examples of this chapter.

Spatial Derivatives
Consider the **grid function** f_i^n defined at each point of D_d. The construction of partial derivatives uses Taylor expansions of f_i^n at near neighbors of (x_i, t^n). The values of the function f at nearest neighbors of D_d along the line $t^n = $ constant are f_{i-1}^n and f_{i+1}^n. In terms of Taylor series expansions around (x_i, t^n) these values are

3.1 Discretization of an IVP

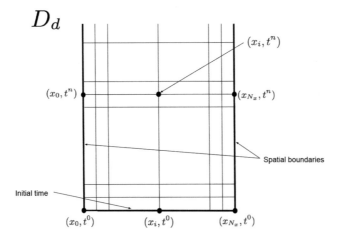

Fig. 3.1 Illustration of the numerical domain D_d

$$f_{i-1}^n = f_i^n - \Delta x \frac{\partial f}{\partial x}\bigg|_{(x_i,t^n)} + \frac{\Delta x^2}{2!} \frac{\partial^2 f}{\partial x^2}\bigg|_{(x_i,t^n)} - \frac{\Delta x^3}{3!} \frac{\partial^3 f}{\partial x^3}\bigg|_{(x_i,t^n)} + \mathcal{O}(\Delta x^4),$$

$$f_i^n = f_i^n,$$

$$f_{i+1}^n = f_i^n + \Delta x \frac{\partial f}{\partial x}\bigg|_{(x_i,t^n)} + \frac{\Delta x^2}{2!} \frac{\partial^2 f}{\partial x^2}\bigg|_{(x_i,t^n)} + \frac{\Delta x^3}{3!} \frac{\partial^3 f}{\partial x^3}\bigg|_{(x_i,t^n)} + \mathcal{O}(\Delta x^4). \quad (3.2)$$

Appropriate combinations of these three expansions isolate the first and second-order derivatives with respect to x.

For the **first-order derivative**, there are various suitable combinations:

$$f_{i+1}^n - f_i^n = \Delta x \frac{\partial f}{\partial x}\bigg|_{(x_i,t^n)} + \mathcal{O}(\Delta x^2)$$

$$\Rightarrow \frac{\partial f}{\partial x}\bigg|_{(x_i,t^n)} = \frac{f_{i+1}^n - f_i^n}{\Delta x} + \mathcal{O}(\Delta x), \quad (3.3)$$

$$f_i^n - f_{i-1}^n = \Delta x \frac{\partial f}{\partial x}\bigg|_{(x_i,t^n)} + \mathcal{O}(\Delta x^2)$$

$$\Rightarrow \frac{\partial f}{\partial x}\bigg|_{(x_i,t^n)} = \frac{f_i^n - f_{i-1}^n}{\Delta x} + \mathcal{O}(\Delta x), \quad (3.4)$$

$$f_{i+1}^n - f_{i-1}^n = 2\Delta x \frac{\partial f}{\partial x}\bigg|_{(x_i,t^n)} + \mathcal{O}(\Delta x^3)$$

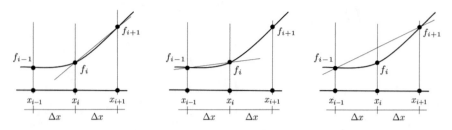

Fig. 3.2 Illustration of formulas (3.3), (3.4) and (3.5). The curve is supposed to represent the continuous $f(x)$ and $f_i = f(x_i)$ for an arbitrary time label n

$$\Rightarrow \left.\frac{\partial f}{\partial x}\right|_{(x_i,t^n)} = \frac{f_{i+1}^n - f_{i-1}^n}{2\Delta x} + \mathcal{O}(\Delta x^2), \tag{3.5}$$

which are three discrete versions of the first-order derivative with respect to x defined at the interior points of D_d. The first two formulas are first-order accurate, whereas the third one is second-order accurate. Their respective names are *forward*, *backward*, and *centered* stencils, or simply discrete versions of the first-order derivative. Illustration of these formulas appears in Fig. 3.2.

These are three expressions that suffice to accurately solve many interesting problems. Nevertheless we want to go slightly further and construct forward and backward second-order accurate operators because they will be needed and, most importantly, to illustrate and reinforce the process of construction of these operators. For this, it is necessary to Taylor expand the values of f at the second nearest neighbors that should accompany expansions (3.2):

$$f_{i-2}^n = f_i^n - 2\Delta x \left.\frac{\partial f}{\partial x}\right|_{(x_i,t^n)} + \frac{2^2 \Delta x^2}{2!} \left.\frac{\partial^2 f}{\partial x^2}\right|_{(x_i,t^n)}$$

$$- \frac{2^3 \Delta x^3}{3!} \left.\frac{\partial^3 f}{\partial x^3}\right|_{(x_i,t^n)} + \mathcal{O}(\Delta x^4),$$

$$f_{i+2}^n = f_i^n + 2\Delta x \left.\frac{\partial f}{\partial x}\right|_{(x_i,t^n)}$$

$$+ \frac{2^2 \Delta x^2}{2!} \left.\frac{\partial^2 f}{\partial x^2}\right|_{(x_i,t^n)} + \frac{2^3 \Delta x^3}{3!} \left.\frac{\partial^3 f}{\partial x^3}\right|_{(x_i,t^n)} + \mathcal{O}(\Delta x^4). \tag{3.6}$$

Combinations of expansions (3.2) and (3.6) for forward and backward versions of the first-order spatial derivative are

$$f_{i+2}^n - 4f_{i+1}^n + 3f_i^n = -2\Delta x \left.\frac{\partial f}{\partial x}\right|_{(x_i,t^n)} + \mathcal{O}(\Delta x^3)$$

3.1 Discretization of an IVP

$$\Rightarrow \left.\frac{\partial f}{\partial x}\right|_{(x_i,t^n)} = -\frac{f^n_{i+2} - 4f^n_{i+1} + 3f^n_i}{2\Delta x} + \mathcal{O}(\Delta x^2), \qquad (3.7)$$

$$f^n_{i-2} - 4f^n_{i-1} + 3f^n_i = 2\Delta x \left.\frac{\partial f}{\partial x}\right|_{(x_i,t^n)} + \mathcal{O}(\Delta x^3)$$

$$\Rightarrow \left.\frac{\partial f}{\partial x}\right|_{(x_i,t^n)} = \frac{f^n_{i-2} - 4f^n_{i-1} + 3f^n_i}{2\Delta x} + \mathcal{O}(\Delta x^2), \qquad (3.8)$$

which are second-order accurate.

Now we construct expressions for the **second-order derivative** with respect to x using combinations of (3.2):

$$f^n_{i-1} - 2f^n_i + f^n_{i+1} = \Delta x^2 \left.\frac{\partial^2 f}{\partial x^2}\right|_{(x_i,t^n)} + \mathcal{O}(\Delta x^4) \Rightarrow$$

$$\left.\frac{\partial^2 f}{\partial x^2}\right|_{(x_i,t^n)} = \frac{f^n_{i-1} - 2f^n_i + f^n_{i+1}}{\Delta x^2} + \mathcal{O}(\Delta x^2), \qquad (3.9)$$

which is a discrete centered version of the second derivative with respect to x defined at interior points of D_d. Like for first-order derivatives, it is possible to construct second-order accurate formulas for second-order derivatives of type forward/backward stencils using only points on one side from the point (x_i, t^n). These formulas are

$$\left.\frac{\partial^2 f}{\partial x^2}\right|_{(x_i,t^n)} = \frac{-f^n_{i+3} + 4f^n_{i+2} - 5f^n_{i+1} + 2f^n_i}{\Delta x^2} + \mathcal{O}(\Delta x^2), \qquad (3.10)$$

$$\left.\frac{\partial^2 f}{\partial x^2}\right|_{(x_i,t^n)} = \frac{-f^n_{i-3} + 4f^n_{i-2} - 5f^n_{i-1} + 2f^n_i}{\Delta x^2} + \mathcal{O}(\Delta x^2). \qquad (3.11)$$

> **Exercise.** Demonstrate these two formulas using the appropriate combinations of Taylor series expansions.

Summarizing, the approximate expressions (3.5), (3.7), (3.8) for first-order derivatives and (3.9), (3.10), (3.11) for second-order derivatives constitute a kit that will be very useful to solve the forthcoming problems.

Time Derivatives

The construction of discrete versions of time derivatives is pretty similar to that of spatial derivatives. In this case we use Taylor expansions of f at neighboring points of (x_i, t^n) along the line of $x_i = $ constant as follows:

$$f_i^{n-2} = f_i^n - 2\Delta t \frac{\partial f}{\partial t}\bigg|_{(x_i,t^n)} + \frac{2^2 \Delta t^2}{2!} \frac{\partial^2 f}{\partial t^2}\bigg|_{(x_i,t^n)} - \frac{2^3 \Delta t^3}{3!} \frac{\partial^3 f}{\partial t^3}\bigg|_{(x_i,t^n)} + \mathcal{O}(\Delta t^4),$$

$$f_i^{n-1} = f_i^n - \Delta t \frac{\partial f}{\partial t}\bigg|_{(x_i,t^n)} + \frac{\Delta t^2}{2!} \frac{\partial^2 f}{\partial t^2}\bigg|_{(x_i,t^n)} - \frac{\Delta t^3}{3!} \frac{\partial^3 f}{\partial t^3}\bigg|_{(x_i,t^n)} + \mathcal{O}(\Delta t^4),$$

$$f_i^n = f_i^n, \tag{3.12}$$

$$f_i^{n+1} = f_i^n + \Delta t \frac{\partial f}{\partial t}\bigg|_{(x_i,t^n)} + \frac{\Delta t^2}{2!} \frac{\partial^2 f}{\partial t^2}\bigg|_{(x_i,t^n)} + \frac{\Delta t^3}{3!} \frac{\partial^3 f}{\partial t^3}\bigg|_{(x_i,t^n)} + \mathcal{O}(\Delta t^4),$$

$$f_i^{n+2} = f_i^n + 2\Delta t \frac{\partial f}{\partial t}\bigg|_{(x_i,t^n)} + \frac{2^2 \Delta t^2}{2!} \frac{\partial^2 f}{\partial t^2}\bigg|_{(x_i,t^n)} + \frac{2^3 \Delta t^3}{3!} \frac{\partial^3 f}{\partial t^3}\bigg|_{(x_i,t^n)} + \mathcal{O}(\Delta t^4).$$

With these expansions one can construct the following set of expressions for the **first-order time derivative** of f:

$$f_i^{n+1} - f_i^n = \Delta t \frac{\partial f}{\partial t}\bigg|_{(x_i,t^n)} + \mathcal{O}(\Delta t^2)$$

$$\Rightarrow \frac{\partial f}{\partial t}\bigg|_{(x_i,t^n)} = \frac{f_i^{n+1} - f_i^n}{\Delta t} + \mathcal{O}(\Delta t), \tag{3.13}$$

$$f_i^n - f_i^{n-1} = \Delta t \frac{\partial f}{\partial t}\bigg|_{(x_i,t^n)} + \mathcal{O}(\Delta t^2)$$

$$\Rightarrow \frac{\partial f}{\partial t}\bigg|_{(x_i,t^n)} = \frac{f_i^n - f_i^{n-1}}{\Delta t} + \mathcal{O}(\Delta t), \tag{3.14}$$

$$f_i^{n+1} - f_i^{n-1} = 2\Delta t \frac{\partial f}{\partial t}\bigg|_{(x_i,t^n)} + \mathcal{O}(\Delta t^3)$$

$$\Rightarrow \frac{\partial f}{\partial t}\bigg|_{(x_i,t^n)} = \frac{f_i^{n+1} - f_i^{n-1}}{2\Delta t} + \mathcal{O}(\Delta t^2), \tag{3.15}$$

$$f_i^{n+2} - 4f_i^{n+1} + 3f_i^n = -2\Delta t \frac{\partial f}{\partial t}\bigg|_{(x_i,t^n)} + \mathcal{O}(\Delta t^3)$$

$$\Rightarrow \frac{\partial f}{\partial t}\bigg|_{(x_i,t^n)} = -\frac{f_i^{n+2} - 4f_i^{n+1} + 3f_i^n}{2\Delta t} + \mathcal{O}(\Delta t^2), \tag{3.16}$$

$$f_i^{n-2} - 4f_i^{n-1} + 3f_i^n = 2\Delta t \frac{\partial f}{\partial t}\bigg|_{(x_i,t^n)} + \mathcal{O}(\Delta t^3)$$

$$\Rightarrow \frac{\partial f}{\partial t}\bigg|_{(x_i,t^n)} = \frac{f_i^{n-2} - 4f_i^{n-1} + 3f_i^n}{2\Delta t} + \mathcal{O}(\Delta t^2). \tag{3.17}$$

Most of these expressions will be very useful in various of the examples presented later on.

3.1 Discretization of an IVP

Finally, an appropriate combination of expressions (3.12) leads to a formula for the **second-order time derivative**:

$$f_i^{n-1} - 2f_i^n + f_i^{n+1} = \Delta t^2 \left.\frac{\partial^2 f}{\partial t^2}\right|_{(x_i, t^n)} + \mathcal{O}(\Delta t^4) \Rightarrow$$

$$\left.\frac{\partial^2 f}{\partial t^2}\right|_{(x_i, t^n)} = \frac{f_i^{n-1} - 2f_i^n + f_i^{n+1}}{\Delta t^2} + \mathcal{O}(\Delta t^2), \quad (3.18)$$

that will be useful as well.

Error Theory and Convergence

As before, the analysis concentrates in the analysis of **truncation errors** due to numerical methods. The error analysis of the numerical solutions of IVPs associated to PDEs is very similar to the theory presented in Sect. 2.7 for ODEs. The key is that convergence has to be satisfied at each point of the two-dimensional numerical domain D_d. This means that at every time t^n the solution should converge to the continuous solution, and the solution along the time direction should converge for each x_i as well.

We define the error of the numerical solution f_i^n of an IVP in our two-dimensional domain, at each point (x_i, t^n) by

$$e_i^n = f_i^n - f_i^{en}, \quad (3.19)$$

where f_i^{en} is the exact solution at the point (x_i, t^n) in the continuous domain. Since convergence tests need the calculation of numerical solutions in domains with different resolutions, it is important to set a hierarchy of numerical domains D_d^m for such purpose, because one needs to make sure that the numerical solutions calculated with different resolutions are compared at the same position and time.

A practical recipe based on that used for the solution of ODEs is described next:

- Calculate a first numerical solution f^{1n}_i using the numerical domain D_d^1, discretized using N_x cells or equivalently $N_x + 1$ points, spatial and time resolutions Δx_1 and $\Delta t_1 = C\Delta x_1$, where C is a constant Courant-Friedrichs-Lewy factor.
- Calculate a second numerical solution f^{2n}_i on a new numerical domain D_d^2 using $2N_x$ cells or $2N_x + 1$ points, with space and time resolutions $\Delta x_2 = \Delta x_1/2$ and $\Delta t_2 = \Delta t_1/2$.
- Calculate a third numerical solution f^{3n}_i on a third numerical domain D_d^3 defined with $2^2 N_x$ cells or $2^2 N_x + 1$ points, with space and time resolutions $\Delta x_3 = \Delta x_2/2 = \Delta x_1/2^2$ and $\Delta t_3 = \Delta t_2/2 = \Delta t_1/2^2$.

Notice that the $N_x + 1$ points of D_d^1 are a subset of the $2N_x + 1$ points of D_d^2 and that these two are subsets of the $2^2 N_x + 1$ points of D_d^3. Then numerical solutions f^{1n}_i, f^{2n}_i, and f^{3n}_i can be compared at points of the coarsest numerical domain D_d^1.

In practice, comparison of solutions on D_d^1 needs the downsampling of the output in space and time of solutions in domains D_d^m, $m > 1$. For this, it is useful to define an integer parameter m that in the code is called `resolution_label` as we did for ODEs. This parameter will help as follows:

- Given D_d^1 is constructed with N_x cells, more refined numerical domains D_d^m can be defined using $2^{m-1} N_x$ cells along the spatial direction.
- Since time resolution will reduce the time step to $\Delta t_m / 2^{m-1}$, the number of time steps for the evolution has to be increased by a factor 2^{m-1}, in order to cover the time domain $[0, t_f]$.
- Then the parameter m will help to output data every 2^{m-1} iterations and every 2^{m-1} points.

The *theory of convergence* is practically the same as for ODEs in Sect. 2.7. The reason is that we assume $\Delta t = C \Delta x$, that is, the time resolution is proportional to the space resolution. This assumption will hold for various of the problems below, not for all of them, in which we describe the appropriate modifications. Then if the error in the approximation of the equation of a PDE is of order k, for instance, an equation approximated with second-order accurate versions of first-order derivatives (3.5), (3.7), (3.8), the PDE will be approximated with an error of order $\mathcal{O}(\Delta t^2, \Delta x^2)$ which is equivalent to $\mathcal{O}(\Delta x^2)$ since $\Delta t = C \Delta x$. Then numerical solutions at time t^n, f^{1n}_i, f^{2n}_i and f^{3n}_i can be written as

$$f^{1n}_i = f^{en}_i + E \Delta x^k,$$

$$f^{2n}_i = f^{en}_i + E \left(\frac{\Delta x}{2}\right)^k,$$

$$f^{3n}_i = f^{en}_i + E \left(\frac{\Delta x}{4}\right)^k, \tag{3.20}$$

where f^{en}_i is the exact solution. If the exact solution is known, the numerical solution converges to the exact solution at point (x_i, t^n) if the following **convergence factor** fulfills the relation:

$$\boxed{CF := \frac{f^{1n}_i - f^{en}_i}{f^{2n}_i - f^{en}_i} \simeq \frac{\Delta x^k}{\left(\frac{\Delta x}{2}\right)^k} = 2^k} \tag{3.21}$$

at time t^n. Analogously, if the exact solution is unknown, one can define **self-convergence**. Specifically, the numerical solution self-converges at point (x_i, t^n) if the following relation between numerical solution holds

$$\boxed{SCF := \frac{f^{1n}_i - f^{2n}_i}{f^{2n}_i - f^{3n}_i} \simeq \frac{\Delta x^k - \left(\frac{\Delta x}{2}\right)^k}{\left(\frac{\Delta x}{2}\right)^k - \left(\frac{\Delta x}{4}\right)^k} = \frac{1 - \frac{1}{2^k}}{\frac{1}{2^k} - \frac{1}{4^k}} = 2^k} \tag{3.22}$$

3.1 Discretization of an IVP

where SFC of the **self-convergence factor**.

Convergence and self-convergence have to be verified at each point of D_d. If one wants to verify convergence at a given time t^n, one has to verify relations (3.21) or (3.22) for all i. If one wants to verify these relations on the whole D_d, it would be necessary to verify these relations at every time t^n for all n, which is difficult due to the possibly great number of time slices n of D_d. Instead, calculating a norm of the error at time t^n allows an easy presentation of the error analysis.

The formal analysis of the convergence of the error of the norm is associated to the **Lax Theorem**, whose construction and foundations can be followed in reference [1]. Commonly used norms are the $\mathcal{L}_1$ and $\mathcal{L}_2$, which for a function w defined along the continuous spatial coordinate x read

$$\mathcal{L}_1(w) = \int_{x_{min}}^{x_{max}} |w| dx ,$$

$$\mathcal{L}_2(w) = \sqrt{\int_{x_{min}}^{x_{max}} w^2 dx} .$$

Since in our case the domain is discrete, it is necessary to evaluate the integrals accordingly. The trapezoidal rule for the integral of a function w_i defined along the discrete spatial domain with points $x_i = x_{min} + i \Delta x$, $i = 0, \ldots, N_x$ reads

$$\int_{x_{min}}^{x_{max}} w dx = \sum_{i=1}^{N_x} \frac{1}{2}(w_{i-1} + w_i) \Delta x + \mathcal{O}(\Delta x^2), \tag{3.23}$$

which is nothing but the sum of the area of rectangles of base Δx and height equal to the average of the value of w at two neighboring points. Then finally, using this method of integration, the discrete expressions L_1 and L_2 of norms $\mathcal{L}_1$ and $\mathcal{L}_2$ of the error (3.19) are

$$\boxed{\begin{array}{c} L_1(e^n) = \sum_{i=1}^{N_x} \frac{1}{2}(|e_{i-1}^n| + |e_i^n|) \Delta x \\ \\ L_2(e^n) = \sqrt{\sum_{i=1}^{N_x} \frac{1}{2}((e_{i-1}^n)^2 + (e_i^n)^2) \Delta x} \end{array}} \tag{3.24}$$

These scalar functions serve to monitor the error in time.

With these concepts it will be easy to construct convergence tests for the numerical solutions of IVPs that will be tackled next.

3.2   1 + 1 Wave Equation: The Paradigm of Hyperbolic Equations

We use the wave equation to illustrate the solution of an IVP associated to a PDE using the Finite Differences approach. The $1 + 1$ name is inherited from Relativity, where space and time coordinates are usually distinguished. The IVP for the $1 + 1$ wave equation can be defined as

$$
\begin{array}{ll}
\dfrac{\partial^2 \phi}{\partial t^2} - \dfrac{\partial^2 \phi}{\partial x^2} = 0 & \phi = \phi(x,t) \\
D = x \in [x_{min}, x_{max}] \times t \in [0, t_f] & Domain \\
\phi(x,0) = \phi_0(x) & Initial\ Conditions \\
\dot{\phi}(x,0) = \dot{\phi}_0(x) & \\
\phi(x_{min}, t),\ \phi(x_{max}, t) & Boundary\ Conditions
\end{array}
\tag{3.25}
$$

1. We define D_d as the set of points $\{(x_i, t^n)\}$ such that $x_i = x_{min} + i \Delta x$ where $i = 0, \ldots, N_x$, $\Delta x = (x_{max} - x_{min})/N_x$ and $t^n = n \Delta t$ with $\Delta t = C \Delta x$.
2. The *discrete version of the PDE* is constructed using expressions (3.9) and (3.18) together

$$\frac{\phi_i^{n-1} - 2\phi_i^n + \phi_i^{n+1}}{\Delta t^2} - \frac{\phi_{i-1}^n - 2\phi_i^n + \phi_{i+1}^n}{\Delta x^2} = \mathcal{O}(\Delta t^2, \Delta x^2). \tag{3.26}$$

From this relation it is possible to solve for ϕ_i^{n+1}:

$$\phi_i^{n+1} = \left(\frac{\Delta t}{\Delta x}\right)^2 [\phi_{i+1}^n - 2\phi_i^n + \phi_{i-1}^n] + 2\phi_i^n - \phi_i^{n-1} + \mathcal{O}(\Delta x^2, \Delta t^2). \tag{3.27}$$

With this expression one can calculate the value of the wave function at point x_i and time t^{n+1} in terms of the values of ϕ at previous times t^{n-1}, t^n. By using (3.27) for all i one has the solution of the wave function at time t^{n+1}, that is, *one can construct the values of the wave function at future times in terms of its values in the past.*

The Courant-Friedrichs-Lewy factor $C = \Delta t / \Delta x$. In practice, for the discretization we first define the spatial resolution Δx, then provide a value for C, and then define the time resolution $\Delta t = C \Delta x$. Fixing C allows to study the stability of evolution schemes like that in (3.27) and practice convergence tests.

The scheme (3.27) is sometimes represented by a molecule, like that in Fig. 3.3, where dots correspond to points in D_d and white dots represent positions where the function ϕ is known and the black dots the position where ϕ is unknown.

3.2 1 + 1 Wave Equation: The Paradigm of Hyperbolic Equations

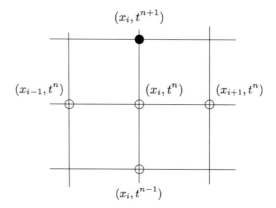

Fig. 3.3 Molecule associated to the evolution scheme (3.27) for the construction of the solution at time t^{n+1} in terms of data at times t^n y t^{n-1}. Black dots correspond to positions where the values of ϕ are unknown, whereas white dots to positions where it is assumed to be known

How to code this scheme. For the solution of the PDE, there is no need to define ϕ_i^n as a two-dimensional array of size $N_x \times N_t$. Instead, it suffices to define three one-dimensional arrays for the wave function, associated to the three time levels involved in (3.27):

phi, contains the entries of ϕ_i^{n+1}, $i = 0, \ldots, N_x$,
phi_p, contains the entries of ϕ_i^n $i = 0, \ldots, N_x$,
phi_pp, contains the entries of ϕ_i^{n-1} $i = 0, \ldots, N_x$.

The names _p and _pp are inspired by the standards of the Cactus code [2]. These arrays can be recycled after using (3.27) for a given time label n by reassigning the values in their entries

$$\phi_i^n \to \phi_i^{n-1},$$
$$\phi_i^{n+1} \to \phi_i^n,$$

and then release the memory assigned to the array phi associated to ϕ_i^{n+1}. The evolution loop in a program looks as follows:

```
do k=1,Nt

  print *, k,t  ! Send some info to the screen
  t = t + dt    ! Evolve the label of time

  ! Recycle variables
  phi_pp = phi_p
  phi_p  = phi

  ! Use eq (2.27)
  do i=1,Nx-1
     phi(i) = CFL**2 * (phi_p(i+1) - 2.0d0 * phi_p(i) + phi_p(i-1))
    &         + 2.0d0 * phi_p(i) - phi_pp(i)
  end do
```

```
! *Apply some boundary conditions

! Save data

end do
```

Unfortunately expression (3.27) works fine only for points that are not at the boundary of D_d. At initial time, for the calculation of ϕ_i^1, the scheme (3.27) needs the information of ϕ_i^{-1}, which is not defined within D_d; see Fig. 3.1. Analogously, at the spatial boundaries (x_0, t) and (x_{N_x}, t), for the calculation of ϕ_0^{n+1} and $\phi_{N_x}^{n+1}$, values of ϕ_{-1}^n and $\phi_{N_x+1}^n$ are required and not defined.

These are situations common to IVPs associated to PDEs, and we now explain their solution for this particular discretization of the wave equation.

Problem at initial time. In the present case, initial conditions are specified by $\phi(x, 0)$ and $\dot{\phi}(x, 0)$ because the wave equation is second order in time. Formally these are **Cauchy** boundary conditions applied to the wave function at the initial time side of the domain $(x, 0)$ (see Fig. 3.1). There is a universe of interesting Cauchy initial time boundary conditions for this equation, and for illustration we will choose the time-symmetric initial conditions which set $\dot{\phi}(x, 0) = \dot{\phi}_0(x) = 0$ and $\phi(x, 0) = \phi_0(x)$. This condition reflects the fact that the evolution has a symmetry between past and future at $t = 0$, an appealing case as we will see.

Setting initial conditions for $\phi_i^0 = \phi_0(x_i, t^0) = \phi_0(x_i, 0)$ is straightforward because one only needs to prescribe a profile for $\phi_0(x)$ and fill in the corresponding array with values. However the time derivative in the discrete domain has to be declared in terms of values of the wave function in at least two time levels according to (3.27). For this we define a virtual time slice t^{-1} and associate values for the wave function ϕ_i^{-1} there. The implementation is straightforward by associating `phi_p` with ϕ_i^0 and `phi_pp` with ϕ_i^{-1}, after which it is possible to use (3.27) to start the evolution.

For the initial wave function profile, we use a Gaussian $\phi_0(x) = Ae^{-x^2/\sigma^2}$ which on D_d reads

$$\phi_i^0 = Ae^{-x_i^2/\sigma^2}. \tag{3.28}$$

The wave function at the virtual past time t^{-1} is obtained by Taylor expanding the wave function backward in time $\phi_i^{-1} = \phi_i^0 - \Delta t \partial_t(\phi_i^0) + \frac{\Delta t^2}{2}\partial_{tt}(\phi_i^0) + O(\Delta t^3)$. By assuming time-symmetry $\dot{\phi}(x, 0) = \dot{\phi}_0(x) = 0$, which on D_d reads $\partial_t(\phi_i^0)=0$, which leads to the third order accurate formula:

$$\phi_i^{-1} = \phi_i^0 + \frac{\Delta t^2}{2}\partial_{tt}(\phi_i^0) + O(\Delta t^3) = \phi_i^0 + \frac{\Delta t^2}{2}\partial_{xx}(\phi_i^0) + O(\Delta t^3), \tag{3.29}$$

where the last equality uses the wave equation to switch time by space derivative, which we can calculate using the discrete formulas (3.9), (3.10), and (3.11) on ϕ_i^0

3.2   1 + 1 Wave Equation: The Paradigm of Hyperbolic Equations

and substituting into (3.29). Finally, start the evolution of the wave function with Eq. (3.27) starting with the initial values (3.28) for phi_p and (3.29) for phi_pp.

Problem at left and right boundaries. Now we focus on the implementation of boundary conditions at boundaries (x_0, t^n) and (x_{N_x}, t^n) for all n as indicated in Fig. 3.1 that we illustrate using various physically motivated examples.

For illustration we have to specify the numerical parameters of the IVP. We set the domain to $[x_{min}, x_{max}] = [-1, 1]$ and $t_f = 2$ that we discretize using $N_x = 200$, or equivalently a spatial resolution $\Delta x = 0.01$ for D_d^1. Time resolution is set to $\Delta t = C \Delta x$ with $C = 0.5$. The amplitude and width of the Gaussian are set to $A = 1$ and $\sigma = 0.1$, respectively.

Next we describe particular boundary conditions and their implementation.

Explicit Boundary Values. For the IVP (3.25) it is possible to impose **Dirichlet** boundary conditions and give the values of the wave function at the boundaries $x_0 = -1$ and $x_{N_x} = 1$ at all times. Therefore, one can simply set the values of the wave function at the boundaries, say $\phi(x_{min}, t) = \phi_0^n = 0$ and $\phi(x_{max}, t) = \phi_{N_x}^n = 0$ at all times.

In this case one avoids the problem of missing values of the wave function at positions with $i = -1$ and $i = N_x + 1$ as needed by the scheme (3.27) in the construction of ϕ_i^{n+1} for $i = 0$ and $i = N_x$. The implementation consists in using (3.27) in a loop with $i = 1, \ldots, N_x - 1$ and afterward apply the boundary conditions by setting $\phi_0^{n+1} = \phi_{N_x}^{n+1} = 0$, as follows:

```
! *Apply some boundary conditions
! Dirichlet boundary conditions
phi(0)  = 0.0d0
phi(Nx) = 0.0d0
```

which is a particular and interesting type of boundary conditions. The result of the solution is shown in Fig. 3.4. Immediately after initial time, the Gaussian pulse splits into two smaller pulses with half the initial amplitude, but the same width, moving at speed one toward the boundaries.

In the same Figure, we see that the wave function bounces from the boundaries with negative amplitude because the boundary conditions force the wave function to be zero at the edges $\phi(-1, t) = \phi(1, t) = 0$. Then at time $t = 2$, the two pulses reflected from the boundaries superpose and acquire the profile of the initial conditions with negative amplitude.

Periodic Boundary Conditions. These boundary conditions are useful for the evolution of a wave that propagates on a periodic domain, for example, a wave propagating through a crystal lattice. Numerically, the domain is such that the right end of the domain is identified with the left end, in such a way that the topology of the domain changes from being a segment of a line from x_0 to x_{N_x}, to be a circle of perimeter $x_{N_x} - x_0$ as illustrated in Fig. 3.5. Thus, more than a boundary condition, it is a change of the domain topology, and the only boundary is that at the initial time surface where Cauchy boundary conditions are applied.

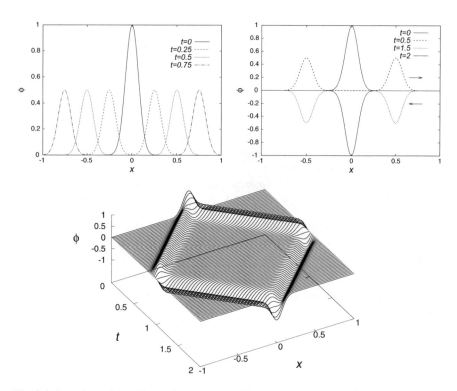

Fig. 3.4 Snapshots of the solution for the case of boundary conditions $\phi(-1,t) = \phi(1,t) = 0$. On the left we show the solution before the pulses reach the boundaries. At the right the pulses after reflection from the boundaries. At the bottom a bundle of snapshots together on D_d

The implementation is based on (3.27) where $i = 1,\ldots,N_x - 1$. For the calculation of ϕ_0^{n+1} in (3.27), the index -1 must be replaced by $N_x - 1$, whereas for the calculation of $\phi_{N_x}^{n+1}$, the index $N_x + 1$ must be substituted by 1. Using this information, the code of formula (3.27) for $i = 0$ and $i = N_x$ for these boundary conditions is as follows:

```
! *Apply some boundary conditions
! Periodic boundary conditions
phi(0) = CFL**2 * (phi_p(1) - 2.0d0 * phi_p(0) + phi_p(Nx-1))
         + 2.0d0 * phi_p(0) - phi_pp(0)
phi(Nx) = CFL**2 * (phi_p(1) - 2.0d0 * phi_p(Nx) + phi_p(Nx-1))
         + 2.0d0 * phi_p(Nx) - phi_pp(Nx)
```

The evolution is shown in Fig. 3.6. By $t = 0.5$ we see the splitting of the initial Gaussian pulse and by $t = 1$ the pulse moving to the left and the one moving to the right add up exactly at the boundaries, since the domain is topologically a circle. At $t = 2$ the two pulses superpose again and resemble the initial Gaussian. These properties are those expected for periodic boundary conditions.

3.2 1+1 Wave Equation: The Paradigm of Hyperbolic Equations

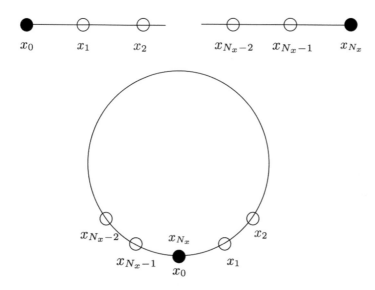

Fig. 3.5 Identification of the spatial domain from being a line segment to being a circle. The boundary condition consists in the appropriate assignation of the wave function values ϕ_i^n for $i = 0, 1, N_x - 1, N_x$ that produces this effect

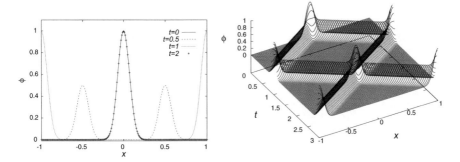

Fig. 3.6 Snapshots of the solution for the case of periodic boundary conditions. Notice that the solution at $t = 2$ is nearly the same as that at initial time $t = 0$. The plot on the right shows snapshots for the time domain with $t_f = 3$ on D_d

Outgoing Wave Boundary Conditions. This is a useful condition that serves to model open boundaries that allow the wave to leave the domain. These are appropriate to model radiative processes of isolated systems that generate signals propagating out of the domain.

The implementation of these conditions needs the following elements. The general solution of the wave equation is of the form $\phi(t, x) = g(x + t) + h(x - t)$ with g and h arbitrary, describing a displacement toward the left and right of signals respectively, along the characteristic lines $t/x =$ constant. These functions g, h are solutions of the equations

$$\left(\frac{\partial}{\partial t} - \frac{\partial}{\partial x}\right) g = 0, \tag{3.30}$$

$$\left(\frac{\partial}{\partial t} + \frac{\partial}{\partial x}\right) h = 0, \tag{3.31}$$

respectively. Therefore the outgoing wave boundary conditions reduce to impose Eq. (3.30) at $x = x_0 = -1$ and (3.31) at $x = x_{N_x} = 1$, for $g, h = \phi$. Since these are conditions imposed on the spatial derivatives of ϕ at each time, they are **Neumann** conditions.

An **important** issue is that in order to impose the condition at the left and right boundaries according to (3.27), one only counts with information from inside of the domain, while one needs discrete versions of (3.30) and (3.31).

Fortunately we have constructed the discrete version of first-order derivatives using forward and backward stencils with second-order accuracy. For the time derivative, in the two equations (3.30) and (3.31), one only has the information in the past time levels t^n and t^{n-1}; thus the only usable operator is the backward formula (3.17). For the spatial derivative in (3.30), interior points are to the right of $i = 0$, and then the forward formula (3.7) is adequate, whereas the space derivative for (3.31) needs the backward formula (3.8) for $i = N_x$. The stencils at boundary points are shown in Fig. 3.7.

With this in mind, the discrete version of conditions (3.30)–(3.31) read

$$\frac{\phi_0^{n-1} - 4\phi_0^n + 3\phi_0^{n+1}}{2\Delta t} - \frac{-\phi_2^{n+1} + 4\phi_1^{n+1} - 3\phi_0^{n+1}}{2\Delta x} = \mathcal{O}(\Delta x^2, \Delta t^2),$$

$$\frac{\phi_N^{n-1} - 4\phi_N^n + 3\phi_N^{n+1}}{2\Delta t} + \frac{\phi_{N-2}^{n+1} - 4\phi_{N-1}^{n+1} + 3\phi_N^{n+1}}{2\Delta x} = \mathcal{O}(\Delta x^2, \Delta t^2). \tag{3.32}$$

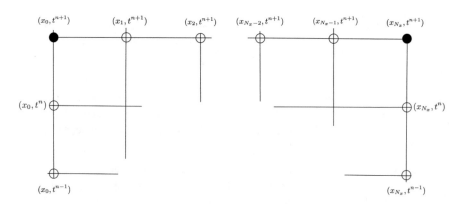

Fig. 3.7 Stencils on D_d for the implementation of the boundary conditions (3.33) and (3.34)

3.2 1 + 1 Wave Equation: The Paradigm of Hyperbolic Equations

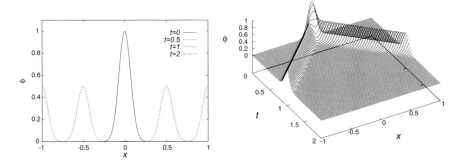

Fig. 3.8 Numerical solution using outgoing wave boundary conditions. By $t = 1$ the wave arrives to the boundary and abandons the domain. The figure at the right shows the solution in the whole domain D_d

From these expressions, it is possible to solve for ϕ_0^{n+1} and $\phi_{N_x}^{n+1}$ in terms of known information:

$$\phi_0^{n+1} = \frac{\frac{\Delta t}{\Delta x}(-\phi_2^{n+1} + 4\phi_1^{n+1}) + 4\phi_0^n - \phi_0^{n-1}}{3(1 + \frac{\Delta t}{\Delta x})} + \mathcal{O}(\Delta x^2, \Delta t^2), \qquad (3.33)$$

$$\phi_{N_x}^{n+1} = \frac{\frac{\Delta t}{\Delta x}(-\phi_{N_x-2}^{n+1} + 4\phi_{N_x-1}^{n+1}) + 4\phi_{N_x}^n - \phi_{N_x}^{n-1}}{3(1 + \frac{\Delta t}{\Delta x})} + \mathcal{O}(\Delta x^2, \Delta t^2). \qquad (3.34)$$

Therefore, using (3.27) for $i = 1, \ldots, N_x - 1$ together with these two formulas for $i = 0$ and $i = N_x$ allows the computation of the solution at time t^{n+1}, and the results are in Fig. 3.8. Notice that after $t = 1$ the wave leaves the numerical domain through the boundary as desired. The implementation follows exactly the formulas (3.33) and (3.34) and reads

```
! *Apply some boundary conditions
! Neumann-type, outgoing wave boundary conditions
phi(0)  = ( CFL * ( -phi(2) + 4.0d0 * phi(1) ) + 4.0d0 * phi_p(0)
          - phi_pp(0) ) / 3.0d0 / ( 1.0d0 + CFL )
phi(Nx) = ( CFL * ( -phi(Nx-2) + 4.0d0 * phi(Nx-1) ) + 4.0d0 *
phi_p(Nx)
          - phi_pp(Nx) ) / 3.0d0 / ( 1.0d0 + CFL )
```

Error and Convergence of the Solution

We illustrate the convergence of the solution corresponding to the case of outgoing wave boundary conditions using the exact solution. The exact solution of the IVP (3.25), with initial conditions $\phi_0(x) = Ae^{-x^2/\sigma^2}$ and $\dot{\phi}_0(x) = 0$, reads

$$\phi^e = \frac{A}{2}\exp[-(x-t)^2/\sigma^2] + \frac{A}{2}\exp[-(x+t)^2/\sigma^2]. \qquad (3.35)$$

In fact this is the solution even if the spatial domain is the whole $\mathbb{R} \times \mathbb{R}$. Notice that the solution at initial time will be a single Gaussian of amplitude A and that time-symmetry holds due to the initial condition $\dot{\phi} = 0$ at $t = 0$. The discrete version of this exact solution for all x_i and t^n in the discrete domain D_d reads

$$\phi^{en}_i = \frac{A}{2} \exp[-(x_i - t^n)^2/\sigma^2] + \frac{A}{2} \exp[-(x_i + t^n)^2/\sigma^2]. \tag{3.36}$$

The error of the numerical solution is defined at each point of D_d by

$$e^n_i = \phi^n_i - \phi^{en}_i. \tag{3.37}$$

Following the hierarchy of numerical domains D^1_d, D^2_d, D^3_d, used to construct solutions with successive resolutions, we calculate the numerical solutions ϕ^{1n}_i, ϕ^{2n}_i and ϕ^{3n}_i using $N_x = 200$, 400 and 800 respectively.

Downsampling in space and time is important for comparison of the solution at points of D^1_d. The code for the output including downsampling in space and time is as follows:

```
resolution_label = 1 ! use 1,2,3 for domains D1,D2,D3
Nx=200
Nx = 2**(resolution_label - 1) * Nx
do n=1,Nt
! Integration stuff
! Output
 if (mod(n,2**resolution_label-1).eq.0) then   ! Write every 2^m
time steps
        do i=0,Nx,2**(resolution_label-1)     ! Write every 2^m points
            write(1,*) x(i),phi(i)
        end do
        write(1,*)
        write(1,*)  ! Two blank spaces are useful for gnuplot
    end if
end do
```

For each numerical solution, the error (3.37) is calculated for a convergence test. In Fig. 3.9 a convergence tests is shown at four different times. Notice that the error with the coarse resolution $N_x = 200$ is four times bigger than that with the middle resolution $N_x = 400$, which in turn is four times bigger than the error using the fine resolution $N_x = 800$, which satisfies the criterion (3.21).

Notice that by $t = 1.5$, the wave has already left the numerical domain; however, there is a finite error propagating backward from the boundaries. Remember that outgoing wave boundary conditions are calculated with second-order accurate formulas, the error is not zero at the boundaries, and part of it is reflected back into the domain but is expected to converge to zero with second order. Therefore, this convergence tests show that the implementation works and converges to the solution in the continuum according to theory in (3.21) for at least four representative snapshots, even after the signal is reflected back from the boundaries.

3.2 1 + 1 Wave Equation: The Paradigm of Hyperbolic Equations

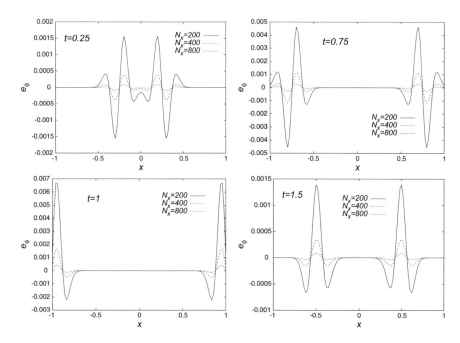

Fig. 3.9 Convergence test of the numerical solution using outgoing wave boundary conditions, at four different times, calculated with $N_x = 200, 400, 800$ cells, or equivalently $\Delta x = 0.01, 0.005, 0.0025$ respectively

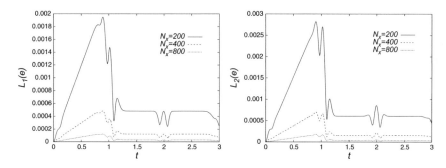

Fig. 3.10 Convergence of $L_1(e)$ and $L_2(e)$ norms of the error (3.37) using outgoing wave boundary conditions and $N_x = 200, 400, 800$ cells, or equivalently $\Delta x = 0.01, 0.005, 0.0025$ respectively

Convergence of the Norms. The convergence during the whole evolution and not only at a few snapshots can be tracked by calculating the convergence of the $L_1(e)$ and $L_2(e)$ norms of the error (3.37), at each time step, using (3.24), for each of the three numerical solutions. The results are shown in Fig. 3.10. The norms of the error using the coarse resolution $N_x = 200$ are four times that of the error using the middle resolution $N_x = 400$, which in turn is four times bigger than the norm of the error using the fine resolution $N_x = 800$. This indicates that $L_1(e)$ and $L_2(e)$ converge with second order to zero.

The Code

Below is the code `WaveSimple.f90` that solves the wave equation using the methods described here. It is equipped with the diagnostics tools to calculate the $L_1(e)$ and $L_2(e)$ norms of the error saved in files with extension `.norms`. It generates output for the one-dimensional plots with extension `.dat` and finally output that allows the generation of the fancy 2D plots in Figs. 3.4, 3.6 and 3.8 with extension `.fancy`.

```
program WaveSimple

implicit none
real(kind=8), allocatable, dimension(:) :: phi,phi_p,phi_pp,x
real(kind=8), allocatable, dimension(:) :: exact, error
real(kind=8) dx,dt,xmin,xmax,CFL,t,amp,sigma,L1,L2
integer i,n,Nx,Nt,resolution_label
resolution_label = 1 ! use 1,2,3 for domains D1,D2,D3
Nx = 200
Nx = 2**(resolution_label-1) * Nx ! Number of cells for the
dicretized
domain
Nt = 2000 * 2**(resolution_label-1)
xmin = -1.0d0
xmax = 1.0d0
CFL = 0.5
amp = 1.0d0
sigma = 0.1

! Alocates memory

allocate(x(0:Nx),phi(0:Nx),phi_p(0:Nx),phi_pp(0:Nx),exact(0:Nx),
error(0:Nx))

! Define domain

dx = (xmax-xmin)/dble(Nx)
dt = CFL * dx
do i=0,Nx
x(i) = xmin + dble(i) * dx
end do

! Init arrays and time
t = 0.0d0
phi = 0.0d0
phi_p = 0.0d0
phi_pp = 0.0d0
error = 0.0d0

! Initial data
```

3.2 1 + 1 Wave Equation: The Paradigm of Hyperbolic Equations   119

```
phi = amp*exp(-x**2/sigma**2) ! Formula (3.28)
do i=1,Nx-1 ! Formula (3.29)
 phi_p(i) = phi(i) + 0.5d0 * CFL**2 * ( phi(i+1) - 2.0*phi(i) +
phi(i-1))
end do
phi_p(0) = 0.0d0
phi_p(Nx) = 0.0d0
exact = amp*exp(-x**2/sigma**2)

open(1,file='WaveSimpleD.dat')    ! File for output of the solution
open(2,file='WaveSimpleD.fancy')  ! File for the fancy 2D plots
open(3,file='WaveSimpleD.norms')  ! File for the error norms

! Saves data at initial time
do i=0,Nx
   write(1,*) x(i),phi(i)
end do
write(1,*)
write(1,*)
do i=0,Nx
   write(2,*) t,x(i),phi(i)
end do
write(2,*)
do n=1,Nt

   print *, n,t ! Send something to the screen
   t = t + dt ! Updates time

! ***** -- LOOP CORE --

   ! Recycle arrays
   phi_pp = phi_p
   phi_p = phi

   ! Formula for the evolution (3.27)
   do i=1,Nx-1
      phi(i) = CFL**2 * (phi_p(i+1) - 2.0d0 * phi_p(i)
+ phi_p(i-1)) + 2.0d0 * phi_p(i) - phi_pp(i)
   end do

   ! --> Uncomment the one to be used <--
   ! Fixed boundary conditions
   ! phi(0) = 0.0d0
   ! phi(Nx) = 0.0d0

   ! Periodic boundary conditions
   ! phi(0) = CFL**2 * (phi_p(1) - 2.0d0 * phi_p(0)
   !        + phi_p(Nx-1)) & + 2.0d0 * phi_p(0) - phi_pp(0)
   ! phi(Nx) = CFL**2 * (phi_p(1) - 2.0d0 * phi_p(Nx)
   !        + phi_p(Nx-1)) & + 2.0d0 * phi_p(Nx) - phi_pp(Nx)

   ! Outgoing wave boundary conditions
```

120   3 Simple Methods for Initial Value Problems Involving PDEs

```fortran
       phi(0) = ( CFL * ( -phi(2) + 4.0d0 * phi(1) ) + 4.0d0  
              * phi_p(0) & - phi_pp(0) ) / 3.0d0 / ( 1.0d0 + CFL )
       phi(Nx) = ( CFL * ( -phi(Nx-2) + 4.0d0 * phi(Nx-1) ) + 4.0d0
     * phi_p(Nx) &
              - phi_pp(Nx) ) / 3.0d0 / ( 1.0d0 + CFL )

 ! ***** -- Ends Loop Core --

       ! Exact solution only uncomment for the outgoing wave boundary
 conditions case
       exact = 0.5d0 * amp * exp(-( x - t )**2 / sigma**2) &
             + 0.5d0 * amp * exp(-( x + t )**2 / sigma**2)

       error = phi - exact

       ! Save data on the fly

       if (mod(n,2**resolution_label).eq.0) then    ! Regular output
       do i=0,Nx,2**resolution_label
        write(1,*) x(i),phi(i),exact(i),error(i)
       end do
       write(1,*)
       write(1,*)
       end if

       if (mod(n,5).eq.0) then    ! Fancy output
       write(2,*)
       do i=0,Nx
        write(2,*) t,x(i),phi(i)
       end do
       write(2,*)
       end if

       ! Norms
       norm1 = 0.0d0
       L2 = 0.0d0
       do i=1,Nx
        L1 = L1 + 0.5D0*(dabs(error(i-1)) + dabs(error(i)))*dx
        L2 = L2 + 0.5D0*(error(i-1)**2 + error(i)**2)*dx
       end do
       L2 = sqrt(L2)
       write(3,*) t,L1,L2

 end do

 close(3)
 close(2)
 close(1)

 end program
```

Stability

Before finishing this section it is important to show that scheme (3.27) has a stability range.

> **Exercise.** Find empirically, using the code, that $C = 1$ is a threshold of stability. This means that using $C > 1$ the amplitude can grow exponentially, whereas using $C < 1$ gives a solution with a bounded amplitude.

A proof that shows that $C \leq 1$ is a sufficient condition for stability of the scheme (3.27) appears in Appendix A.

3.2.1 Dissipation and Dispersion

Aside of the **truncation error**, whose theory is in Sect. 3.1 and its convergence has been illustrated for the wave equation in Figs. 3.9 and 3.10, there are other types of error that accumulate and produce important effects, manifest particularly in large time domains.

This can be exemplified with the solution of the wave equation using periodic boundary conditions during a number of **crossing times**, defined as the time it takes a signal to travel across the domain; in the case of our wave equation, since the speed is one and the domain is two units wide, the crossing time is two units of time. For the illustration we use the same spatial resolution defined to produce Fig. 3.6, the only difference being the time domain, which is set to $t \in [0, 50]$. The results are in Fig. 3.11, where we show snapshots of the numerical solution every two units of time, so that the pulse is captured with its center at $x = 0$. In the continuum limit the figure should be the Gaussian pulse of the initial conditions superposed 25 times.

The numerical solution presents two effects to be noticed. First, the amplitude of the Gaussian pulse decreases, an effect associated to **dissipation**. Second, the pulse is distorted with time, an effect associated with **dispersion** of the wave. The results are shown for three resolutions in order to see how the magnitude is in each case.

Dissipation can be measured with the central value of ϕ as function of time, registered every time the pulse passes through the center of the domain. This is shown in the fourth plot of Fig. 3.11, and its smaller deviation for higher resolution indicates consistency.

Different evolution schemes have different **dissipation** and **dispersion** properties. All depends on how similar to the diffusion equation the scheme is and how different modes depend on propagation velocities. For example, later on in Sect. 4.4, we use a different integration scheme, the Method of Lines to solve the same wave equation as done here, and the effects of dissipation and dispersion are different

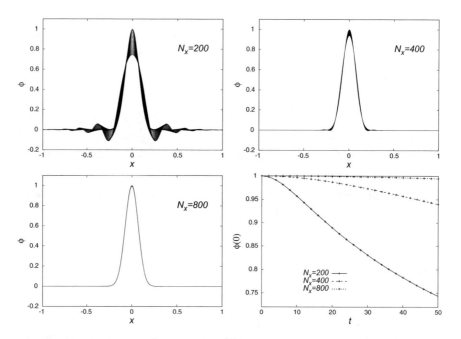

Fig. 3.11 Snapshots of the solution using periodic boundary conditions in the time domain [0, 50], every two units of time using three resolutions with $N_x = 200, 400, 800$. At the right-bottom we show the central value of ϕ as function of time for the three runs, which illustrates that diffusion is resolution dependent

and can be seen in Figs. 4.7 and 4.8. The later is an example of how dissipation is important for the stability of a scheme.

> **Exercise.** Results in Fig. 3.11 are constructed using the Courant-Friedrichs-Lewy factor $CFL = 0.5$, which illustrates the dependency on spatial resolution. Investigate if there is a type of dependency on time resolution by using $CFL = 0.25$.

The code `WaveSimple.f90` is available as Electronic Supplementary Material described in Appendix B and can be used to reproduce the results in this section.

3.3 1 + 1 Wave Equation Using Implicit Methods

We recall the IVP for the 1 + 1 wave equation from (3.25):

3.3 1 + 1 Wave Equation Using Implicit Methods

$$\begin{aligned}
&\frac{\partial^2 \phi}{\partial t^2} - \frac{\partial^2 \phi}{\partial x^2} = 0 \qquad \phi = \phi(x,t) \\
&D = x \in [x_{min}, x_{max}] \times t \in [0, t_f] \quad \text{Domain} \\
&\phi(x,0) = \phi_0(x) \qquad \qquad \qquad \text{Initial Conditions} \\
&\dot{\phi}(x,0) = \dot{\phi}_0(x) \\
&\phi(x_{min}, t), \; \phi(x_{max}, t) \qquad \qquad \text{Boundary Conditions}
\end{aligned} \qquad (3.38)$$

1. For the *discretization* of D_d, we use exactly the construction described in the previous section.
2. The *discrete version of the PDE* using this new method is different from (3.26). In this case it is assumed that the discrete version of the equation around the point $(x_i, t^n) \in D_d$ is a time average of the spatial part at times t^{n-1} and t^{n+1}, a method called Implicit Crank-Nicolson, and the discretized version of the equation is

$$\frac{\phi_i^{n+1} - 2\phi_i^n + \phi_i^{n-1}}{\Delta t^2} = \frac{1}{2}\left[\frac{\phi_{i+1}^{n+1} - 2\phi_i^{n+1} + \phi_{i-1}^{n+1}}{\Delta x^2} + \frac{\phi_{i+1}^{n-1} - 2\phi_i^{n-1} + \phi_{i-1}^{n-1}}{\Delta x^2}\right]$$
$$+ \mathcal{O}(\Delta x^2, \Delta t^2). \qquad (3.39)$$

The wave equation is discretized using the centered version of the second derivative in space (3.9) and time (3.18). Unlike the simple discrete version (3.26), one cannot isolate ϕ_i^{n+1} like in (3.27), in terms of known information at times t^n and t^{n-1}. This is the reason why the method (3.27) is called an **explicit method**, whereas (3.39) is called **implicit method**.

The strategy to solve the IVP has to be different. Let $\alpha = \frac{1}{2}\left(\frac{\Delta t}{\Delta x}\right)^2$ and rearrange (3.39) as follows:

$$(-\alpha)\phi_{i-1}^{n+1} + (1+2\alpha)\phi_i^{n+1} +|, (-\alpha)\phi_{i+1}^{n+1} = (\alpha)\phi_{i-1}^{n-1} + (-1-2\alpha)\phi_i^{n-1} + (\alpha)\phi_{i+1}^{n-1}$$
$$+ 2\phi_i^n + \mathcal{O}(\Delta x^2, \Delta t^2), \qquad (3.40)$$

which defines a linear system of equations for $\phi_{i-1}^{n+1}, \phi_i^{n+1}, \phi_{i+1}^{n+1}$ at the interior points of D_d along the spatial direction $i = 1, \ldots, N_x - 1$. The information in formula (3.40) is represented by the molecule in Fig. 3.12, to be compared with that in Fig. 3.3. Notice that there are now three unknowns simultaneously at time t^{n+1}. Different boundary conditions define different linear systems that we illustrate here, because we will need to construct an appropriate evolution equation similar to (3.40) for $i = 0$ and $i = N_x$, using only information at points of D_d.

Dirichlet Boundary Conditions. This boundary condition sets values of ϕ at boundaries (x_0, t^n) and (x_{N_x}, t^n) in Fig. 3.1. In the example below, we use $\phi(x_0, t) =$

Fig. 3.12 Molecule associated to the evolution scheme (3.40) at points of D_d. Black dots correspond to points where the value of ϕ is unknown, whereas white dots to points where ϕ is known

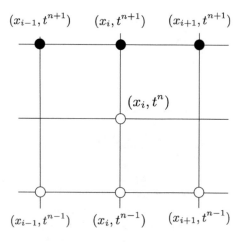

$\phi(x_{N_x}, t) = 0$, and first we deal with the boundary at (x_0, t^n). The condition means that $\phi_0^n = 0$ for all n. Substituting $i = 0$ in formula (3.40), and assuming the virtual values of ϕ_{-1}^k for $k = n-1, n+1$ at x_{-1} exist, one obtains

$$(-\alpha)\phi_{-1}^{n+1} + (1+2\alpha)\phi_0^{n+1} + (-\alpha)\phi_1^{n+1} = (\alpha)\phi_{-1}^{n-1} + (-1-2\alpha)\phi_0^{n-1} + (\alpha)\phi_1^{n-1} + 2\phi_0^n.$$

Now, setting $\phi_0^k = 0$, constructing a line in the $x\phi$–plane that passes through the points (x_0, ϕ_0^k) and (x_1, ϕ_1^k), one extrapolates the wave function at the virtual point and find that $\phi_{-1}^k = -\phi_1^k$. This can be substituted back into the above equation and obtain

$$(1 + 2\alpha)\phi_0^{n+1} = (-1 - 2\alpha)\phi_0^{n-1} + 2\phi_0^n \qquad (3.41)$$

which includes only values of ϕ defined at points of D_d.

At the right boundary (x_{N_x}, t^n), assuming that the information at x_{N_x+1} exists, Eq. (3.40) reads

$$(-\alpha)\phi_{N_x-1}^{n+1} + (1 + 2\alpha)\phi_{N_x}^{n+1} + (-\alpha)\phi_{N_x+1}^{n+1} = (\alpha)\phi_{N_x-1}^{n-1} + (-1 - 2\alpha)\phi_{N_x}^{n-1}$$
$$+ (\alpha)\phi_{N_x+1}^{n-1} + 2\phi_{N_x}^n.$$

The boundary condition is $\phi_{N_x}^k = 0$ for $k = n-1, n, n+1$. Constructing the line that passes through the points $(x_{N_x-1}, \phi_{N_x-1}^k)$ and $(x_{N_x}, \phi_{N_x}^k)$ where $\phi_{N_x}^k = 0$, one extrapolates the virtual value $\phi_{N_x+1}^k = -\phi_{N_x-1}^k$. Then the previous formula reduces to

$$(1 + 2\alpha)\phi_{N_x}^{n+1} = (-1 - 2\alpha)\phi_{N_x}^{n-1} + 2\phi_{N_x}^n. \qquad (3.42)$$

3.3 1 + 1 Wave Equation Using Implicit Methods

Finally, formulas (3.40), (3.41), and (3.42) define the following linear tridiagonal system of equations:

$$\begin{bmatrix} 1+2\alpha & 0 & 0 & & & & & \\ -\alpha & 1+2\alpha & -\alpha & 0 & & & & \\ 0 & -\alpha & 1+2\alpha & -\alpha & 0 & & & \\ & & \ldots & & & & & \\ & & \ldots & & & & & \\ & & \ldots & & & & & \\ & & & 0 & -\alpha & 1+2\alpha & -\alpha & 0 \\ & & & & 0 & -\alpha & 1+2\alpha & -\alpha \\ & & & & & 0 & 0 & 1+2\alpha \end{bmatrix} \begin{bmatrix} \phi_0^{n+1} \\ \phi_1^{n+1} \\ \phi_2^{n+1} \\ \cdot \\ \cdot \\ \cdot \\ \phi_{N_x-2}^{n+1} \\ \phi_{N_x-1}^{n+1} \\ \phi_{N_x}^{n+1} \end{bmatrix}$$

$$= \begin{bmatrix} (-1-2\alpha)\phi_0^{n-1} + 2\phi_0^n \\ (\alpha)\phi_0^{n-1} + (-1-2\alpha)\phi_1^{n-1} + (\alpha)\phi_2^{n-1} + 2\phi_1^n \\ (\alpha)\phi_1^{n-1} + (-1-2\alpha)\phi_2^{n-1} + (\alpha)\phi_3^{n-1} + 2\phi_2^n \\ \cdot \\ \cdot \\ \cdot \\ (\alpha)\phi_{N_x-3}^{n-1} + (-1-2\alpha)\phi_{N_x-2}^{n-1} + (\alpha)\phi_{N_x-1}^{n-1} + 2\phi_{N_x-2}^n \\ (\alpha)\phi_{N_x-2}^{n-1} + (-1-2\alpha)\phi_{N_x-1}^{n-1} + (\alpha)\phi_{N_x}^{n-1} + 2\phi_{N_x-1}^n \\ (-1-2\alpha)\phi_{N_x}^{n-1} + 2\phi_{N_x}^n \end{bmatrix}, \quad (3.43)$$

that has to be solved for ϕ_i^{n+1} in terms of ϕ_i^n and ϕ_i^{n-1} for $i = 0, \ldots, N_x$.

A parenthesis on the solution of tridiagonal systems is in turn. Assume the system of equations to be solved is $A\mathbf{x} = \mathbf{d}$ with A tridiagonal as follows:

$$\begin{bmatrix} b_0 & c_0 & 0 & & & & \\ a_1 & b_1 & c_1 & 0 & & & \\ 0 & a_2 & b_2 & c_2 & 0 & & \\ & & \ldots & & & & \\ & & \ldots & & & & \\ & & \ldots & & & & \\ & & & 0 & a_{N_x-2} & b_{N_x-2} & c_{N_x-2} & 0 \\ & & & & 0 & a_{N_x-1} & b_{N_x-1} & c_{N_x-1} \\ & & & & & 0 & a_{N_x-1} & b_{N_x} \end{bmatrix} \begin{bmatrix} x_0 \\ x_1 \\ x_2 \\ \cdot \\ \cdot \\ \cdot \\ x_{N_x-2} \\ x_{N_x-1} \\ x_{N_x} \end{bmatrix} = \begin{bmatrix} d_0 \\ d_1 \\ d_2 \\ \cdot \\ \cdot \\ \cdot \\ d_{N_x-2} \\ d_{N_x-1} \\ d_{N_x} \end{bmatrix}, \quad (3.44)$$

The system can be solved by forward and then backward substitution [3]. For this, it is simple to store the nonzero entries of the matrix and the right hand side vector a_i, b_i, c_i, d_i as four one-dimensional arrays a, b, c, d. Since we want to make the

126   3 Simple Methods for Initial Value Problems Involving PDEs

code easy, we will use arrays with $i = 0, \ldots, N_x$ entries, even if we waste storage for the entries a_0 and c_{N_x}.

In terms of the arrays a,b,c,d, the auxiliary number tmp and the auxiliary array aux, given that $b_0 \neq 0$, literally following the recipe in the Numerical Recipes book [3], the solution of the system can be coded as

```
! Forward substitution
tmp = b(0)
phi(0) = d(0)/tmp
do k=1,Nx
    aux(k) = c(k-1)/tmp
    tmp = b(k)-a(k)*aux(k)
    phi(k) = (d(k)-a(k)*phi(k-1))/tmp
end do
! Backward substitution
do k = Nx-1,0,-1
    phi(k) = phi(k) - aux(k+1)*phi(k+1)
end do
```

This is the method used to solve the system (3.43). The essential loop of the program that solves this problem and evolves the information using data at times t^{n-1} and t^n to construct the solution at time t^{n+1} appears below. Here we assume, like in the previous method, that ϕ_i^{n-1}, ϕ_i^n, ϕ_i^{n+1} are associated to the arrays phi_pp, phi_p and phi respectively:

```
! ***** --> LOOP CORE <--

alpha = 0.5d0*dt**2/dx**2
! Recycle arrays
phi_pp = phi_p
phi_p = phi


a(0) = 0.0d0 ! waste this entry
c(Nx) = 0.0d0 ! waste this entry
! ZERO boundary conditions
b(0) = (2.0d0*alpha + 1.0d0)
c(0) = 0.0d0
a(Nx) = 0.0d0
b(Nx) = (2.0d0*alpha + 1.0d0)
d(0) = - (2.0d0*alpha + 1.0d0)*phi_pp(0) + 2.0d0*phi_p(0)
d(Nx) = - (2.0d0*alpha + 1.0d0)*phi_pp(Nx) + 2.0d0*phi_p(Nx)
do i=1,Nx-1
    a(i) = -alpha
    b(i) = (2.0d0*alpha + 1.0d0)
    c(i) = -alpha
    d(i) = alpha*phi_pp(i-1) -(2.0d0*alpha + 1.0d0)*phi_pp(i) &
        + alpha*phi_pp(i+1) + 2.0d0*phi_p(i)
end do
```

3.3 1 + 1 Wave Equation Using Implicit Methods

```
! Three diagonal system solution
! Forward substitution
tmp = b(0)
phi(0) = d(0)/tmp
do i=1,Nx
    aux(i) = c(i-1)/tmp
    tmp = b(i)-a(i)*aux(i)
    phi(i) = (d(i)-a(i)*phi(i-1))/tmp
end do
! Backward substitution
do i = Nx-1,0,-1
    phi(i) = phi(i) - aux(i+1)*phi(i+1)
end do

! ***** --> Ends Loop Core <--
```

Entries of the arrays phi(i), phi_p(i), phi_pp(i) are the entries of the wave functions ϕ_i^{n+1}, ϕ_i^n, and ϕ_i^{n-1} respectively, that appear in the tridiagonal system.

For the implementation of the method, we set the numerical parameters to the following. The numerical domain defined by $[x_{min}, x_{max}] = [-1, 1]$, $t \in [0, t_f]$, with $D_d = \{(x_i, t^n)\}$ such that $x_i = x_{min} + i\Delta x$, with $\Delta x = (x_{max} - x_{min})/N_x$ where $N_x = 200$, which implies $\Delta x = 0.01$ as resolution for D_d^1. The time discretization is such that $\Delta t = C \Delta x$ with $C = 0.5$. The evolution time is set to $t_f = 2$.

The initial conditions are, like in the previous section, **Cauchy** boundary conditions consisting of a time-symmetric Gaussian pulse with $\phi_0(x) = A e^{-x^2/\sigma^2}$ and $\dot{\phi}_0(x) = 0$, where the amplitude and width are set to $A = 1$ and $\sigma = 0.1$. These initial conditions are implemented at time t^0 using (3.28) for ϕ_i^0 and at the virtual time t^{-1} using (3.29) for ϕ_i^{-1}.

The results are shown in Fig. 3.13, which can be compared visually with those in Fig. 3.4.

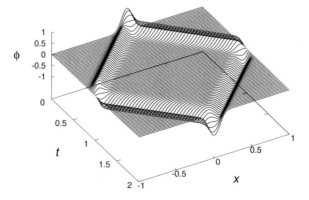

Fig. 3.13 Solution for the boundary conditions $\phi(-1, t) = \phi(1, t) = 0$. Pretty similar to Fig. 3.4, but this time solved using the implicit method

Outgoing Wave Boundary Conditions. These are **Neumann** boundary conditions at boundaries (x_0, t^n) and (x_{N_x}, t^n) in Fig. 3.1, imposed on the space derivatives of ϕ, that have to be translated into the first and last of equations that complete formula (3.40) for $i = 0$ and $i = N_x$.

First, at the left boundary (x_0, t^n), we impose the condition $\partial_t \phi = \partial_x \phi$ as seen from (3.30) for $g = \phi$. Notice that formula (3.40) for $i = 0$ needs the unavailable values ϕ_{-1}^{n+1} and ϕ_{-1}^{n-1} that we now construct in terms of known information. A second-order accurate discretization of this boundary condition at point (x_0, t^{n+1}), centered along the spatial direction (3.5) and backward along the time direction (3.17), reads

$$\frac{\phi_0^{n-1} - 4\phi_0^n + 3\phi_0^{n+1}}{2\Delta t} = \frac{\phi_1^{n+1} - \phi_{-1}^{n+1}}{2\Delta x} + \mathcal{O}(\Delta x^2, \Delta t^2),$$

whereas the discretization at point (x_0, t^{n-1}), centered along the spatial direction (3.5) and forward along the time direction (3.16) is

$$-\frac{\phi_0^{n+1} - 4\phi_0^n + 3\phi_0^{n-1}}{2\Delta t} = \frac{\phi_1^{n-1} - \phi_{-1}^{n-1}}{2\Delta x} + \mathcal{O}(\Delta x^2, \Delta t^2).$$

The stencils used for these two formulas are shown in Fig. 3.14, where circles indicate the position of points belonging to D_d, whereas triangles indicate positions out of D_d where the values of ϕ are unavailable. From these two expressions, dropping the error terms and keeping in mind these are second order, one finds

$$\phi_{-1}^{n+1} = -\frac{\Delta x}{\Delta t} \left(\phi_0^{n-1} - 4\phi_0^n + 3\phi_0^{n+1} \right) + \phi_1^{n+1},$$

$$\phi_{-1}^{n-1} = \frac{\Delta x}{\Delta t} \left(\phi_0^{n+1} - 4\phi_0^n + 3\phi_0^{n-1} \right) + \phi_1^{n-1}, \tag{3.45}$$

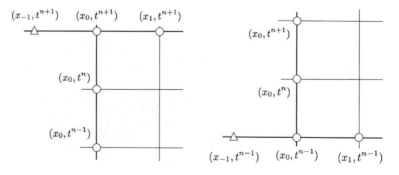

Fig. 3.14 Stencils for the construction of outgoing wave boundary conditions at the left boundary (x_0, t^n). The circles indicate the position of points in D_d, whereas triangles indicate the location of the virtual values ϕ_{-1}^{n+1} and ϕ_{-1}^{n-1} outside D_d

3.3 1 + 1 Wave Equation Using Implicit Methods

values that can be substituted back into the general discrete equation (3.40) for $i = 0$ as follows:

$$(-\alpha)\phi_{-1}^{n+1} + (1 + 2\alpha)\phi_0^{n+1} + (-\alpha)\phi_1^{n+1} = (\alpha)\phi_{-1}^{n-1} + (-1 - 2\alpha)\phi_0^{n-1} + (\alpha)\phi_1^{n-1} + 2\phi_0^n.$$

Defining $\kappa = \frac{\Delta x}{\Delta t}$ and substituting (3.45) into this expression, one finds

$$(2\kappa\alpha + 2\alpha + 1)\phi_0^{n+1} + (-2\alpha)\phi_1^{n+1} = (2\kappa\alpha - 2\alpha - 1)\phi_0^{n-1} + (2\alpha)\phi_1^{n-1} + 2\phi_0^n, \quad (3.46)$$

which will be the first equation of the tridiagonal linear system (3.40) using outgoing wave boundary conditions.

Now let us see the problem at the right boundary (x_{N_x}, t^n) in Fig. 3.1. In this case we impose the condition $\partial_t\phi = -\partial_x\phi$ from (3.31) for $h = \phi$. For this we construct a second-order accurate discretization of this condition at point (x_{N_x}, t^{n+1}), centered along the spatial direction (3.5) and backward along the time direction (3.17):

$$\frac{\phi_{N_x}^{n-1} - 4\phi_{N_x}^n + 3\phi_{N_x}^{n+1}}{2\Delta t} = -\frac{\phi_{N_x+1}^{n+1} - \phi_{N_x-1}^{n+1}}{2\Delta x},$$

and the discretization at point (x_{N_x}, t^{n-1}), centered along the spatial direction (3.5) and forward along the time direction (3.16):

$$\frac{\phi_{N_x}^{n+1} - 4\phi_{N_x}^n + 3\phi_{N_x}^{n-1}}{2\Delta t} = \frac{\phi_{N_x+1}^{n-1} - \phi_{N_x-1}^{n-1}}{2\Delta x}.$$

The stencils used for these two formulas are shown in Fig. 3.15. From these expressions it is possible to isolate the virtual values:

$$\phi_{N_x+1}^{n+1} = -\frac{\Delta x}{\Delta t}\left(\phi_{N_x}^{n-1} - 4\phi_{N_x}^n + 3\phi_{N_x}^{n+1}\right) + \phi_{N_x-1}^{n+1},$$

$$\phi_{N_x+1}^{n-1} = \frac{\Delta x}{\Delta t}\left(\phi_{N_x}^{n+1} - 4\phi_{N_x}^n + 3\phi_{N_x}^{n-1}\right) + \phi_{N_x-1}^{n-1}. \quad (3.47)$$

Substituting these expressions into the general discrete formula (3.40) leads to

$$(-2\alpha)\phi_{N_x-1}^{n+1} + (2\kappa\alpha + 2\alpha + 1)\phi_{N_x}^{n+1} = (2\alpha)\phi_{N_x-1}^{n-1} + (2\kappa\alpha - 2\alpha - 1)\phi_{N_x}^{n-1} + 2\phi_{N_x}^n, \quad (3.48)$$

which is the N_x-th equation of the linear system (3.40) using these boundary conditions. In this way, Eqs. (3.46) and (3.48) complete the tridiagonal system (3.40).

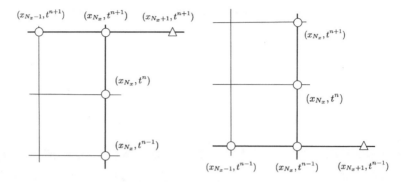

Fig. 3.15 Stencils for the construction of outgoing boundary conditions at the right boundary (x_{N_x}, t^n). Circles indicate the points of D_d, whereas triangles indicate the location of points out of D_d, for the virtual values $\phi_{N_x+1}^{n+1}$ and $\phi_{N_x+1}^{n-1}$

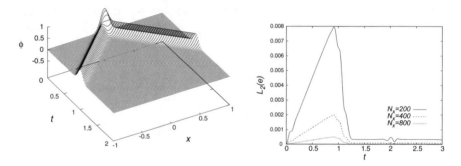

Fig. 3.16 Solution for the case of outgoing wave boundary conditions. Similar to Fig. 3.8, but this time solved using the implicit method. At the right we show the $L_2(e)$ norm of the error that can be seen to converge to zero with second order, pretty similar to that in Fig. 3.10, with a different amplitude

Now the tridiagonal system is complete.

The solution using these boundary conditions is shown in Fig. 3.16, which can be visually compared with the solution in Fig. 3.8.

This ends the explanation of how the *Implicit Crank-Nicolson* method works at solving the wave equation. Nevertheless the **most important property of this method** is the stability. We remind the reader that the method from the previous section provides a stable solution when the condition $C = \Delta t / \Delta x \leq 1$ is fulfilled.

The Crank-Nicolson method is *unconditionally stable*, that is, the solution is stable despite the value of $C = \Delta t / \Delta x$. It is possible to use a big time step Δt and illustrate the stability property using the case of outgoing wave boundary conditions, keeping in mind that we have used $C = 0.5$ so far. We now show how the method performs for $C = 1$ and $C = 2$, for which we will only change the CFL parameter in the code, which will produce time steps two and four times bigger than in the construction of Fig. 3.16. The results are in Fig. 3.17.

3.3 1 + 1 Wave Equation Using Implicit Methods

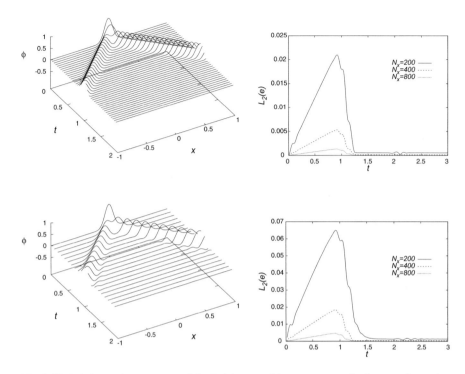

Fig. 3.17 Solution and convergence of the $L_2(e)$ norm of the error to zero, for the case of outgoing wave boundary conditions using $C = 1$ at the top and $C = 2$ at the bottom

The solution is stable and convergent, since the error converges to zero with second order according to the plots in Fig. 3.17. However, the error is bigger for bigger value of C and from the deformation of the solution near the pulses in the solution for $C = 2$. This is only a lesson showing that stability is a desirable property of an evolution scheme, but that the price to pay is accuracy.

Notice that the discrete version in (3.39) and the boundary conditions imposed here are accurate to order $\mathcal{O}(\Delta x^2, \Delta t^2)$, or equivalently $\mathcal{O}(\Delta t^2)$ given that $\Delta t = C\Delta x$. Then using a bigger C for a fixed resolution Δx implies a bigger Δt, which increases the **truncation error** quadratically. It can also be noticed that the solution with $C = 2$ presents a distorted wave function, a consequence of the **dispersion error** of the scheme, that becomes more evident for bigger Δt. Again, this is a price to pay for using big time steps.

> **Exercise.** Following the theory and examples in Appendix A, show that the implicit scheme (3.40) is unconditionally stable.

> **Exercise.** Produce plots equivalent to those in Fig. 3.11, this time using the Crank-Nicolson scheme.

The code `WaveImplicit.f90` is available as Electronic Supplementary Material described in Appendix B and can be used to reproduce all the results in this section.

3.4 Diffusion Equation: The Paradigm of Parabolic Equations

The IVP for the $1 + 1$ diffusion equation can be defined as

$$
\begin{array}{ll}
\dfrac{\partial u}{\partial t} = \kappa \dfrac{\partial^2 u}{\partial x^2} & u = u(x,t) \\
D = x \in [x_{min}, x_{max}] \times t \in [0, t_f] & Domain \\
u(x,0) = u(x) & Initial\ Conditions \\
u(x_{min}, t),\ u(x_{max}, t) & Boundary\ Conditions
\end{array}
\tag{3.49}
$$

where κ is the diffusion coefficient. We illustrate the numerical solution in the domain $[x_{min}, x_{max}] = [0, 1]$ with $t_f = 1$. The discrete domain is $D_d = \{(x_i, t^n)\}$, $x_i = i \Delta x, i = 0, \ldots, N_x$ which implies the spatial resolution $\Delta x = 1/N_x$. We use Dirichlet boundary conditions, $u(x_0, t^n) = u(x_{N_x}, t^n) = 0$.

This IVP has exact solution. Let us assume $\kappa > 0$, and separation of variables assume $u(x,t) = X(x)T(t)$, which substituted into Eq. (3.49) implies

$$\frac{1}{T}\frac{dT}{dt} = \kappa \frac{1}{X}\frac{d^2 X}{dx^2} = -K,$$

with K a constant that we assume positive. Integration of the equations for T and X gives

$$T = T_0 e^{-K(t-t_0)}, \quad X = A \cos\left(\sqrt{\frac{K}{\kappa}} x\right) + B \sin\left(\sqrt{\frac{K}{\kappa}} x\right),$$

where $T_0 = T(0)$ and A, B are constants. Imposing the boundary conditions $X(0) = X(1) = 0$, one finds that $A = 0$ and $K = \kappa n^2 \pi^2$, where n is an integer. Setting initial time to $t_0 = 0$, the exact solution reduces to

3.4 Diffusion Equation: The Paradigm of Parabolic Equations

$$u^e(x,t) = X(x)T(t) = B\sin(n\pi x)e^{-\kappa n^2 \pi^2 t} \tag{3.50}$$

where B is an arbitrary amplitude. This is the exact solution we will use to define initial conditions in numerical examples with $B = 1$ and to compare the numerical solutions with.

At this point we use the two methods previously explored in this chapter and illustrate how the solutions behave.

3.4.1 Simple Discretization Method

A straightforward discrete version of the PDE in (3.49) is constructed using the expressions (3.9) and (3.13) together:

$$\frac{u_i^{n+1} - u_i^n}{\Delta t} = \kappa \frac{u_{i+1}^n - 2u_i^n + u_{i-1}^n}{\Delta x^2} + \mathcal{O}(\Delta t, \Delta x^2), \tag{3.51}$$

which is accurate to first order in time and second order in space if the Taylor expansions are developed around the point (x_i, t^n). From this relation it is possible to solve for u_i^{n+1} in terms of past values

$$u_i^{n+1} = \kappa \left(\frac{\Delta t}{\Delta x^2}\right) [u_{i+1}^n - 2u_i^n + u_{i-1}^n] + u_i^n + \mathcal{O}(\Delta t, \Delta x^2). \tag{3.52}$$

With this expression one can calculate the value of the unknown u at point x_i and time t^{n+1} in terms of the values of u at previous time t^n. The molecule associated to this expression appears in Fig. 3.18. One can define time resolution through a factor similar to the CFL factor $C := \frac{\Delta t}{\Delta x^2}$. Notice the quadratic dependence of Δt on Δx, which is a subtlety to take into account when setting time resolution.

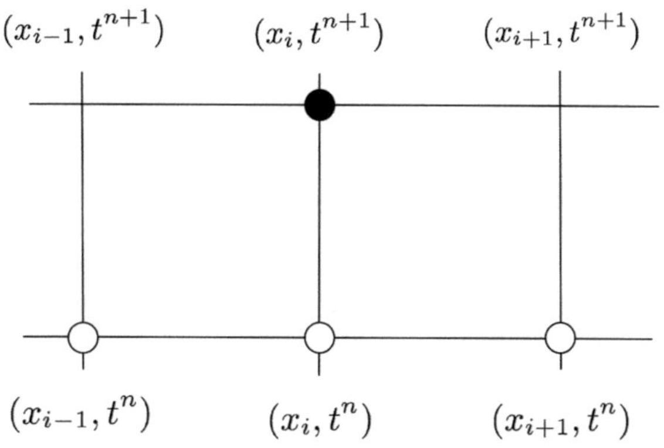

Fig. 3.18 Molecule associated to the evolution scheme (3.52) at interior points of D_d. Black dots correspond to positions where u is unknown, whereas white dots indicate the position where u is known and needed

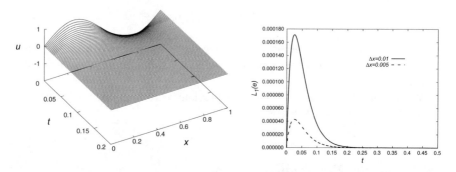

Fig. 3.19 Solution using the simple discretization scheme for initial conditions with $B = 1, n = 2$ and $\kappa = 1$. At the left, snapshots of the numerical solution u on D_d with resolution $\Delta x = 0.01$ and $C = 0.5$. At the right we show the L_1 norm of the error, using resolutions $\Delta x_1 = 0.01$ and $\Delta x_2 = 0.005$ keeping $C = 0.5$, which implies time resolution is $\Delta t_1 = 5 \times 10^{-5}$ and $\Delta t_2 = 1.25 \times 10^{-5}$

Initial Conditions are boundary conditions at $t = 0$ as seen in Fig. 3.1, and since the PDE is first order in time, we impose Dirichlet conditions $u(x, 0) = u_0(x)$ using the exact solution for $t = 0$, $u_0(x) = B \sin(n\pi x)$.

Boundary conditions at (x_0, t^n) and (x_{N_x}, t^n) are also Dirichlet conditions $u(x_0, t^n) = u(x_{N_x}, t^n) = 0$.

The numerical solution for $n = 2$ is shown in Fig. 3.19 for $\kappa = 1$. The numerical parameters are $N_x = 100$ with $C = 0.5$, which implies space and time resolutions are $\Delta x = 0.01$ and $\Delta t = 0.5 \Delta x^2 = 5 \times 10^{-5}$. As expected, the diffusion equation diffuses the initial data exponentially to zero.

Convergence. A convergence test becomes a little bit subtle. Assuming a base resolution Δx_1 of D_d^1 and $\Delta t_1 = C \Delta x_1^2$, when increasing resolution $\Delta x_2 = \Delta x_1/2$ in the construction of D_d^2 keeping $C = \Delta t / \Delta x^2$ constant, $\Delta t_2 = C \Delta x_2^2 = C \Delta x_1^2/4 = \Delta t_1/4$. This means that time resolution of D_d^2 is not halved, but divided by four.

Since the approximation is accurate with first order in time and second order in space as seen in (3.51), it should converge to first order in time and second order in space. In Fig. 3.19 we show the L_1 norm of the error of the numerical solution, calculated as $e_i = u_i - u_i^e$, with u^e the exact solution (3.50). The factor between $L_1(e)$ calculated with two resolutions is four. Let us analyze this part. The convergence factor is easily justified for the accuracy of the spatial operator, which is second-order accurate. If time resolution is doubled the convergence factor should be two, however, time resolution is not being increased by a factor of two but by a factor of four. Thus the result in Fig. 3.19 shows second-order convergence in space and first-order convergence in time.

Stability. Time resolution is defined for an adequate evolution of initial data. In the case of this IVP the method is stable as long as the condition

3.4 Diffusion Equation: The Paradigm of Parabolic Equations

$$\Delta t \leq \frac{1}{2}\frac{\Delta x^2}{\kappa} \tag{3.53}$$

if satisfied. A proof is offered in Appendix A.

The code `DiffSimple.f90` available as Electronic Supplementary Material, solves this problem is pretty similar to that used for the wave equation in Sect. 3.2. The main loop of the evolution code is below:

```
CFL = 0.5
! ***** --> LOOP CORE <--
! Recycle arrays
u_p = u
! Formula for the evolution
do i=1,Nx-1
    u(i) = CFL * (u_p(i+1) - 2.0d0 * u_p(i) + u_p(i-1)) + u_p(i)
end do
! Zero boundary conditions
u(0) = 0.0d0
u(Nx) = 0.0d0
! ***** --> Ends Loop Core <--
```

3.4.2 Implicit Crank-Nicolson Method

This method is based on the average of the spatial part of the differential equation between two consecutive time slices for the calculation of the time derivative. The discrete version of (3.49) for this method reads

$$\frac{u_i^{n+1} - u_i^n}{\Delta t} = \frac{1}{2}\kappa\left[\frac{u_{i+1}^{n+1} - 2u_i^{n+1} + u_{i-1}^{n+1}}{\Delta x^2} + \frac{u_{i+1}^n - 2u_i^n + u_{i-1}^n}{\Delta x^2}\right]$$

$$+ \mathcal{O}(\Delta t^2, \Delta x^2). \tag{3.54}$$

This discrete version is obtained from the Taylor expansion of u around the virtual point $(x_i, t^{n+1/2}) = (x_i, (t^{n+1} + t^n)/2)$. This is the reason why the time derivative of u is second-order accurate here, unlike the scheme (3.52). On the other hand, the spatial derivative operator is averaged using its values at t^{n+1} and t^n.

Defining $\alpha = \frac{1}{2}\kappa\left(\frac{\Delta t}{\Delta x^2}\right)$, one can rearrange (3.54) as

$$(-\alpha)u_{i-1}^{n+1} + (1+2\alpha)u_i^{n+1} + (-\alpha)u_{i+1}^{n+1} = (\alpha)u_{i-1}^n + (1-2\alpha)u_i^n + (\alpha)u_{i+1}^{n-1} \tag{3.55}$$

that defines a linear system for u_{i-1}^{n+1}, u_i^{n+1}, u_{i+1}^{n+1} at points of D_d along the spatial direction for $i = 1, \ldots, N_x - 1$. The information in formula (3.55) is represented by the molecule in Fig. 3.20, to be compared with Figs. 3.12 and 3.18. Notice the

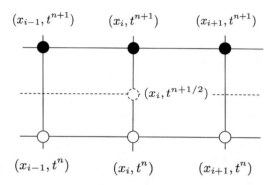

Fig. 3.20 Molecule associated to the evolution scheme (3.55) at interior points of D_d. Black dots indicate the positions where u is unknown, whereas white dots indicate the position where u is known and needed. The dotted line helps to illustrate formula (3.54), obtained from Taylor series expansion around the point $(x_i, t^{n+1/2})$ that does not belong to D_d

three unknowns simultaneously at time t^{n+1}. The equation at boundary points will be defined in terms of the boundary conditions used.

Initial conditions are the same as before.

Boundary conditions are also Dirichlet with $u(x_0, t^n) = u(x_{N_x}, t^n) = 0$. For the implementation let us first see the boundary (x_0, t^n). The condition means that $u_0^n = 0$ for all n. Substituting $i = 0$ in formula (3.55) and assuming the information at x_{-1} is available for u_{-1}^k for $k = n-1, n+1$, one obtains

$$(-\alpha)u_{-1}^{n+1} + (1+2\alpha)u_0^{n+1} + (-\alpha)u_1^{n+1} = (\alpha)u_{-1}^n + (1-2\alpha)u_0^n + (\alpha)u_1^n.$$

Now, setting $u_0^k = 0$, constructing a line in the xu-plane, from (x_0, u_0^k) to (x_1, u_1^k), one extrapolates the virtual value $u_{-1}^k = -u_1^k$, which substituted into the above equation implies the first equation of the linear system (3.55):

$$(1+2\alpha)u_0^{n+1} = (1-2\alpha)u_0^n, \tag{3.56}$$

which includes values of u defined on D_d only.

At the right boundary (x_{N_x}, t^n), Eq. (3.55) for $i = N_x$, assuming the information at x_{N_x+1} is available, becomes

$$(-\alpha)u_{N_x-1}^{n+1} + (1+2\alpha)u_{N_x}^{n+1} + (-\alpha)u_{N_x+1}^{n+1} = (\alpha)u_{N_x-1}^n + (1-2\alpha)u_{N_x}^n + (\alpha)u_{N_x+1}^n.$$

The boundary condition is $u_{N_x}^k = 0$ for $k = n, n+1$. Constructing the line from $(x_{N_x-1}, u_{N_x-1}^k)$ to $(x_{N_x}, u_{N_x}^k)$ where $u_{N_x}^k = 0$, one can extrapolate the virtual value $u_{N_x+1}^k = -u_{N_x-1}^k$. Then the previous formula reduces to

3.4 Diffusion Equation: The Paradigm of Parabolic Equations

$$(1+2\alpha)u_{N_x}^{n+1} = (1-2\alpha)u_{N_x}^n, \tag{3.57}$$

which is the N_x-th equation of system (3.55).
Finally, formulas (3.55), (3.56), and (3.57) define the following linear tridiagonal system:

$$\begin{bmatrix} 1+2\alpha & 0 & 0 & & & & & \\ -\alpha & 1+2\alpha & -\alpha & 0 & & & & \\ 0 & -\alpha & 1+2\alpha & -\alpha & 0 & & & \\ & & \cdots & & & & & \\ & & \cdots & & & & & \\ & & \cdots & & & & & \\ & & & 0 & -\alpha & 1+2\alpha & -\alpha & 0 \\ & & & & 0 & -\alpha & 1+2\alpha & -\alpha \\ & & & & & 0 & 0 & 1+2\alpha \end{bmatrix} \begin{bmatrix} u_0^{n+1} \\ u_1^{n+1} \\ u_2^{n+1} \\ \cdot \\ \cdot \\ \cdot \\ u_{N_x-2}^{n+1} \\ u_{N_x-1}^{n+1} \\ u_{N_x}^{n+1} \end{bmatrix}$$

$$= \begin{bmatrix} (1-2\alpha)u_0^n \\ (\alpha)u_0^n + (1-2\alpha)u_1^n + (\alpha)u_2^n \\ (\alpha)u_1^n + (1-2\alpha)u_2^n + (\alpha)u_3^n \\ \cdot \\ \cdot \\ \cdot \\ (\alpha)u_{N_x-3}^n + (1-2\alpha)u_{N_x-2}^n + (\alpha)u_{N_x-1}^n \\ (\alpha)u_{N_x-2}^n + (1-2\alpha)u_{N_x-1}^n + (\alpha)u_{N_x}^n \\ (1-2\alpha)u_{N_x}^n \end{bmatrix}, \tag{3.58}$$

that has to be solved for u_i^{n+1} in terms of u_i^n for $i = 0, \ldots, N_x$.

The implementation of this method is illustrated for the initial conditions $u_0(x) = B\sin(n\pi x)$ with $B = 1$ and $n = 3$ in Fig. 3.21, where we also show the convergence of $L_1(e)$ to zero.

The code DiffusionImplicit.f90 solves the diffusion equation using the Crank-Nicolson method and is available as Electronic Supplementary Material described in Appendix B. The essential evolution loop is the following:

```
! ***** --> LOOP CORE <--
! Recycle arrays
u_p = u

! ZERO boundary conditions
b(0)  = 1.0d0 + 2.0d0*alpha
c(0)  = 0.0d0
a(Nx) = 0.0d0
b(Nx) = 1.0d0 + 2.0d0*alpha
d(0)  = (1.0d0 - 2.0d0*alpha)*u_p(0)
```

```
d(Nx) = (1.0d0 - 2.0d0*alpha)*u_p(Nx)

! Entries of Matrix A and vector b for inner points do i=1,Nx-1
   a(i) = -alpha
   b(i) = 1.0d0 + 2.0d0*alpha
   c(i) = -alpha
   d(i) = alpha*u_p(i-1) + (1.0d0 - 2.0d0*alpha)*u_p(i)
   + alpha*u_p(i+1)
end do

! Forward substitution
tmp = b(0)
u(0) = d(0)/tmp
do i=1,Nx
   aux(i) = c(i-1)/tmp
   tmp = b(i)-a(i)*aux(i)
   u(i) = (d(i)-a(i)*u(i-1))/tmp
end do
! Backward substitution
do i = Nx-1,0,-1
   u(i) = u(i) - aux(i+1)*u(i+1)
end do
! ***** --> Ends Loop Core <--
```

Exercise. Solve the Diffusion equation using $\kappa = 2$ and a different number of nodes n.

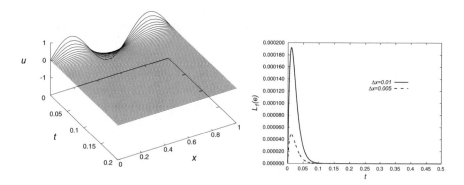

Fig. 3.21 Solution using the implicit Crank-Nicolson scheme with $n = 3$ and $\kappa = 1$. At the left, the numerical solution u on D_d with $\Delta x = 0.01$ and $C = 0.5$. At the right we show the norm $L_1(e)$ of the error, using resolutions $\Delta x_1 = 0.01$ and $\Delta x_2 = 0.005$ keeping $C = 0.5$, which implies time resolutions $\Delta t_1 = 5 \times 10^{-5}$ and $\Delta t_2 = 1.25 \times 10^{-5}$

Exercise. Solve the Diffusion equation using $\kappa = 2$ and $CFL = 2$. Show the explicit method is unstable whereas the Implicit method is stable.

3.4.3 Dissipation and Dispersion Again

In Sect. 3.2.1 above, we showed numerical solutions of the wave equation that illustrate the effects of dispersion and dissipation, which become evident when errors have more time to accumulate in long time domains.

After solving the Diffusion equation is a good opportunity to understand the origin of diffusion. The formal description of these processes and applications to various evolution schemes can be found in Sections 7.2 and 7.3 of [1], while here we follow only a schematic description useful in this book.

Consider u is a function that can be expanded as a combination of plane waves of the type $u = \hat{u} e^{i\omega t} e^{ikx}$. If u is to be a solution of the wave equation $\frac{\partial^2 u}{\partial t^2} - \frac{\partial^2 u}{\partial x^2} = 0$, the dispersion relation $\omega = \pm k$ between ω and k must be fulfilled. This would imply that the solution for mode k would be written as $u = \hat{u} e^{ik(x \pm t)}$ that propagates at the k-independent velocity ± 1. This expression is in fact the superposition of two plane waves whose amplitude should be strictly time-independent.

Consider now that u is intended to be solution of the diffusion equation $\frac{\partial u}{\partial t} = \kappa \frac{\partial^2 u}{\partial t^2}$. Then the dispersion-diffusion relation $\omega = i\kappa k^2$ must be fulfilled, and the solution will be the superposition of waves of type $u = \hat{u} e^{-\kappa k^2 t} e^{ikx}$, which are plane waves modulated by the decaying exponential of time, very similar to the exact solution in (3.50) for the diffusion equation. Notice moreover that for each mode k the amplitude decay is different, all originated from the dispersion relation for the diffusion equation. The result is that the amplitude of different modes of the solution decay with different rates.

Results in Fig. 3.11 for the wave equation illustrate that there are some modes (that in theory should not be there) that spread away from the main pulse, modes that can only be due to the discretization and evolution scheme used to solve the equation. These modes also appear in Fig. 3.17 for the wave equation using the Crank-Nicolson method with a big CFL factor.

3.5 1 + 1 Schrödinger Equation

The general form of an IVP associated to the one-dimensional Schrödinger equation reads

$$\begin{aligned}
&i\frac{\partial \Psi}{\partial t} = -\frac{1}{2}\frac{\partial^2 \Psi}{\partial x^2} + V(x,t)\Psi \qquad &\Psi = \Psi(x,t)& \\
&D = x \in [x_{min}, x_{max}] \times t \in [0, t_f] \qquad &Domain& \\
&\Psi(x, 0) = \Psi_0(x) \qquad &Initial\ Conditions& \\
&\Psi(x_{min}, t),\ \Psi(x_{max}, t) \qquad &Boundary\ Conditions&
\end{aligned} \qquad (3.59)$$

where $V(x,t)$ is a general potential. In this section we illustrate the solution for two particular cases, the particle in a box, and the particle subject to a harmonic oscillator potential using the Implicit Crank-Nicolson method.

3.5.1 Particle in a 1D Box

This case corresponds to a particle subject to the potential

$$V(x) = \begin{cases} 0 & x \in [0,1] \\ \infty & \text{otherwise} \end{cases} \qquad (3.60)$$

that is, the particle is free in the domain $x \in [0, 1]$, and the infinity potential emulates the solid walls of a box, which means that there $\Psi = 0$. With this in mind, we consider the spatial domain $[x_{min}, x_{max}] = [0, 1]$ and the continuity conditions on the wave function forces the boundary conditions to be Dirichlet type $\Psi(0, t) = \Psi(1, t) = 0$. This is the problem we will solve numerically, but first let us construct the exact solution.

Using separation of variables in the equation of (3.59) such that $\Psi(x,t) = \psi(x)T(t)$ and considering a time-independent potential $V(x,t) = V(x)$, which is our case, the equation becomes

$$i\frac{1}{T}\frac{dT}{dt} = -\frac{1}{2\psi}\frac{d^2\psi}{dx^2} + V(x) = E, \qquad (3.61)$$

where the separation constant is the energy E. The time dependence of the wave function $T = T_0 e^{-iE(t-t_0)}$ and the spatial wave function $\psi(x)$ must be solution of the **stationary Schrödinger equation**:

$$-\frac{1}{2}\frac{d^2\psi}{dx^2} + V(x)\psi = E\psi. \qquad (3.62)$$

If $\Psi(x,t)$ is to be eventually normalized, one can set $T_0 = 1$ and $t_0 = 0$ without loss of generality and leave the normalization to the spatial wave function $\psi(x)$. Then the time-dependent wave function would be

3.5 1 + 1 Schrödinger Equation

$$\Psi(x,t) = e^{-iEt}\psi(x). \tag{3.63}$$

In the particular case of the potential (3.60), the wave function is nonzero only in the domain $x \in [0, 1]$. There $V(x) = 0$, and (3.62) is reduced to

$$\frac{\partial^2 \psi}{\partial x^2} + 2E\psi = 0, \tag{3.64}$$

which is the Helmholtz equation we already solved in Chapter 1. Applying the Dirichlet boundary conditions $\Psi(0, t) = \Psi(1, t) = 0$, the solution is

$$\psi_{n_x}(x) = A_{n_x} \sin(n_x \pi x),$$
$$E_{n_x} = \frac{(n_x \pi)^2}{2}, \tag{3.65}$$

where $n_x = 1, 2, \ldots$ is the quantum number labeling the permitted energy values E_{n_x}.

If one demands the wave function to be normalized, then ψ must satisfy $\int_0^1 \psi^* \psi dx = \int_0^1 A_{n_x}^2 \sin^2(n_x \pi x) dx = 1$, which implies $\frac{A_{n_x}^2}{2} = 1$, and the amplitude of the wave function reduces to $A_{n_x} = \sqrt{2}$. This finally implies $\psi_{n_x}(x) = \sqrt{2} \sin(n_x \pi x)$. Therefore, the complete time-dependent solution for a given n_x is the following (3.63):

$$\Psi_{n_x}(x, t) = \sqrt{2} e^{-iE_{n_x} t} \sin(n_x \pi x). \tag{3.66}$$

The density of probability of the solution (3.66) is $\rho = \Psi^* \Psi = 2 \sin^2(n_x \pi x)$, which is *time-independent*. Time-independence is **key to test** the numerical solution, that is, the solution has to show the harmonic time dependence of Ψ with angular frequency E_{n_x} whereas ρ has to remain constant in time.

In Fig. 3.22 we show the exact solution at $t = 0$ for various values of n_x, as well as the density of probability. The solutions can be distinguished because Re(Ψ) for $n_x = 1$ has no zeroes within the domain, for $n_x = 2$ it has one zero and so on.

3.5.2 Particle in a Harmonic Oscillator Potential

In this case the potential is $V(x) = \frac{1}{2}kx^2$. Since the potential is time-independent, Schrödinger equation separates according to (3.61) and has harmonic time dependence $\Psi(x, t) = e^{-iEt}\psi(x)$, in agreement with (3.63). For this potential the resulting stationary Schrödinger equation is

$$\frac{d^2 \psi}{dx^2} + [\beta - \alpha^2 x^2]\psi = 0, \tag{3.67}$$

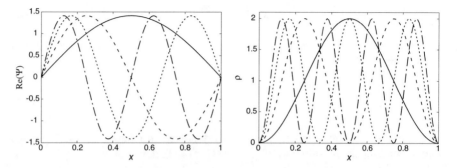

Fig. 3.22 Exact solution for the particle in the one-dimensional box [0, 1]. On the left Re($\Psi(x, 0)$) for $n_x = 0, 1, 2, 3, 4$ at $t = 0$. At the right the corresponding densities of probability $\rho = \Psi^*(x, 0)\Psi(x, 0)$

where $\alpha = \sqrt{k}$ and $\beta = 2E$. Defining the new variable $\xi = \sqrt{\alpha}x$, one obtains

$$\frac{d^2\psi}{d\xi^2} + \left(\frac{\beta}{\alpha} - \xi^2\right)\psi = 0. \tag{3.68}$$

Bounded solutions for $\xi \to \pm\infty$ are such that

$$\psi(\xi) = e^{-\xi^2/2} H(\xi),$$

where the equation for the H is Hermite's equation

$$\frac{d^2H}{d\xi^2} - 2\frac{dH}{d\xi} + \left(\frac{\beta}{\alpha} - 1\right)H = 0, \tag{3.69}$$

whose solutions are Hermite polynomials

$$\begin{aligned} H_0(\xi) &= 1, \\ H_1(\xi) &= 2\xi, \\ H_2(\xi) &= 2 - 4\xi^2, \\ H_3(\xi) &= 12\xi - 8\xi^3, \\ H_4(\xi) &= 12 - 48\xi^2 + 16\xi^4, \\ &\ldots \text{etcetera.} \end{aligned} \tag{3.70}$$

The normalized solution for $\psi(x)$ is obtained from the condition $\int_{-\infty}^{\infty} \psi^*\psi \, dx = 1$ and reads

$$\psi_{n_x}(\xi) = \sqrt{\frac{1}{\sqrt{\pi}2^{n_x}n_x!}} e^{-\xi^2/2} H_{n_x}(\xi),$$

3.5 1 + 1 Schrödinger Equation

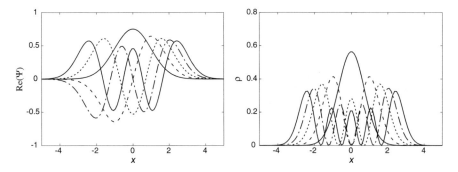

Fig. 3.23 Exact solution for the particle in a harmonic oscillator potential. On the left Re($\Psi(x,0)$) for $n_x = 0, 1, 2, 3, 4$ at $t = 0$. At the right the corresponding densities of probability $\rho = \Psi^*(x,0)\Psi(x,0)$

with $n_x = 0, 1, 2, \ldots$. Recovering the original variables and the time dependence of Ψ, the wave-function solution is

$$\Psi_{n_x}(x,t) = \sqrt{\frac{1}{\sqrt{\pi} 2^{n_x} n_x!}} e^{-iE_{n_x} t} e^{-x^2/2} H_{n_x}(x), \quad (3.71)$$

where the energy takes values $E_{n_x} = n_x + \frac{1}{2}$, restricted for Hermite equation to have bounded solutions (see, e.g., [4]). We show the exact solution at $t = 0$ for various values of n_x and also the density of probability in Fig. 3.23.

3.5.3 Implicit Crank-Nicolson Method Applied to Schrödinger Equation

We solve the IVP on the discrete domain $D_d = \{(x_i, t^n)\}$ with $x_i = x_{min} + i\Delta x$, where $\Delta x = (x_{max} - x_{min})/N_x$ and $t^n = n\Delta t$, with $\Delta t = C\Delta x^2$. Consider one needs to evolve the wave function from time t^n to t^{n+1} at all space positions x_i in the discrete domain D_d. This can be achieved with the application of the evolution operator to the wave function $\Psi_i^n = \Psi(x_i, t^n)$ as follows:

$$\Psi_i^{n+1} = e^{-i\hat{H}\Delta t}\Psi_i^n, \quad (3.72)$$

where $\hat{H}$ is the Hamiltonian operator in (3.59)

$$i\frac{\partial \Psi}{\partial t} = \hat{H}\Psi, \quad \text{and} \quad \hat{H} = -\frac{1}{2}\frac{d^2}{dx^2} + V(x). \quad (3.73)$$

Expression (3.72) can be written and Taylor expanded for small Δt as follows:

$$\Psi_i^{n+1} = \frac{e^{-i\frac{1}{2}\hat{H}\Delta t}}{e^{i\frac{1}{2}\hat{H}\Delta t}} \Psi_i^n \quad \Rightarrow$$

$$\Psi_i^{n+1} = \frac{(1 - \frac{1}{2}i\hat{H}\Delta t)}{(1 + \frac{1}{2}i\hat{H}\Delta t)} \Psi_i^n \quad \Rightarrow$$

$$\left(1 + \frac{1}{2}i\hat{H}\Delta t\right)\Psi_i^{n+1} = \left(1 - \frac{1}{2}i\hat{H}\Delta t\right)\Psi_i^n.$$

Rearranging the terms in such a way that the discrete first-order time derivative is isolated one finds that

$$\Psi_i^{n+1} + \frac{1}{2}i\hat{H}\Psi_i^{n+1}\Delta t = \Psi_i^n - \frac{1}{2}i\hat{H}\Psi_i^n \Delta t \quad \Rightarrow$$

$$i\frac{\Psi_i^{n+1} - \Psi_i^n}{\Delta t} = \frac{1}{2}\left[\hat{H}\Psi_i^{n+1} + \hat{H}\Psi_i^n\right]. \tag{3.74}$$

Notice that this formula is Schrödinger equation with the spatial part averaged at times t^n and t^{n+1}, which is the essence of the Crank-Nicolson method, as illustrated with the wave and diffusion equations before. Notice also that (3.74) is the second-order accurate discrete version of Eq. (3.59) centered at the virtual point $(x_i, t^{n+1/2})$, as illustrated in Fig. 3.20 for the Diffusion equation.

Now, let us expand (3.74) including the operator terms of $\hat{H}$ as follows:

$$\Psi_i^{n+1} + \frac{1}{2}i\left[-\frac{1}{2}\frac{\Psi_{i+1}^{n+1} - 2\Psi_i^{n+1} + \Psi_{i-1}^{n+1}}{\Delta x^2} + V_i\Psi_i^{n+1}\right]\Delta t$$

$$= \Psi_i^n - \frac{1}{2}i\left[-\frac{1}{2}\frac{\Psi_{i+1}^n - 2\Psi_i^n + \Psi_{i-1}^n}{\Delta x^2} + V_i\Psi_i^n\right]\Delta t$$

$$+ \mathcal{O}(\Delta t^2, \Delta x^2), \tag{3.75}$$

which defines a tridiagonal linear system for the wave function at time t^{n+1}. Defining $\alpha = \frac{1}{4}i\frac{\Delta t}{\Delta x^2}$, $\beta = \frac{1}{2}i\Delta t$ and dropping the error term, the system reads

$$(-\alpha)\Psi_{i-1}^{n+1} + (1 + 2\alpha + \beta V_i)\Psi_i^{n+1} + (-\alpha)\Psi_{i+1}^{n+1}$$
$$= (\alpha)\Psi_{i-1}^n + (1 - 2\alpha - \beta V_i)\Psi_i^n + (\alpha)\Psi_{i+1}^n, \tag{3.76}$$

for inner points $i = 1, \ldots, N_x - 1$. Boundary conditions at $x_{min} = x_0$ and $x_{max} = x_{N_x}$ will determine the first equation for $i = 0$ and the last equation of the tridiagonal system for $i = N_x$.

3.5 1 + 1 Schrödinger Equation

3.5.3.1 Numerical Solution for the Particle in a Box

For comparison with the exact solution, we use the spatial domain $[x_{min}, x_{max}] = [0, 1]$. The numerical solution is constructed on the discrete domain D_d, using $\Delta t = C \Delta x^2$. Notice that Δt is not proportional to Δx, instead Δx^2. We set $N = 100$ or equivalently the resolution to $\Delta x = 0.01$ of D_d^1, $C = 0.125$ and the time domain $t \in [0, 10]$.

In this case the Dirichlet type of boundary conditions are appropriate on the wave function $\psi(0, t) = \Psi(1, t) = 0$, that is, we require $\Psi_0^k = \Psi_{N_x}^k = 0$ for all k. Assuming there are virtual values for Ψ_{-1}^k and $\Psi_{N_x+1}^k$, which take values

$$\Psi_{1-}^k = \Psi_1^k \text{ and } \Psi_{N_x+1}^k = \Psi_{N_x-1}^k,$$

to second order, which results from connecting the point $(x_0, \Psi_0^k = 0)$ and (x_1, Ψ_1^k) and extrapolate with a line the value Ψ_{-1}^k, and $(x_{N_x}, \Psi_{N_x}^k = 0)$ with $(x_{N_x-1}, \Psi_{N_x-1}^k)$ allows the extrapolation of $\Psi_{N_x+1}^k$. These virtual values can be used in Eq. (3.76) with $i = 0$ and $i = N_x$ to construct the first and last equations of the tridiagonal system, which are

$$(1 + 2\alpha + \beta V_0)\Psi_0^{n+1} = (1 - 2\alpha - \beta V_0)\Psi_0^n, \quad \text{at the left boundary } (x_0, t^n),$$

$$(1 + 2\alpha + \beta V_{N_x})\Psi_{N_x}^{n+1} = (1 - 2\alpha - \beta V_{N_x})\Psi_{N_x}^n, \quad \text{at the right boundary } (x_{N_x}, t^n).$$

These conditions are constructed for any time-independent potential V. For the case of a particle in a box $V = 0$ in the numerical domain $x \in [0, 1]$, therefore the system simplifies. Using the later two expressions together with (3.76), the linear system to be solved for the evolution reads

$$\begin{bmatrix} 1+2\alpha & 0 & 0 & & & & & \\ -\alpha & 1+2\alpha & -\alpha & 0 & & & & \\ 0 & -\alpha & 1+2\alpha & -\alpha & 0 & & & \\ & & \cdots & & & & & \\ & & \cdots & & & & & \\ & & \cdots & & & & & \\ & & & 0 & -\alpha & 1+2\alpha & -\alpha & 0 \\ & & & & 0 & -\alpha & 1+2\alpha & -\alpha \\ & & & & & 0 & 0 & 1+2\alpha \end{bmatrix} \begin{bmatrix} \Psi_0^{n+1} \\ \Psi_1^{n+1} \\ \Psi_2^{n+1} \\ \cdot \\ \cdot \\ \cdot \\ \Psi_{N_x-2}^{n+1} \\ \Psi_{N_x-1}^{n+1} \\ \Psi_{N_x}^{n+1} \end{bmatrix}$$

$$= \begin{bmatrix} (1-2\alpha)\Psi_0^n \\ (\alpha)\Psi_0^n + (1-2\alpha)\Psi_1^n + (\alpha)\Psi_2^n \\ (\alpha)\Psi_1^n + (1-2\alpha)\Psi_2^n + (\alpha)\Psi_3^n \\ \cdot \\ \cdot \\ \cdot \\ (\alpha)\Psi_{N_x-3}^n + (1-2\alpha)\Psi_{N_x-2}^n + (\alpha)\Psi_{N_x-1}^n \\ (\alpha)\Psi_{N_x-2}^n + (1-2\alpha)\Psi_{N_x-1}^n + (\alpha)\Psi_{N_x}^n \\ (1-2\alpha)\Psi_{N_x}^n \end{bmatrix} \qquad (3.77)$$

In order to solve this system, we use the method described for the **generic tridiagonal case** (3.44), which determines the evolution of Ψ_i^n from t^n to t^{n+1}. Consider now that the entries of the system are *complex*, which requires the definition of arrays of complex entries `psi` and `psi_p` for Ψ_i^{n+1} and Ψ_i^n, respectively, and for the matrix and vector elements `a`, `b`, `c`, `d`.

The workhorse example will be the case $n_x = 2$. As initial conditions we use the exact solution (3.66) at $t = 0$, $\Psi_2 = \sqrt{2}\sin(2\pi x)$, that is, we fill in the array of the wave function at initial time as follows:

$$\Psi_i^0 = \sqrt{2}\sin(2\pi x_i) + 0\,\mathrm{i},$$

and these initial data are evolved using the scheme (3.77) at each time step.

Diagnostics of the Solution. The first diagnostics consists in showing that the wave function Ψ oscillates, whereas the density of probability $\rho = \Psi^*\Psi$ remains time-independent. This is shown in Fig. 3.24.

The second physical diagnostics is the unitarity of the evolution. This means that the number of particles in the domain, namely, $N = \int_0^1 \rho dx$, has to remain constant in time. The result is calculated using the trapezoidal rule (3.23) and is shown in Fig. 3.24. The value of N in this solution deviates from one only by 10^{-8} over the time window used.

Finally, the error of the numerical solution is estimated for the real part of the wave function at every point of $(x_i, t^n) \in D_d$, as the numerical solution minus the exact solution:

$$e_i^n = \mathrm{Re}(\Psi_i^n) - \mathrm{Re}(\Psi_i^{e,n}). \qquad (3.78)$$

In order to show the numerical solution is convergent, we calculate $L_2(e_i^n)$ using (3.24), as function of time for $N_x = 100$, 200, and 400 or, equivalently, resolutions $\Delta x = 0.01$, 0.005, and 0.0025. The result appears in Fig. 3.24. The factor between the errors using consecutive resolutions is nearly four, although we decided to present the error using a log plot which is common in long runs. In a

3.5 1 + 1 Schrödinger Equation

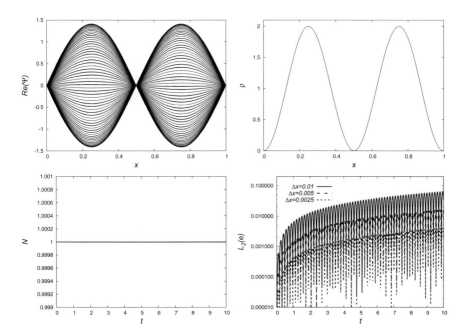

Fig. 3.24 Numerical solution for the particle in a box with $n_x = 2$, using $\Delta x = 0.01$ and $C = 0.125$. At the top we show snapshots of $\text{Re}(\Psi)$ and ρ at various times. This shows the wave function oscillates, whereas the density remains nearly stationary. At the bottom we show that the integral $N = \int_0^1 \rho \, dx$ remains nearly constant in time. Finally we show the convergence of the $L_2(e)$ norm of the error (3.78) to zero, using the numerical solution calculated with various resolutions

semi-log plot, the difference between the solutions in D_d^1 and D_d^2, should be nearly the same as the difference between the solutions in D_d^2 and D_d^3.

All these diagnostics indicate that the numerical solution for the particle in a box is reliable.

Exercise. Enhance the code below, to calculate $\text{Re}(\Psi)$ at $x = 0.25$ as a function of time. The result will be a time series with the value of the wave function at this position that oscillates in time. Then calculate the Fourier Transform with an external software and show that the dominant peak is found approximately at the angular frequency $E_2 = (2\pi)^2/2$.

Exercise. Adapt the proof for the stability of the Crank-Nicolson equation applied to the Diffusion Equation in Appendix A, and show that the scheme (3.76) is unconditionally stable.

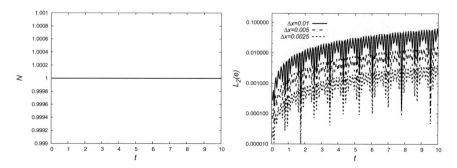

Fig. 3.25 Numerical solution for the particle in a box with $n_x = 2$, using various resolutions and $C = 1$. We show that the integral N remains nearly constant in time during the evolution. Finally we show the convergence of the $L_2(e)$ norm of the error (3.78) to zero

An effect of doubling resolution, halving Δx, is that in order to cover the time domain the number of time steps if four times bigger if we keep C constant in the relation $\Delta t = C \Delta x^2$. Since this method is unconditionally stable for any value of Δt, let us use $C = 1$, which implies the code will execute 1/8 of the time-steps than with the value $C = 0.125$ used before. The evolution of Ψ and ρ is shown in Fig. 3.25, and the results are very similar to those in 3.24. We show the preservation of N and the convergence of the error. Notice that the L_2 norm of the error is of the same order as that using $C = 0.125$, which indicates the solution is nearly as good as when using a small time step.

Results in Figs. 3.24 and 3.25 are obtained using `CFL=0.125` and `CFL=1` respectively, in the code provided at the end of this section.

> **Challenge.** Using $C = 2$, estimate t_f such that the change in the number of particles deviates more than 1%.

3.5.3.2 Numerical Solution for the Particle in a Harmonic Oscillator Potential

Boundary conditions in this case are different from those for the particle in the box. The reason is that within the numerical domain, the wave function decays with a Gaussian profile very fast, but is not zero. Therefore a more appropriate condition is that the space derivative of Ψ at the boundaries vanishes, which is Neumann type.

Assuming there are virtual values for Ψ^k_{-1} and $\Psi^k_{N_x+1}$, for $k = n, n+1$, one can use the centered formula for the first-order spatial derivative at the boundary points and find the relations:

3.5 1 + 1 Schrödinger Equation

$$\frac{\Psi_1^k - \Psi_{-1}^k}{2\Delta x} = 0 \quad \Rightarrow \quad \Psi_{-1}^k = \Psi_1^k,$$

$$\frac{\Psi_{N_x+1}^k - \Psi_{N_x-1}^k}{2\Delta x} = 0 \quad \Rightarrow \quad \Psi_{N_x+1}^k = \Psi_{N_x-1}^k.$$

Using these expressions into (3.76), we find the equations to be fulfilled at the boundaries:

$$(1 + 2\alpha + \beta V_0)\Psi_0^{n+1} + (-2\alpha)\Psi_1^{n+1} = (1 - 2\alpha - \beta V_0)\Psi_0^n + (2\alpha)\Psi_1^n,$$

$$(-2\alpha)\Psi_{N_x-1}^{n+1} + (1 + 2\alpha + \beta V_{N_x})\Psi_{N_x}^{n+1} = (2\alpha)\Psi_{N_x-1}^n + (1 - 2\alpha - \beta V_{N_x})\Psi_{N_x}^n.$$

In this case the potential is $V = \frac{1}{2}x^2$, defined in the numerical domain for each x_i as $V_i = \frac{1}{2}x_i^2$. Using these two expressions together with (3.76), the linear system to be solved for the evolution reads:

$$\begin{bmatrix} 1+2\alpha+\beta V_0 & -2\alpha & 0 & & & & & \\ -\alpha & 1+2\alpha+\beta V_1 & -\alpha & 0 & & & & \\ 0 & -\alpha & 1+2\alpha+\beta V_2 & -\alpha & 0 & & & \\ & & \ldots & & & & & \\ & & \ldots & & & & & \\ & & \ldots & & & & & \\ & & & 0 & -\alpha & 1+2\alpha+\beta V_{N_x-2} & -\alpha & 0 \\ & & & & 0 & -\alpha & 1+2\alpha+\beta V_{N_x-1} & -\alpha \\ & & & & & 0 & -2\alpha & 1+2\alpha+\beta V_{N_x} \end{bmatrix}$$

$$\begin{bmatrix} \psi_0^{n+1} \\ \psi_1^{n+1} \\ \psi_2^{n+1} \\ \cdot \\ \cdot \\ \cdot \\ \psi_{N_x-2}^{n+1} \\ \psi_{N_x-1}^{n+1} \\ \psi_{N_x}^{n+1} \end{bmatrix} = \begin{bmatrix} (1-2\alpha-\beta V_0)\psi_0^n + 2\alpha\psi_1^n \\ (\alpha)\psi_0^n + (1-2\alpha-\beta V_1)\psi_1^n + (\alpha)\psi_2^n \\ (\alpha)\psi_1^n + (1-2\alpha-\beta V_2)\psi_2^n + (\alpha)\psi_3^n \\ \cdot \\ \cdot \\ \cdot \\ (\alpha)\psi_{N_x-3}^n + (1-2\alpha-\beta V_{N_x-2})\psi_{N_x-2}^n + (\alpha)\psi_{N_x-1}^n \\ (\alpha)\psi_{N_x-2}^n + (1-2\alpha-\beta V_{N_x-1})\psi_{N_x-1}^n + (\alpha)\psi_{N_x}^n \\ 2\alpha\psi_{N_x-1}^n + (1-2\alpha-\beta V_{N_x})\psi_{N_x}^n \end{bmatrix} \quad (3.79)$$

The numerical solution is constructed on the discrete domain $D_d = \{(x_i, t^n)\}$, with $x_i = x_{min} + i\Delta x$ where $\Delta x = (x_{max} - x_{min})/N_x$ and $\Delta t = C\Delta x^2$. In this case we set $[x_{min}, x_{max}] = [-10, 10]$, for the wave function to have space to vanish due to the Gaussian envelope. We set $N_x = 2000$ or equivalently $\Delta x = 0.01$ as the resolution for D_d, $C = 0.125$ and the time domain to $t \in [0, 10]$.

150 3 Simple Methods for Initial Value Problems Involving PDEs

The workhorse example will be *the case* $n_x = 3$. As initial conditions we use the exact solution (3.71) at $t = 0$:

$$\Psi_{n_x}(x, 0) = \sqrt{\frac{1}{\sqrt{\pi} 2^3 3!}} e^{-x^2/2} H_3(x) = \sqrt{\frac{1}{\sqrt{\pi} 2^3 3!}} e^{-x^2/2} (12x - 8x^3), \qquad (3.80)$$

and the array of the wave function at initial time is filled with values

$$\Psi_i^0 = \sqrt{\frac{1}{\sqrt{\pi} 2^3 3!}} e^{-x_i^2/2} (12x_i - 8x_i^3), \qquad (3.81)$$

which is evolved over the time domain using the scheme (3.79).

Diagnostics of the solution. Like for the particle in the box, the first diagnostics consists in showing that the wave function Ψ oscillates, whereas the density of probability $\rho = \Psi^* \Psi$ remains time-independent. This is shown in Fig. 3.26. Secondly, the unitarity of the solution is shown through the conservation of $N = \int_{x_{min}}^{x_{max}} \rho dx$, also shown in Fig. 3.26.

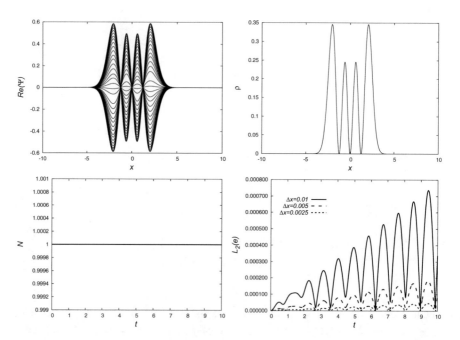

Fig. 3.26 Numerical solution for the particle in a harmonic oscillator potential with $n_x = 3$ using $\Delta x = 0.01$ and $C = 0.125$. At the top we show snapshots of Re(Ψ) and the density of probability ρ at various times. This shows that the wave function oscillates, whereas the density remains nearly time-independent. At the bottom we show that the integral N remains nearly constant in time. Finally we show the convergence of the $L_2(e)$ norm of the error to zero with second order

The error of the numerical solution is estimated for the real part of the wave function at every point $(x_i, t^n) \in D_d$ using Eq. (3.78). Finally, in order to show that the numerical solution converges to the exact solution, we calculate $L_2(e_i^n)$ as function of time for $N = 2000, 4000, 8000$ or equivalently resolutions $\Delta x = 0.01, 0.005, 0.0025$. The resulting value of $L_2(e_i^n)$ appears also in Fig. 3.26, showing second-order convergence of the error to zero. Unlike the particle in the box, for completeness, here we show a typical plot in a linear escale of the L_2 norm of the error, where the convergence factor close to 2^2 is consistent with the second-order accuracy of the methods.

Exercise. Enhance the code to calculate $\text{Re}(\Psi)$ at the point where this function acquires its maximum. Then carry out a Fourier Transform of the resulting time series with an external software and show the peak frequency is approximately $E_3 = 3 + 1/2$.

Exercise. Results in Fig. 3.26 can be reproduced with CFL=0.125 using the code below. Consider experiments with bigger values of C in order to measure execution time, error amplitudes, and convergence. Also it is possible to evolve initial conditions with $n_x = 1, 2, 3, 4$.

The code `SchroedingerImplicit.f90` can be used to solve Schrödinger equation for the cases in this section and is listed in Appendix B.

3.6 How to Improve the Methods in this Chapter

We defined and used the basic methods for the solution of IVPs in $1+1$ dimensions. The codes provided are a template where a garden variety of methods can be implemented for another universe of problems. We recommend to study the methods and examples in reference [1].

3.7 Projects

With the methods described in this chapter, it is possible to carry out the following projects.

3.7.1 Bosonic Dark Matter

This model proposes that dark matter is an ultralight spinless particle, whose dynamics is ruled by the Gross-Pitaevskii equation suitable for a Bose gas, which mathematically is Schrödinger equation with a macroscopic interpretation. The potential in this equation is the gravitational potential sourced by the density of the bosonic gas itself that obeys Poisson equation.

The problems described in this chapter are defined in $1+1$ dimensions, yet when a three-dimensional problem has spherical symmetry, it can be translated into a $1+1$ problem in spherical coordinates. The equations that rule the dynamics of bosonic dark matter in spherical symmetry and spherical coordinates are

$$i\frac{\partial \Psi}{\partial t} = -\frac{1}{2r}\frac{\partial^2(r\Psi)}{\partial r^2} + V\Psi,$$

$$\frac{\partial^2(rV)}{\partial r^2} = r|\Psi|^2, \tag{3.82}$$

respectively, Schrödinger and Poisson equations. This is an IVP with constrained evolution, that is, the wave function evolves according to Schrödinger equation, whereas the potential is sourced by the density $|\Psi|^2$ itself, and V has to be updated during the evolution because Ψ is time-dependent.

The collapse of fluctuations that eventually may form galaxies can be studied by solving these equations. With some modifications of the code provided here to solve Schrödinger equation, the evolution, collapse, and relaxation of a fluctuation can be studied as indicated in [5]. For the evolution of the system, the same Crank-Nicolson method described here is used in [5], whereas Poisson equation can be solved at each time step using a Runge-Kutta method just like the one used for the solution of TOV equations in Chapter 1.

3.7.2 Newtonian Boson Stars

The problem above has a special set of solutions that are stationary. Their construction assumes that the time dependence of Ψ is harmonic in time, and therefore the source of Poisson equation is time-independent, which leads to a stationary potential. The resulting stationary Schrödinger equation is solved using the Shooting method described in Sect. 2.11.8. A detailed description to follow for the construction of these solutions can be found in [6].

Once these solutions are constructed, they can be used as initial conditions for the solution of the system (3.82). Then Boson Stars can be perturbed, and it is possible to study their stability and oscillation modes.

References

1. J. W. Thomas, in *Numerical Partial Differential Equations: Finite Difference Methods*, (Springer-Verlag, New York, 1998). Texts in Applied Mathematics 22
2. Cactus Einstein Toolkit http://einsteintoolkit.org/
3. W. H. Press, S. A Teukolsky, W. T. Vettering, B. P. Flannery, in *Numerical Recipes in Fortran 77: The Art of Scientific Computing*, 2nd edn. (Cambridge University Press, Cambridge, 2003)
4. C. Cohen-Tanoudji, B. Diu, F. Laloë, in *Quantum Mechanics*, vol. one (Wiley-VCH, London, 1977)
5. F. S. Guzmán, L. A. Ureña-López, in Newtonian collapse of scalar field dark matter. Phys. Rev. D **68**, 024023 (2003). arXiv:astro-ph/0303440 https://doi.org/10.1103/PhysRevD.68.024023
6. F. S. Guzmán, L. A. Ureña-López, Evolution of the Schrödinger-Newton system for a self-gravitating scalar-field. Phys. Rev. D **69**, 124033 (2004). arXiv:gr-qc/0404014 https://doi.org/10.1103/PhysRevD.69.124033

Chapter 4
Method of Lines for Initial Value Problems Involving PDEs

Abstract In this chapter, we describe the method of lines (MoL) and apply it to various of the problems solved above in order to compare properties. The first example used is the advection equation, which is adequate to the methods since it is a first-order equation in space and time. Later on, we solve the wave equation, which needs a little work to bring it from second to first order. We illustrate the power of the MoL with the wave equation on a general Minkowski space-time, which also helps illustrating some concepts of relativistic physics. We also solve Schrödinger equation with this method and compare the solutions with those constructed previously. We finalize the chapter with an advanced problem of relativity, the accretion of a spherical wave by a black hole.

Keywords Partial differential equations · Method of lines · Wave equation · Schrödinger equation · Diffusion equation · Error theory · Convergence

In this chapter, we describe the method of lines (MoL) and apply it to various of the problems solved above in order to compare properties. The first example used is the advection equation, which is adequate to the methods since it is a first-order equation in space and time. Later on, we solve the wave equation, which needs a little work to bring it from second to first order. We illustrate the power of the MoL with the wave equation on a general Minkowski space-time, which also helps illustrating some concepts of relativistic physics. We also solve Schrödinger equation with this method and compare the solutions with those constructed previously. We finalize the chapter with an advanced problem of relativity, the accretion of a spherical wave by a black hole.

4.1 Method of Lines

Assume that the IVP of interest is associated with a first order in time PDE defined on a two-dimensional domain, described with independent variables (x, t), where t

represents time and x a spatial coordinate. A sufficiently general case that contains all the examples developed in this chapter has the form

$$\boxed{\begin{aligned} &\text{combination of }\left(x, t, f, \frac{\partial f}{\partial t}, \frac{\partial f}{\partial x}, \frac{\partial^2 f}{\partial x^2}\right) = g(x,t) \quad f = f(x,t) \\ &D = x \in [x_{min}, x_{max}] \times t \in [0, t_f] \quad\quad\quad\quad Domain \\ &f(x, 0) = f_0(x) \quad\quad\quad\quad\quad\quad\quad\quad\quad\quad Initial\ Conditions \\ &f(x_{min}, t),\ f(x_{max}, t) \quad\quad\quad\quad\quad\quad\quad Boundary\ Conditions \end{aligned}}$$

(4.1)

and is such that the PDE can be written as

$$\frac{\partial f}{\partial t} = \text{rhs}(f) \tag{4.2}$$

where the right-hand side (rhs) defines the evolution of f.

The idea is to construct a numerical solution to the above IVP on the numerical domain D_d defined as the set of points (x_i, t^n), with $x_i = x_{min} + i\Delta x$, $i = 0, \ldots, N_x$, $\Delta x = (x_{max} - x_{min})/N_x$ and $t^n = n\Delta t$. Assuming that for each point $(x_i, t^n) \in D_d$, it is possible to define an evolution equation of the form

$$\frac{\partial f}{\partial t} = \text{rhs}\left(x_i, t^n, f\Big|_{(x_i, t^n)}, g\Big|_{(x_i, t^n)}, \frac{\partial f}{\partial x}\Big|_{(x_i, t^n)}, \frac{\partial^2 f}{\partial x^2}\Big|_{(x_i, t^n)}\right), \tag{4.3}$$

which is **an ODE for f at each spatial point** x_i, to be integrated in time from t^n to t^{n+1}. In fact, this is an IVP in the time domain $t \in [0, t_f]$ for each x_i. This approach is called the method of lines (MoL), because the solution to the IVP associated with a PDE will be the solution of $N_x + 1$ ODEs integrated along lines of constant spatial position x_i for $i = 0, \ldots, N_x$.

The objective of the MoL is the construction of a solution to the IVP by solving the $N_x + 1$ ODEs using the integrators described in Chap. 2.

4.2 Time Integrators

Equation (4.3) is called the **semi discrete version of the PDE** when the right-hand side is written in terms of the discrete version of f and its spatial derivatives.

The time integration of f from (x_i, t^n) to (x_i, t^{n+1}) can use an integrator of ODEs. We use the Heun flavor of the RK2 method from Sect. 2.8 to illustrate how. The integration of $\frac{\partial f}{\partial t} = \text{rhs}(x_i, t^n, f_i^n)$ would be, according to Eqs. (2.32) for each x_i, as follows: Identifying from (2.32) and (2.33) the unknown u at time t_{i-1} with the unknown f at point (x_i, t^n) as seen in Fig. 4.1, $u_{i-1} \to f_i^n$ and the slope $k_1 \to \text{rhs}(f_i^n)$, the argument of k_2 is $u_{i-1} + k_1 \Delta t \to f_i^n + \text{rhs}(f_i^n)\Delta t := f^*$ and finally

4.2 Time Integrators

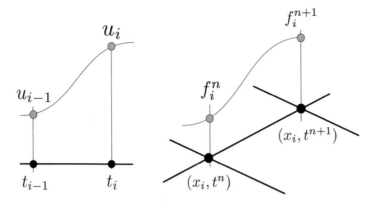

Fig. 4.1 Left, illustration of the integration of the unknown u of an ODE at the left, from point t_{i-1} to t_i of the numerical domain, as seen in Figs. 2.2 and 2.4. Right, integration of f of a PDE from space-time point (x_i, t^n) to (x_i, t^{n+1}) of the numerical domain

$k_2 \to \text{rhs}(f^*)$. With this, the integration of f from (x_i, t^n) to (x_i, t^{n+1}) can be expressed as

$$f_i^* = f_i^n + \text{rhs}\left(f_i^n\right) \Delta t = f_i^n + k_1 \Delta t \quad \Rightarrow k_1 = \frac{f_i^* - f_i^n}{\Delta t},$$

$$f_i^{n+1} = f_i^n + \left(\frac{1}{2}k_1 + \frac{1}{2}k_2\right)\Delta t$$

$$= f_i^n + \frac{1}{2}k_1 \Delta t + \frac{1}{2}\text{rhs}\left(f_i^*\right) \Delta t$$

$$= f_i^n + \frac{1}{2}\frac{f_i^* - f_i^n}{\Delta t}\Delta t + \frac{1}{2}\text{rhs}\left(f_i^*\right) \Delta t$$

$$= \frac{1}{2}\left(f_i^n + f_i^* + \text{rhs}\left(f_i^*\right)\Delta t\right). \quad (4.4)$$

This method can be implemented in two steps, for each x_i, as follows:

```
! Heun RK2
do rk=1,2
  calculate rhs_f  ! With a discrete version of the rhs
  if (rk.eq.1) then
    f = f_p + rhs_f * dt
  else
    f = 0.5d0 * (f_p + f + rhs_f * dt)
  end if
end do
```

Other flavors of RK2 or other ODE integrators can be implemented as well. For example, a method that we will use frequently is a third-order Runge-Kutta (RK3) method described in [1]:

$$f_i^* = f_i^n + \text{rhs}(f_i^n)\Delta t,$$
$$f_i^{**} = \frac{3}{4}f_i^n + \frac{1}{4}f_i^* + \frac{1}{4}\text{rhs}(f_i^*)\Delta t,$$
$$f_i^{n+1} = \frac{1}{3}f_i^n + \frac{2}{3}f_i^{**} + \frac{2}{3}\text{rhs}(f_i^{**})\Delta t, \qquad (4.5)$$

which can be programmed as follows:

```
! RK3
do rk=1,3
  calculate rhs_f   ! With a discretization on D_d
  if (rk.eq.1) then
    dt_temp = dt
    f = f_p + dt_temp * rhs_f
  else if (rk.eq.2) then
    dt_temp = 0.25 * dt
    f = 0.75 * f_p + 0.25 * f + dt_temp * rhs_f
  else
    dt_temp = 2.0D0 * dt / 3.0D0
    f = f_p / 3.0D0 + 2.0D0 * f / 3.0D0 + dt_temp * rhs_f
  end if
end do
```

For the implementation of these two methods, three basic one-dimensional arrays have to be defined:

- One for entries of f_i^n
- One for entries of f_i^{n+1}
- One for the right-hand side of the semi-discrete Eq. (4.3) rhs_f_i^n

with $N_x + 1$ entries $i = 0, \ldots, N_x$. With these arrays, it is possible to integrate the semi-discrete expression (4.3) using only two time levels of f stored. The arrays can be recycled after each time step as follows:

$$f_i^{n+1} \rightarrow f_i^n,$$

and then release the memory assigned to the array associated with f_i^{n+1}. Integrators (4.4) and (4.5) will be very used in the forthcoming examples; sometimes, we use Heun RK2 and some others RK3.

When instead of a single equation the IVP involves a set of k coupled equations $\partial_t u_1 = \text{rhs}(u_1),\ldots,\partial_t u_k = \text{rhs}(u_k)$, the evolution method has to be applied to each of the unknowns $u_1, \ldots, u_k$.

4.3 Advection Equation

The advection equation is a good first example of IVP because the PDE is first order in time and space. The problem is formulated as follows:

$$\boxed{\begin{aligned} &\frac{\partial u}{\partial t} + a \frac{\partial u}{\partial x} = 0 & & u = u(x,t) \\ & D = x \in [x_{min}, x_{max}] \times t \in [0, t_f] & & Domain \\ & u(x, 0) = u_0(x) & & Initial\ Conditions \\ & u(x_{min}, t),\ u(x_{max}, t) & & Boundary\ Conditions \end{aligned}} \quad (4.6)$$

We construct a discrete domain D_d as done previously, with points $(x_i, t^n) = (x_{min} + i\Delta x, n\Delta t)$, where $\Delta x = (x_{max} - x_{min})/N_x$ and $\Delta t = C\ \Delta x$.

The spatial derivative of the advection equation is written using the second-order accurate approximation (3.5). With these elements, the **semi-discrete** version of the advection equation at the interior point $(x_i, t^n) \in D_d$ reads

$$\left. \frac{\partial u}{\partial t} \right|_{(x_i, t^n)} = -a \frac{u_{i+1}^n - u_{i-1}^n}{2\Delta x} + \mathcal{O}(\Delta x^2). \quad (4.7)$$

This illustrates how the rhs of the time derivative of u is a number, namely, the discrete version of the spatial derivative of u in (4.6). The molecule that illustrates this method is shown at the left in Fig. 4.2.

For the example of numerical solution, we use the domain D_d with $[x_{min}, x_{max}] = [-1, 1]$ and $t_f = 2$, as well as $N_x = 200$, which implies a resolution $\Delta x = 0.01$, and $C = 0.5$, which implies $\Delta t = 0.005$.

Initial conditions are Dirichlet boundary conditions at the boundary $t = 0$ in Fig. 3.1, given by a Gaussian $u(x, 0) = u_0(x) = A \exp(-x^2/\sigma^2)$ with $A = 1$ and $\sigma = 0.1$.

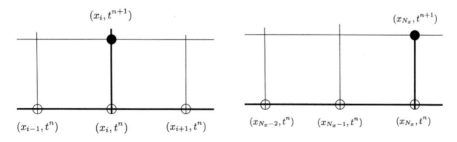

Fig. 4.2 Left: molecule illustrating scheme for the advection equation in Eq. (4.7). Right: molecule for the implementation of the boundary condition at the $x = x_{N_x}$, using the one-sided backward discrete derivative in Eq. (4.8)

Boundary Conditions. The advection equation in (4.6) drives the initial conditions along the characteristic line $t = x$ across the domain, and the pulse will move toward the right side of the spatial domain. Knowing this, we use the following set of boundary conditions:

- Impose a Dirichlet condition $u(x_0, t^n) = 0$ at the left boundary (x_0, t^n) of Fig. 3.1.
- Use the Neumann outgoing wave boundary condition at the right boundary (x_{N_x}, t^n). For this condition, one has to solve Eq. (3.31), $\partial_t u + \partial_x u = 0$, this time using the MoL. The semi-discrete version of this equation at the boundary is

$$\left.\frac{\partial u}{\partial t}\right|_{(x_{N_x}, t^n)} = -a \frac{u^n_{N_x-2} - 4u^n_{N_x-1} + 3u^n_{N_x}}{2\Delta x} + \mathcal{O}(\Delta x^2), \qquad (4.8)$$

where the second-order accurate backward derivative (3.8) is used at $i = N_x$.

The code that solves this problem is MoLAdvection.f90. Here, we show the core of the code, that is, the time integration loop using the Heun RK2 integrator.

```
do n=1,Nt
t = t + dt  ! Updates time

! *****  --> LOOP CORE <--
! Recycle arrays
u_p = u
! Heun RK2
do rk=1,2
  ! Calculate the RHS
  do i=1,Nx-1
    rhs_u(i) = - 0.5d0 * a * ( u(i+1) - u(i-1) ) / dx
  end do
  ! Now at the boundaries
  rhs_u(0) = 0.0d0
  rhs_u(Nx) = -0.5d0*a*( u(Nx-2) - 4.0d0 * u(Nx-1) + 3.0d0
  * u(Nx) ) / dx
  if (rk.eq.1) then
    u = u_p + rhs_u * dt
  else
    u = 0.5d0 * (u_p + u + rhs_u * dt)
  end if
end do

! *****  --> Ends Loop Core <--

! Calculate the exact solution
! Calculate the error
! Calculate norms L1 and L2 of errors
! Save data in ordinary and fancy ways

end do
```

4.3 Advection Equation

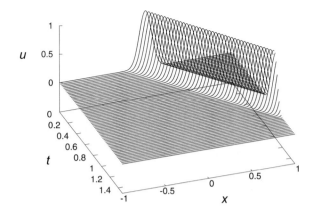

Fig. 4.3 Numerical solution of the advection equation for $a = 1$, using $N_x = 200$, or equivalently $\Delta x = 0.01$ and $C = 0.5$

Notice that we indicate a `Loop Core`, an essential piece of code that will be used and adapted in the further examples in this chapter.

The resulting numerical solution is shown for $a = 1$ in Fig. 4.3. The pulse evolves, and the boundary conditions do as expected, fixed at the left and outgoing wave at the right. Notice that the solution corresponds to one of the pulses of the wave equation in Figs. 3.8 or 3.16.

The exact solution to this IVP for the initial Gaussian pulse is $u^e = A \exp[-(x - t)^2/\sigma^2]$, which we use to calculate the error of the numerical solution $e = u - u^e$, and its convergence to zero. The time integrator RK2 is second-order accurate, the discrete version of the spatial derivative in (4.7) is second-order accurate as well, and the semi-discrete version of the boundary condition at the right end (4.8) is also second-order accurate. In summary, the error is second order, and second-order convergence is expected as long as the evolution is stable according to Lax Theorem [1].

In Fig. 4.4, we show the L_1 and L_2 norms of the error, calculated with formulas (3.24), for three resolutions $N_x = 200, 400$, and 800 or equivalently $\Delta x_1 = 0.01$, $\Delta x_2 = \Delta x_1/2$, and $\Delta x_3 = \Delta x_2/2$. The convergence of the error is second order, which indicates that the implementation is correct and that the resolution used is within the convergent regime.

Let us now implement the RK3 method (4.5). The loop core corresponding to the implementation of this method for the advection equation reads

```
! *****  --> LOOP CORE <--
! RK3
do rk=1,3
   ! Calculate the RHS
   do i=1,Nx-1
      rhs_u(i) = - 0.5d0 * a * ( u(i+1) - u(i-1) ) / dx
   end do
   ! Now at the boundaries
```

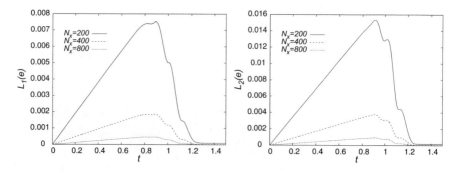

Fig. 4.4 Norms $L_1(e)$ and $L_2(e)$ of the error. Notice that the approximate factor is four between the error calculated using successive resolutions which indicates second-order convergence of the two norms to zero

```
  rhs_u(0) = 0.0d0
  rhs_u(Nx) = -0.5d0 * a * ( u(Nx-2) - 4.0d0 * u(Nx-1)
+ 3.0d0 * u(Nx) ) / dx
  if (rk.eq.1) then
    dt_temp = dt
    u = u_p + dt_temp * rhs_u
  else if (rk.eq.2) then
    dt_temp = 0.25 * dt
    u = 0.75 * u_p + 0.25 * u + dt_temp * rhs_u
  else
    dt_temp = 2.0D0 * dt / 3.0D0
    u = u_p / 3.0D0 + 2.0D0 * u / 3.0D0 + dt_temp
* rhs_u
    end if
end do
! ***** --> Ends Loop Core <--
```

The result of a convergence test using the RK3 integrator would be very similar to that in Fig. 4.4, including the second-order convergence expected for the RK2 method. It means that despite of using a third-order accurate method for time integration, the convergence would still be second order. The reason is that the accuracy of the solution will be dominated by the leading order error. In this case, the time integration is third-order accurate; however, the discretization of the rhs(u) is second order, and then the overall error is expected to be second order.

Even though the accuracy is second order, we will use the RK3 integrator in various of the following examples because it is more stable than the RK2 and is equally easy to implement:

Exercise. Solve the advection equation with $a = 0.5$, and verify that it takes the Gaussian pulse twice the time to arrive at the boundary compared to the case when $a = 1$.

4.4 1+1 Wave Equation

The advection equation serves as a guide to implement the solution of first order in time PDEs using the method of lines. We would like to apply this method to solve the wave equation, and part of the work consists in converting a second-order equation in space and time, into a system of first-order equations.

This can be done by defining the variables:

$$\psi := \frac{\partial \phi}{\partial x}, \quad \pi := \frac{\partial \phi}{\partial t},$$

in terms of which the wave equation $\frac{\partial^2 \phi}{\partial t^2} - \frac{\partial^2 \phi}{\partial x^2} = 0$ can be written as $\frac{\partial \pi}{\partial t} = \frac{\partial \psi}{\partial x}$, which is an evolution equation for π. However, the definition of ψ is a constraint that must be satisfied at all times. Therefore, it is necessary to construct an evolution equation for ψ, which can be done as follows: $\frac{\partial \psi}{\partial t} = \frac{\partial}{\partial t}\frac{\partial \phi}{\partial x} = \frac{\partial}{\partial x}\frac{\partial \phi}{\partial t} = \frac{\partial \pi}{\partial x}$, provided ϕ is at least of class C^2.

For ease, from now on, we will use indistinctly the notation $\partial_t f = \partial f / \partial t$ and $\partial_x f = \partial f / \partial x$. Considering this, the wave equation is converted into a system of two first-order equations:

$$\partial_t \psi = \partial_x \pi,$$
$$\partial_t \pi = \partial_x \psi.$$

This system of equations can be solved using the MoL; however, the solution would be for the auxiliary functions ψ and π, while the original goal is the construction of ϕ. The wave function can be integrated at the same time as ψ and π by adding a third equation that is actually the definition of π:

$$\partial_t \phi = \pi.$$

With these three equations, one can formulate the following IVP:

$$\partial_t \psi = \partial_x \pi$$
$$\partial_t \pi = \partial_x \psi$$
$$\partial_t \phi = \pi \qquad \phi = \phi(x,t),\ \psi = \psi(x,t),\ \pi = \pi(x,t)$$
$$D = x \in [x_{min}, x_{max}] \times t \in [0, t_f] \qquad Domain$$
$$\phi(x,0) = \phi_0(x),\ \psi(x,0) = \psi_0(x),$$
$$\pi(x,0) = \pi_0(x) \qquad Initial\ Conditions$$
$$\phi(x_{min}, t),\ \phi(x_{max}, t)$$
$$\psi(x_{min}, t),\ \psi(x_{max}, t)$$
$$\pi(x_{min}, t),\ \pi(x_{max}, t) \qquad Boundary\ Conditions$$
$$(4.9)$$

The IVP will be solved using the MoL on the domain D_d with points $(x_i, t^n) = (x_{min} + i\Delta x, n\Delta t)$, where $\Delta x = (x_{max} - x_{min})/N_x$ and $\Delta t = C\ \Delta x$. The semi-discrete version of the evolution equations in (4.9), at the point (x_i, t^n), using the second-order accurate formula of the spatial derivative, reads

$$\partial_t \psi = \frac{\pi_{i+1}^n - \pi_{i-1}^n}{2\Delta x} + \mathcal{O}(\Delta x^2),$$
$$\partial_t \pi = \frac{\psi_{i+1}^n - \psi_{i-1}^n}{2\Delta x} + \mathcal{O}(\Delta x^2),$$
$$\partial_t \phi = \pi_i^n, \qquad (4.10)$$

valid at the interior points of the domain $i = 1, ..., N_x - 1$. For this system of three equations, one-dimensional arrays are needed for ϕ^{n+1}, ϕ^n, rhs(ϕ), ψ^{n+1}, ψ^n, rhs(ψ), π^{n+1}, π^n, and rhs(π).

Initial conditions are Dirichlet boundary condition for each variable at the side $(x, 0)$ instead of Cauchy conditions as in Sect. 3.2. Nevertheless, we will set the same time-symmetric Gaussian pulse, which in terms of the wave function and the auxiliary variables reads:

$$\phi_0(x) = Ae^{-x^2/\sigma^2},\quad \psi_0(x) = -\frac{2x}{\sigma^2}\phi_0(x),\quad \pi_0(x) = 0, \qquad (4.11)$$

with $A = 1$ and $\sigma = 0.1$. Notice that time symmetry means simply $\pi_0(x) = 0$ and that the initial conditions of the three variables are set at time $t = t^0$, unlike in Sect. 3.2, where information on two time slices was needed in order to set time-symmetric initial conditions.

At boundaries (x_0, t^n) and (x_{N_x}, t^n), we impose *outgoing wave boundary conditions*. The implementation can be done in various ways; here, we show one strategy and will show a more sophisticated one later on when we solve the wave equation on a general Minkowski space-time.

4.4 1+1 Wave Equation

At the boundary at (x_0, t^n), the outgoing wave equation to be satisfied is Eq. (3.30), $\partial_t \phi = \partial_x \phi$, which is Neumann-type and can be semi-discretized using the forward formula (3.7) as follows:

$$\partial_t \phi = -\frac{\phi_2^n - 4\phi_1^n + 3\phi_0^n}{2\Delta x} + \mathcal{O}(\Delta x^2). \tag{4.12}$$

With this semi-discrete version, ϕ can be integrated up to the boundary. Now, the condition on ψ, let ϕ_0^m be the value of the wave function at this boundary at the intermediate iteration m within the RK integrator, it is possible to calculate the spatial derivative of ϕ, ψ_0^m using the forward approximation (3.7):

$$\psi_0^m = -\frac{\phi_2^m - 4\phi_1^m + 3\phi_0^m}{2\Delta x} + \mathcal{O}(\Delta x^2). \tag{4.13}$$

Finally, the outgoing wave condition itself is a relation between ψ and π, that is, $\partial_t \phi = \partial_x \phi$ at this left boundary:

$$\pi_0^m = \psi_0^m. \tag{4.14}$$

Expressions (4.12), (4.13), and (4.14) provide the values for all the three variables at (x_0, t^n).

At the right boundary (x_{N_x}, t^n), the equation to be fulfilled is (3.31), $\partial_t \phi = -\partial_x \phi$. Proceeding analogously, ϕ can be integrated up to the boundary using the appropriate backward formula for the spatial derivative (3.8):

$$\partial_t \phi = -\frac{\phi_{N_x-2}^n - 4\phi_{N_x-1}^n + 3\phi_{N_x}^n}{2\Delta x} + \mathcal{O}(\Delta x^2).$$

Knowing that the value of ϕ_{N_x} allows the calculation of ψ_{N_x} and π_{N_x}, using again the definition of ψ and the boundary condition, the values of the auxiliary variables are:

$$\psi_{N_x}^m = \frac{\phi_{N_x-2}^m - 4\phi_{N_x-1}^m + 3\phi_{N_x}^m}{2\Delta x} + \mathcal{O}(\Delta x^2),$$

$$\pi_{N_x}^m = -\psi_{N_x}^m, \tag{4.15}$$

at all intermediate steps m of the RK integrator. In this strategy, the order of the instructions within the program is important. The Loop Core of the code that implements the evolution using these boundary conditions can be as follows. We show the code for the Heun RK2 method, but the implementation using the RK3 method is analogous:

```
! ***** --> LOOP CORE <--
! Recycle arrays
phi_p = phi
```

```
psi_p = psi
pi_p = pi

! Mid-point RK2
do rk=1,2
  ! Calculate the RHS at the interior
  do i=1,Nx-1
    rhs_psi(i) = 0.5d0 * ( pi(i+1) - pi(i-1) ) / dx
    rhs_pi(i)  = 0.5d0 * ( psi(i+1) - psi(i-1) ) / dx
    rhs_phi(i) = pi(i)
  end do
  ! The rhs only for phi at the boundaries
  rhs_phi(0) = - 0.5d0 * ( phi(2) - 4.0d0*phi(1) + 3.0d0 * phi(0)
) / dx
  rhs_phi(Nx) = - 0.5d0 * ( phi(Nx-2) - 4.0d0*phi(Nx-1) + 3.0d0 *
phi(Nx) ) / dx
  if (rk.eq.1) then
    psi = psi_p + rhs_psi * dt
    pi  = pi_p + rhs_pi * dt
    phi = phi_p + rhs_phi * dt
  else
    psi = 0.5d0 * ( psi_p + psi + rhs_psi * dt)
    pi  = 0.5d0 * ( pi_p + pi + rhs_pi * dt)
    phi = 0.5d0 * ( phi_p + phi + rhs_phi * dt)
  end if
  psi(0) = - 0.5d0 * ( phi(2) - 4.0d0*phi(1) + 3.0d0*phi(0)) / dx
  pi(0) = psi(0)
  psi(Nx) = 0.5d0 * ( phi(Nx-2) - 4.0d0*phi(Nx-1)
+ 3.0d0 * phi(Nx) ) / dx
  pi(Nx) = -psi(Nx)
end do
! ***** --> Ends Loop Core <--
```

The results are shown in Fig. 4.5, where the wave function ϕ behaves as we have already seen in the previous chapter. New in this case is the behavior of the auxiliary variables ψ and π. The wave function can be compared with the results obtained with the simple discretization method in Fig. 3.8 and the Crank-Nicolson method in Fig. 3.16.

Periodic Boundary Conditions. The implementation consists in identifying the values of $\phi, \psi,$ and π at the left boundary (x_0, t^n) with those at the right boundary (x_{N_x}, t^n), which produce the effect of topology change of the domain as illustrated in Fig. 3.5. In practice, one has to specify the appropriate right-hand sides of $\phi, \psi,$ and π at (x_0, t^n) and (x_{N_x}, t^n). At the left, boundary equations (4.10) are written as

$$\partial_t \psi = \frac{\pi_1^n - \pi_{N_x-1}^n}{2\Delta x} + \mathcal{O}(\Delta x^2),$$

$$\partial_t \pi = \frac{\psi_1^n - \psi_{N_x-1}^n}{2\Delta x} + \mathcal{O}(\Delta x^2), \qquad (4.16)$$

4.4 1+1 Wave Equation

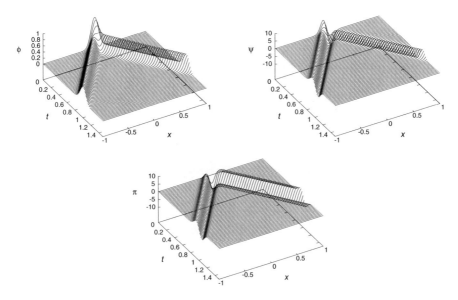

Fig. 4.5 Numerical solution using the MoL for the wave equation with outgoing wave boundary conditions. Shown are the wave function ϕ and the auxiliary variables ψ and π

$$\partial_t \phi = \pi_0^n,$$

whereas at the right boundary equations (4.10) become

$$\partial_t \psi = \frac{\pi_1^n - \pi_{N_x-1}^n}{2\Delta x} + \mathcal{O}(\Delta x^2),$$

$$\partial_t \pi = \frac{\psi_1^n - \psi_{N_x-1}^n}{2\Delta x} + \mathcal{O}(\Delta x^2), \quad (4.17)$$

$$\partial_t \phi = \pi_{N_x}^n.$$

For the numerical solution, we define the domain D_d with $[x_{min}, x_{max}] = [-1, 1]$, $t_f = 2$, using $C = 0.5$, and resolution with $N_x = 200$, equivalent to $\Delta x_1 = 0.01$. The results are shown in Fig. 4.6, where we notice that all the variables get off through one of the boundaries $((x_0, t^n)$ or $(x_{N_x}, t^n))$ and reenter the domain through the other one. These boundary conditions are implemented on the right-hand side of the equation for each variable, which is handy. The Loop Core code would read as follows:

```
! ***** --> LOOP CORE <--
! Recycle arrays
phi_p = phi
psi_p = psi
pi_p = pi
```

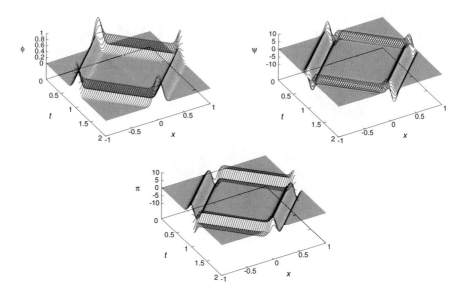

Fig. 4.6 Numerical solution using MoL for the wave equation with periodic boundary conditions. Shown are the wave function ϕ and the auxiliary variables ψ and π. The wave function can be compared with that of Fig. 3.6

```
! Mid-point RK2
do rk=1,2
  ! Calculate the RHS at the interior
  do i=1,Nx-1
    rhs_psi(i) = 0.5d0 * ( pi(i+1) - pi(i-1) ) / dx
    rhs_pi(i) = 0.5d0 * ( psi(i+1) - psi(i-1) ) / dx
    rhs_phi(i) = pi(i)
  end do
  ! RHS for periodic boundary conditions
  rhs_psi(0) = 0.5d0 * ( pi(1) - pi(Nx-1) ) / dx
  rhs_pi(0) = 0.5d0 * ( psi(1) - psi(Nx-1) ) / dx
  rhs_phi(0) = pi(Nx)
  rhs_psi(Nx) = 0.5d0 * ( pi(1) - pi(Nx-1) ) / dx
  rhs_pi(Nx) = 0.5d0 * ( psi(1) - psi(Nx-1) ) / dx
  if (rk.eq.1) then
    psi = psi_p + rhs_psi * dt
    pi = pi_p + rhs_pi * dt
    phi = phi_p + rhs_phi * dt
  else
    psi = 0.5d0 * ( psi_p + psi + rhs_psi * dt)
    pi = 0.5d0 * ( pi_p + pi + rhs_pi * dt)
    phi = 0.5d0 * ( phi_p + phi + rhs_phi * dt)
  end if
end do
! ***** --> Ends Loop Core <--
```

4.4 1+1 Wave Equation

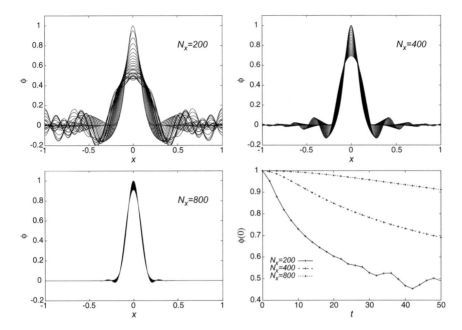

Fig. 4.7 Snapshots of the solution using periodic boundary conditions in the time domain [0, 50] and the RK3 integrator, every two units of time using three resolutions with $N_x = 200, 400,$ and 800. At the right bottom, we show the central value of ϕ as function of time for the three runs, which illustrates that diffusion is resolution-dependent. These results are to be compared with those in Fig. 3.11

We finally show the effects of **diffusion** and **dispersion** as done in Fig. 3.11, this time using the method of lines with the third-order Runge-Kutta integrator in Fig. 4.7 and with the second-order Runge-Kutta in Fig. 4.8. In the later case, the evolution is unstable after a few crossing times since errors accumulate and trigger high-frequency modes that destroy the solution.

Continuing with the expression of solutions in terms of plane waves whose velocities may be different for different modes in Sect. 3.4.3, a summary of effects from [1] reads as follows: A scheme is **unstable** if at least one Fourier mode grows without bound in the sense explained in Appendix A, as happens with the results using RK2 method in Fig. 4.8. It is **dissipative** if no Fourier mode grows and at least one decays, like in the case of the diffusion equation in Sect. 3.4. It is **non-dissipative** if Fourier modes neither grow or decay, nearly like the simple method used to solve the wave equation with results in Fig. 3.11 with high resolution. It is dispersive if different modes propagate at different speeds, like the solution of the wave equation using the Crank-Nicolson method with $C = 2$ in Fig. 3.17 or using the MoL with the RK3 integrator in Fig. 4.7.

Finally, the code that solves the wave equation is MoLWave.f90, listed in Appendix B.

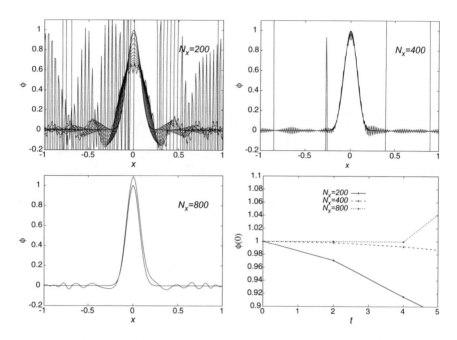

Fig. 4.8 Snapshots of the solution using periodic boundary conditions in the time domain [0, 50] and the RK2 integrator, every two units of time using three resolutions with $N_x = 200, 400,$ and 800. At the right bottom, we show the central value of ϕ as function of time for the three runs, which illustrates that diffusion is resolution-dependent. These results are to be compared with those in Figs. 3.11 and 4.7

Exercise. The **crossing time** is defined as the time it takes a signal to travel across the domain; in the case of our wave equation, since the speed is one and the domain is two units wide, the crossing time is two units of time. Run the code using periodic boundary conditions, and determine the number of crossing times at which the amplitude of ϕ changes by 1%. This will tell about the dissipation of the method.

4.5 Wave Equation in a General 1+1 Minkowski Space-Time: Example of Characteristic Analysis

This is an advanced application in relativistic physics. With this example, we intend to illustrate the use of gauge in the solution of relativistic problems. We choose the wave equation because it is familiar from previous sections.

4.5 Wave Equation in a General 1+1 Minkowski Space-Time: Example of...

We now use a sufficiently general description of Minkowski space-time. Let us assume that the space-time is described with coordinates $(\tilde{x}, \tilde{t})$. Then the line element of this space-time is given by $ds^2 = -d\tilde{t}^2 + d\tilde{x}^2$ or equivalently $ds^2 = g_{\mu\nu} dx^\mu dx^\nu$ where $g_{\mu\nu}$ are the components of the metric tensor in the Minkowski space-time.

This line element indicates the space-time is foliated with straight lines of constant $\tilde{t}$ and constant $\tilde{x}$, and the coordinates of events are points with coordinates $(\tilde{x}, \tilde{t})$.

Nevertheless, the space-time is the same despite how it is foliated; it is independent of the coordinates we use to label its points. A more general way to label the space-time points using lines other than straight lines of constant time and space can be done using a coordinate transformation. For example, instead of foliating time with lines of constant time, one can define a time coordinate t such that $d\tilde{t} = \alpha(x) dt$. The spatial coordinate x can also be defined in coordinates such that $d\tilde{x} = dx - \beta dt$, where β is a velocity at which the x coordinate displaces. These two generalizations for time and space coordinates can be combined to produce very general descriptions of the space-time.

Within this general description, the curves of constant space and time coordinates are not necessarily horizontal and vertical straight lines anymore. Illustrations of various foliations appear in Fig. 4.9. First we show the standard foliation with $\alpha = 1$, $\beta = 0$ corresponding to horizontal and vertical lines in the space-time. A second case uses a nontrivial $\alpha = \frac{1}{4} \tanh(10x) + \frac{3}{4}$ with $\beta = 0$. A third case uses a trivial $\alpha = 1$ and a constant $\beta = 1/2$. Finally, a fourth case uses $\alpha = 1$ and a nonconstant velocity $\beta = x$.

Notice that events corresponding to a given value of coordinate time $\tilde{t}$ lie along continuous lines and events corresponding to a constant coordinate position $\tilde{x}$ are located along dashed lines. The most common case corresponds to $\alpha = 1$ and $\beta = 0$, with $t = \tilde{t}$ and $\tilde{x} = x$, where lines of constant time are horizontal and those of constant position are vertical.

We present one case where lines of constant time are not straight lines and another case where lines of constant position are not straight lines either. We will illustrate in this section the utility of using these coordinates.

In relativistic problems, the use of such kind of general coordinates allows the control of the evolution in terms of space and time. For example, in the second case of Fig. 4.9, the evolution of a process at the left half of the domain will be slower than in the right half, whereas in the third plot the spatial coordinate moves at constant velocity at all times.

Therefore, the election of functions α and β depends on the type of process we would like to study or the type of effect one wants to capture in the solution. In these very general coordinates, the line element ds^2, which is invariant under coordinate changes, reads

$$ds^2 = -d\tilde{t}^2 + d\tilde{x}^2$$
$$= (-\alpha^2 + \beta^2) dt^2 + 2\beta dt dx + dx^2.$$

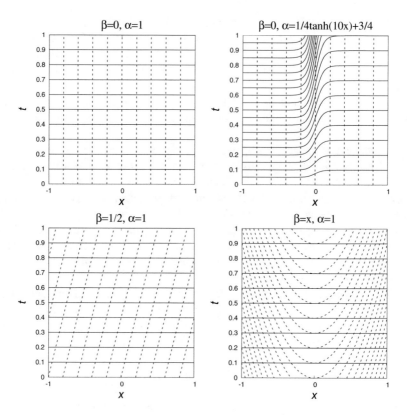

Fig. 4.9 Four different foliations of the space-time. Lines of constant time are represented by continuous lines, whereas dashed lines indicate lines of constant position

$$= g_{\mu\nu}dx^{\mu}dx^{\nu}, \tag{4.18}$$

where $x^{\mu} = t, x$ and Einstein summation convention holds. Explicitly, the metric tensor and its inverse in matrix form are

$$g_{\mu\nu} = \begin{pmatrix} (-\alpha^2 + \beta^2) & \beta \\ \beta & 1 \end{pmatrix}, \tag{4.19}$$

$$g^{\mu\nu} = \begin{pmatrix} -1/\alpha^2 & \beta/\alpha^2 \\ \beta/\alpha^2 & (1 - \beta^2/\alpha^2) \end{pmatrix}, \tag{4.20}$$

where $\alpha > 0$ is called the **lapse function** and β the **shift vector**, with only one component for the case of the 1+1 space-time.

Now, let us focus on the wave equation. The wave D'Alembertian operator in a general space-time is defined by $\Box \phi = \frac{1}{\sqrt{-g}} \partial_{\mu}[\sqrt{-g} g^{\mu\nu} \partial_{\nu}\phi]$ where $g = \det(g_{\mu\nu})$ and Einstein summation convention holds. For the metric (4.19), one has $\sqrt{-g} = \alpha$, and the wave equation for this general description of the space-time reads

4.5 Wave Equation in a General 1+1 Minkowski Space-Time: Example of...

$$0 = \Box \phi$$
$$= \frac{1}{\sqrt{-g}} \partial_\mu [\sqrt{-g} g^{\mu\nu} \partial_\nu \phi]$$
$$= \frac{1}{\alpha} \partial_t [\alpha g^{t\nu} \partial_\nu \phi] + \frac{1}{\alpha} \partial_x [\alpha g^{x\nu} \partial_\nu \phi]$$
$$= \frac{1}{\alpha} \partial_t [\alpha g^{tt} \partial_t \phi + \alpha g^{tx} \partial_x \phi] + \frac{1}{\alpha} \partial_x [\alpha g^{xt} \partial_t \phi + \alpha g^{xx} \partial_x \phi]$$
$$= \frac{1}{\alpha} \partial_t \left[-\frac{1}{\alpha} \partial_t \phi + \frac{\beta}{\alpha} \partial_x \phi \right] + \frac{1}{\alpha} \partial_x \left[\frac{\beta}{\alpha} \partial_t \phi + \alpha \left(1 - \frac{\beta^2}{\alpha^2} \right) \partial_x \phi \right]. \quad (4.21)$$

Notice that when $\alpha = 1$ y $\beta = 0$, one recovers the usual wave equation $\partial_t^2 \phi - \partial_x^2 \phi = 0$, which holds only in the traditional description of Minkowski space-time. It is desirable to write down this equation as a system of first-order PDEs, which would allow us to use the method of lines for the construction of numerical solutions.

Expression (4.21) suggests the definition of new variables $\psi := \partial_x \phi$ and $\pi := (\partial_t \phi - \beta \partial_x \phi)/\alpha$. Observe that π is simply the argument of the first-order time derivative in (4.21). With these new variables, one can write the wave equation as a system of two first-order equations, one for π:

$$\partial_t \pi = \partial_x (\alpha \psi + \beta \pi), \quad (4.22)$$

and by assuming ϕ is at least of class C^2, the evolution equation for ψ is $\partial_t \psi = \partial_t (\partial_x \phi) = \partial_x (\partial_t \phi)$, which implies an equation for ψ:

$$\partial_t \psi = \partial_x (\alpha \pi + \beta \psi). \quad (4.23)$$

Equations (4.22)–(4.23) constitute the first-order version of the wave equation with the constraint $\psi = \partial_x \phi$. Finally, we construct the wave function ϕ from the definition of π, that is, $\partial_t \phi = \alpha \pi + \beta \psi$. With all these conventions, the IVP to solve reads

$\partial_t \pi = \partial_x (\alpha \psi + \beta \pi),$	
$\partial_t \psi = \partial_x (\alpha \pi + \beta \psi),$	
$\partial_t \phi = \alpha \pi + \beta \psi,$	$\phi = \phi(x,t) \; \psi = \psi(x,t), \; \pi = \pi(x,t)$
$D = x \in [x_{min}, x_{max}] \times t \in [0, t_f]$	Domain
$\phi(x,0) = \phi_0(x), \; \psi(x,0) = \psi_0(x),$	
$\pi(x,0) = \pi_0(x)$	Initial Conditions
$\phi(x_{min},t), \; \phi(x_{max},t)$	
$\psi(x_{min},t), \; \psi(x_{max},t)$	
$\pi(x_{min},t), \; \pi(x_{max},t)$	Boundary Conditions

$$(4.24)$$

The IVP will be solved using the MoL on the domain D_d with points $(x_i, t^n) = (x_{min} + i\Delta x, n\Delta t)$, where $\Delta x = (x_{max} - x_{min})/N_x$ and $\Delta t = C\,\Delta x$. Before dealing with the implementation of the MoL to solve this IVP, we discuss various mathematical aspects of the problem.

Characteristic Analysis. The characteristic analysis of a system of equations is a formal strategy to understand the construction of solutions associated with evolution problems and is also useful for various reasons, for example:

- In some cases, like the wave equation, it allows to write down the system in (4.24) as a set of advection equations.
- The variables obeying these advection equations are the **characteristic variables**. These variables correspond to the modes that carry the information of the wave to the *right* and to the *left*.
- These characteristic variables allow the construction of boundary conditions, even in general conditions like in our case of the wave equation here for $\alpha = \alpha(x)$ y $\beta = \beta(x)$.
- In general, for IVPs associated with less known equations, the analysis allows to determine the well posedness of the problem and stability of the solutions.
- The analysis is essential in the solution of hydrodynamics equations.

Let us then proceed to the analysis of the wave equation in (4.24). Notice that the third equation is only auxiliary for the construction of ϕ and that the wave equation was actually converted into two equations, one for π (4.22) and one for ψ (4.23). Then the analysis centers on these two equations.

We define a state vector $\mathbf{u} = (\pi, \psi)^T$, so that the system of equations can be written as

$$\partial_t \mathbf{u} + \partial_x(\mathbf{A}\mathbf{u}) = 0 \quad \Rightarrow$$
$$\partial_t \mathbf{u} + \mathbf{A}\partial_x \mathbf{u} = -\partial_x(\mathbf{A})\mathbf{u} \tag{4.25}$$

where

$$\mathbf{A} = -\begin{pmatrix} \beta & \alpha \\ \alpha & \beta \end{pmatrix}. \tag{4.26}$$

The characteristic velocities of propagation of signals in the xt plane are the eigenvalues of the matrix $\mathbf{A}$, which are solutions of the equation $\det(\mathbf{A} - I_2 \lambda) = 0$ with λ a local velocity of propagation of data in time. The eigenvalues of $\mathbf{A}$ are

$$\lambda_\pm = -\beta \pm \alpha. \tag{4.27}$$

This means that the propagation velocities depend on the coordinates used to foliate the space-time represented by the gauge functions α and β. Notice from (4.21) that α has to be nonzero, which implies that the two eigenvalues are real and different; thus, the eigenvectors of $\mathbf{A}$ constitute a complete set, and then the system of equations is

4.5 Wave Equation in a General 1+1 Minkowski Space-Time: Example of…

strongly hyperbolic, which is a property that guarantees that the IVP is well posed (see, for example, [2] for a detailed discussion of well-posed IVPs).

In the case $\beta = 0$ y $\alpha = 1$, the characteristic velocities are $\lambda = \pm 1$, constant in the whole space-time, which implies the trajectories of the initial data are straight lines with slopes ± 1 in the space-time, that define the light cones $x = x_0 \pm t$ with $x_0 \in \mathbb{R}$, whose lines correspond to the characteristics of the solution, that is, those curves along which the value of the solution is the same as at initial time [3]. An **illustration of this fact appears in Fig. 4.5**; notice that at initial time there are initial superposed pulses of ϕ, ψ, π that separate and propagate along the straight lines with slope ± 1; these lines are the characteristic lines.

The eigenvectors associated with $\lambda_\pm$ are $u_1 = (1, -1)^T$ y $u_2 = (1, 1)^T$; then the matrix that diagonalizes **A** is

$$\mathbf{P} = \begin{pmatrix} 1 & 1 \\ -1 & 1 \end{pmatrix}, \quad \mathbf{P}^{-1} = \frac{1}{2}\begin{pmatrix} 1 & -1 \\ 1 & 1 \end{pmatrix}. \tag{4.28}$$

From here, **A** can be written as $\mathbf{A} = \mathbf{P}\boldsymbol{\Lambda}\mathbf{P}^{-1}$ with $\boldsymbol{\Lambda} = diag(\lambda_+, \lambda_-)$. Multiplying (4.25) by $\mathbf{P}^{-1}$, one obtains

$$\mathbf{P}^{-1}\partial_t \mathbf{u} + \mathbf{P}^{-1}\mathbf{A}\partial_x \mathbf{u} = -\partial_x(\mathbf{P}^{-1}\mathbf{A})\mathbf{u},$$
$$\partial_t \mathbf{w} + \boldsymbol{\Lambda}\partial_x \mathbf{w} = -\partial_x(\boldsymbol{\Lambda})\mathbf{w}, \tag{4.29}$$

where we have used that **P** has constant entries and that $\mathbf{P}^{-1}\mathbf{A} = \boldsymbol{\Lambda}\mathbf{P}^{-1}$. Notice the new state vector

$$\mathbf{w} = \mathbf{P}^{-1}\mathbf{u} = \frac{1}{2}\begin{pmatrix} \pi - \psi \\ \pi + \psi \end{pmatrix} = \begin{pmatrix} R \\ L \end{pmatrix}, \tag{4.30}$$

whose components are the *characteristic variables* R and L. In this way, Eqs. (4.22) and (4.23) decouple, and the dynamics of the wave is described by a mode traveling to the right $R = \frac{1}{2}(\pi - \psi)$ and another mode traveling to the left $L = \frac{1}{2}(\pi + \psi)$.

Characteristic analysis has translated the system (4.22)–(4.23) into two decoupled **advection equations** for R and L, with sources on the right-hand side:

$$\partial_t R + \lambda_+ \partial_x R = -\partial_x(\lambda_+)R,$$
$$\partial_t L + \lambda_- \partial_x L = -\partial_x(\lambda_-)L, \tag{4.31}$$

equations that in principle are easy to solve as done in Sect. 4.3.

Boundary Conditions
We discuss the implementation of boundary conditions in detail because the implementation is subtle for a general description of Minkowski space-time.

Periodic boundary conditions can be implemented as in the previous examples, whereas outgoing wave boundary conditions are more complex, and their implementation uses the characteristic information of the system of equations developed above and are very important. We focus on these conditions only.

In the case $\alpha = 1$, $\beta = 0$, we know the wave operator $\square$ can be easily factorized $\square = (\partial_t + \partial_x)(\partial_t - \partial_x)$, and as seen before, this property allows to impose outgoing wave boundary conditions by setting $(\partial_t - \partial_x)\phi = 0$ at the left boundary (x_0, t^n) and $(\partial_t + \partial_x)\phi = 0$ at the right boundary (x_{N_x}, t^n), as done in Sects. 3.2 and 3.3.

However, for more general α and β, the operator in (4.21) cannot be easily factorized. This is where the characteristic structure of the system is helpful again.

In terms of the characteristic variables (R, L), the boundary conditions are as follows: At the left boundary (x_0, t^n), we would like L to remain moving toward the left, whereas we do not want information traveling to the right, which means we need to set $R = 0$ there, that is, $R_0 = 0$. At the right boundary (x_{N_x}, t^n), we want the mode moving to the right to move freely, whereas we want to avoid information moving to the left, for which we set $L = 0$ there, that is, $L_{N_x} = 0$. Notice that since $R = \frac{1}{2}(\pi - \psi)$ and $L = \frac{1}{2}(\pi + \psi)$, these are exactly the same conditions, $\pi = \psi$ at the left and $\pi = -\psi$ at the right boundaries, for the simple case $\alpha = 1$ and $\beta = 0$ as described in (4.14) and (4.15). The difference here is just that now π has a more complex definition.

One can choose to solve the system of equations in (4.24) or the system (4.31) plus the evolution equation for ϕ in (4.24). In this case, the semi-discrete version of these equations at the interior points of D_d reads

$$\partial_t R = -\lambda_{+,i} \frac{R_{i+1} - R_{i-1}}{2\Delta x} - \frac{\lambda_{+,i+1} - \lambda_{+,i-1}}{2\Delta x} R_i,$$

$$\partial_t L = -\lambda_{-,i} \frac{L_{i+1} - L_{i-1}}{2\Delta x} - \frac{\lambda_{-,i+1} - \lambda_{-,i-1}}{2\Delta x} L_i,$$

$$\partial_t \phi = \alpha_i \pi_i + \beta_i \psi_i.$$

At the left boundary (x_0, t^n), the semi-discrete equations are

$$\partial_t R = -\lambda_{+,0} \frac{-R_2 + 4R_1 - 3R_0}{2\Delta x} - \frac{-\lambda_{+,2} + 4\lambda_{+,1} - 3\lambda_{+,0}}{2\Delta x} R_0,$$

$$\partial_t L = -\lambda_{-,0} \frac{-L_2 + 4L_1 - 3L_0}{2\Delta x} - \frac{-\lambda_{-,2} + 4\lambda_{-,1} - 3\lambda_{-,0}}{2\Delta x} L_0,$$

$$\partial_t \phi = \alpha_0 \pi_0 + \beta_0 \psi_0, \tag{4.32}$$

and at the right boundary (x_{N_x}, t^n)

$$\partial_t R = -\lambda_{+,N_x} \frac{R_{N_x-2} - 4R_{N_x-1} + 3R_{N_x}}{2\Delta x} - \frac{\lambda_{+,N_x-2} - 4\lambda_{+,N_x-1} + 3\lambda_{+,N_x}}{2\Delta x} R_{N_x},$$

4.5 Wave Equation in a General 1+1 Minkowski Space-Time: Example of...

$$\partial_t L = -\lambda_{-,N_x} \frac{L_{N_x-2} - 4L_{N_x-1} + 3L_{N_x}}{2\Delta x} - \frac{\lambda_{-,N_x-2} - 4\lambda_{-,N_x-1} + 3\lambda_{-,N_x}}{2\Delta x} L_{N_x},$$

$$\partial_t \phi = \alpha_{N_x} \pi_{N_x} + \beta_{N_x} \psi_{N_x}, \qquad (4.33)$$

where we have used the forward (3.7) and backward (3.8) formulas for the spatial derivatives.

Notice that R and L can be integrated up to the boundary points provided values for L, R, $\lambda_\pm$ at the boundaries; however, ϕ can be integrated in the whole spatial domain, as long as we know the values of ψ and π. We know that outgoing wave conditions consist in setting $R_0 = L_{N_x} = 0$ as described above. These values should enter in expressions (4.32) and (4.33) and imply the following set of algebraic equations at the left boundary:

$$\left. \begin{array}{l} \frac{1}{2}(\pi_0^n + \psi_0^n) = L_0 \\ \frac{1}{2}(\pi_0^n - \psi_0^n) = R_0 = 0 \end{array} \right\} \Rightarrow \pi_0^n = \psi_0^n = L_0 \qquad (4.34)$$

which in turn determines the value of ϕ_0 using the third of Eqs. (4.32). On the other hand, at the right boundary, the condition reduces to

$$\left. \begin{array}{l} \frac{1}{2}(\pi_{N_x}^n + \psi_{N_x}^n) = L_{N_x} = 0 \\ \frac{1}{2}(\pi_{N_x}^n - \psi_{N_x}^n) = R_{N_x} \end{array} \right\} \Rightarrow \pi_{N_x}^n = -\psi_{N_x}^n = R_{N_x} \qquad (4.35)$$

which provides the numbers needed to construct ϕ_{N_x} with the help of the third of Eqs. (4.33).

Initial conditions are defined again as a Dirichlet boundary condition at $(x, 0)$ and set the same initial data used in the previous examples of the wave equation, namely, a time-symmetric Gaussian:

$$\phi_0(x) = Ae^{-x^2/\sigma^2}, \quad \psi_0(x) = -\frac{2x}{\sigma^2}\phi_0(x), \quad \pi_0(x) = 0, \qquad (4.36)$$

with $A = 1$ and $\sigma = 0.1$. Finally, for time integration, we use the method of lines with the Heun flavor of the RK2 method in Eq. (4.4).

For the examples below, we define the domain $[x_{min}, x_{max}] = [-1, 1]$, $t_f = 2$ and its discrete version D_d using $N_x = 400$ cells, which implies a resolution $\Delta x_1 = 0.005$ and $\Delta t = C\Delta x$ with $C = 0.1$.

Case 1. We set $\alpha = 1$ and a constant value of $\beta = \beta_0$. In this case, the characteristic velocities are $\lambda_\pm = -1 \pm \beta_0 = dx/dt$. The characteristic lines are curves whose equations are

$$\frac{dt}{dx} = \frac{1}{-\beta_0 \pm 1} = \begin{cases} \frac{1}{-\beta_0+1}, \\ \frac{1}{-\beta_0-1} \end{cases} \quad (4.37)$$

and determine the characteristic lines followed by the initial data on the space-time, which in this case are straight lines. In Fig. 4.10, we show the characteristic lines associated with (4.37) and the numerical solution of the IVP in these coordinates, for two values of $\beta_0 = 0.5$ and 1. Continuous characteristic lines correspond to the eigenvalue λ_+, and dotted lines correspond to λ_-.

For $\beta_0 = 0.5$, the characteristic speeds are $\lambda_+ = 1/2$ and $\lambda_- = -3/2$, which implies that the pulse moving to the left will arrive to the boundary (x_{N_x}, t^n) at $t \sim 2$, whereas the pulse moving to the left arrives at $t \sim 2/3$ to the boundary (x_0, t^n) as can be seen in the plot for the numerical solution of ϕ.

Now, for $\beta_0 = 1$, the characteristic velocities are $\lambda_+ = 0$ and $\lambda_- = -2$, which means that the pulse moving to the right will remain frozen during the evolution, whereas the pulse moving to the left will arrive to the boundary at $t \sim 0.5$.

The stars at initial time of the characteristic lines in Fig. 4.10 illustrate positions of some initial conditions on D_d. From these points, the right pulse will move along

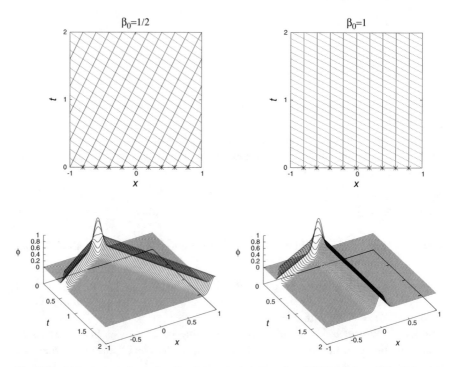

Fig. 4.10 At the top, we show a bundle of characteristic lines from (4.37) for $\beta_0 = 0.5$ and $\beta_0 = 1$. The stars at $t = 0$ represent positions of certain points, where the value of the wave function at initial time $\phi_0(x)$ will propagate with constant profile along the characteristic lines. At the bottom, we show the numerical solution for ϕ

4.5 Wave Equation in a General 1+1 Minkowski Space-Time: Example of...

the continuous characteristic lines, whereas the pulse moving to the left will follow the dashed characteristic lines. This idea is complemented in the plots showing the numerical solution of ϕ for the two different values of β_0 on D_d.

Case 2. Here, we consider $\beta = 0$ and a nonconstant lapse $\alpha = \frac{1}{4}\tanh(10x) + \frac{3}{4}$, which is a smooth version of a step function that brings the lapse from 0.5 for $x < 0$ to 1 for $x > 0$. The characteristics are solutions of the equation

$$\frac{dx}{dt} = -\beta \pm \alpha = \pm\left(\frac{1}{4}\tanh(10x) + \frac{3}{4}\right) \Rightarrow$$

$$-\frac{1}{20}\ln(2 + e^{-20x}) + x + C = \pm(t - t_0) \tag{4.38}$$

where C and t_0 are integration constants. A sample of these characteristics and the numerical solution in this case are shown in Fig. 4.11, where the solution ϕ in this gauge is also shown. The characteristics pass continuously from having slope ± 2 in the region $x < 0$ to ± 1 for $x > 0$.

The solution reflects this fact: on the left part of the domain ($x < 0$), time slices are separated half the time than at the right part of the domain ($x > 0$), because the lapse α has value ~ 0.5 for $x < 0$ and ~ 1 for $x > 0$. This is the reason why the pulse moving to the left takes nearly twice as long to reach the boundary than the pulse moving to the right.

Case 3. In this case, we use a trivial lapse but a nonconstant shift $\alpha = 1$ y $\beta = x$. The characteristics are such that the Gaussian pulses do not reach the boundaries because these boundaries move at the characteristic speeds $\lambda_\pm = \pm 1$, and therefore the wave never reaches the boundaries. The results are shown in Fig. 4.12, where we show the characteristic lines, which tend to focus near the boundaries and the pulses of ϕ squeeze but never arrive at the boundary. The pulses of the solution separate

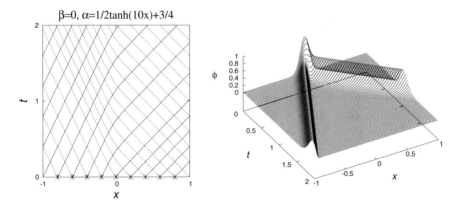

Fig. 4.11 Characteristics and solution for Case 2. Characteristic velocities pass from ± 0.5 for $x < 0$ to ± 1 for $x > 0$. Notice that the pulse moving to the left is twice slower than the pulse moving to the right, which reflects the fact that in these coordinates the time flows slower in the zone where $\alpha \sim 0.5$ than in the zone where $\alpha \sim 1$

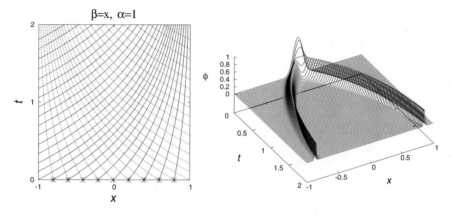

Fig. 4.12 Characteristics and solution for Case 3 with $\alpha = 1$ and $\beta = x$. Notice that near $x = 0$ the characteristics have slopes near ± 1, whereas they tend to focus near the boundaries ($x = \pm 1, t$). Also shown is the numerical solution of ϕ on D_d, which illustrates how the Gaussian pulses squeeze when approaching the boundary

at initial time and later on, since the coordinates are stretching because $\beta = x$, the Gaussian pulses squeeze in these coordinates, and for this case, we use a high-resolution $\Delta x_1 = 0.005$ or equivalently $N_x = 400$ in order to resolve the pulses. Finally, because $\beta = \pm 1$ at the boundaries, the pulses never reach the boundary points.

We use Case 3 to verify self-convergence as indicated in Sect. 3.1. For this, we calculate numerical solutions ϕ^1, ϕ^2, and ϕ^3 using the three resolutions $\Delta x_1 = 0.005$, $\Delta x_2 = 0.0025$, and $\Delta x_3 = 0.00125$ and compare these solutions at the snapshot $t = 1$ shown in Fig. 4.13. On the left, we show the solution for the three resolutions where differences among the results are difficult to see. However, in the right, we show a zoom centered at one of the pulses. In the zoomed plot, one can see the differences.

According to (3.22), the solution self-converges if the difference between + and × points is nearly four times bigger than the difference between × points and circles, that is, $\frac{\phi_i^1 - \phi_i^2}{\phi_i^2 - \phi_i^3} \sim 2^2$ for these few points. This is consistent with the second-order accuracy of spatial stencils and the Heun RK2 used for time integration.

The formal calculation of the self-convergence factor has to be done at each time step along the whole spatial direction, not only at a few points. This requires the post-processing of the numerical solutions ϕ^1, ϕ^2, and ϕ^3 and the construction of the two vectors:

$$\phi_i^1 - \phi_i^2,$$
$$\phi_i^2 - \phi_i^3, \tag{4.39}$$

4.5 Wave Equation in a General 1+1 Minkowski Space-Time: Example of...

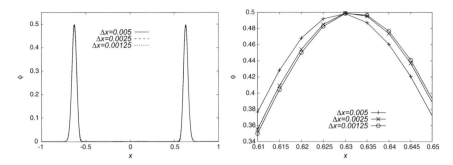

Fig. 4.13 Snapshot of the solution of Case 3 at $t = 1$ using three resolutions. On the left a whole domain view and on the right a zoomed plot centered around one of the peaks. The fact that the difference between the solution using the coarse and medium resolution is nearly four times the difference between the solution using medium and high resolutions, indicates second-order self-convergence of the solution at this snapshot

for $i = 0, ..., N_x$ and all times t^n. Then calculate one norm, say the L_2 norm of these two vectors at each time t^n. According to theory in (3.22), **the solution self-converges if** the following result holds:

$$\frac{L_2(\phi_i^1 - \phi_i^2)}{L_2(\phi_i^2 - \phi_i^3)} \sim 2^2, \qquad (4.40)$$

for the second-order accurate methods used here.

> **Exercise/Challenge.** Program a code that reads the output data files for the three resolutions used to generate Fig. 4.13, define the vectors (4.39), and calculate the norms $L_2(\phi_i^1 - \phi_i^2)$ and $L_2(\phi_i^2 - \phi_i^3)$ for all n. Finally, calculate the left expression in (4.40) as a function of time, and show that the result is a curve that oscillates around 4.

The code that implements this solution is `MoLWaveGauge.f90` with the gauge and boundary conditions described above. Again, the order of instructions in the implementation is important for which we show the following piece of code:

```
! ***** --> LOOP CORE <--

! Recycle arrays
phi_p = phi
psi_p = psi
pi_p = pi
R_p = R
L_p = L

! Heun RK2
do rk=1,2
  ! Calculate the RHS
  do i=1,Nx-1
  rhs_phi(i) = alpha(i)*pi(i) + beta(i)*psi(i)
  rhs_psi(i) = 0.5d0 * ( alpha(i+1)*pi(i+1) + beta(i+1)*psi(i+1) &
     - alpha(i-1)*pi(i-1) - beta(i-1)*psi(i-1) ) / dx
  rhs_pi(i) = 0.5d0 * ( alpha(i+1)*psi(i+1) + beta(i+1)*pi(i+1) &
     - alpha(i-1)*psi(i-1) - beta(i-1)*pi(i-1)) / dx
  rhs_R(i) = -lambda_plus(i) * 0.5d0 * ( R(i+1) - R(i-1) ) / dx &
     - 0.5d0 * ( lambda_plus(i+1) - lambda_plus(i-1) ) / dx * R(i)
  rhs_L(i) = -lambda_minus(i) * 0.5d0 * ( L(i+1) - L(i-1) ) / dx &
     - 0.5d0 * ( lambda_minus(i+1) - lambda_minus(i-1) ) / dx * L(i)
  end do
  ! RHS for outgoing wave boundary conditions at the boundaries
  rhs_R(0) = -lambda_plus(0) * 0.5d0 * ( -R(2) + 4.0d0* R(1) - 3.0d0*R(0) ) / dx &
     - 0.5d0 * ( -lambda_plus(2) + 4.0d0*lambda_plus(1) - 3.0d0*lambda_plus(0)
) / dx * R(0)
  rhs_R(Nx) = -lambda_plus(Nx) * 0.5d0 * ( R(Nx-2) - 4.0d0* R(Nx-1)
+ 3.0d0*R(Nx) ) / dx &
     - 0.5d0 * ( lambda_plus(Nx-2) - 4.0d0*lambda_plus(Nx-1)
+ 3.0d0*lambda_plus(Nx) ) / dx * R(Nx)
  rhs_L(0) = -lambda_minus(0) * 0.5d0 * ( -L(2) + 4.0d0*L(1) - 3.0d0*L(0) ) / dx &
     - 0.5d0 * ( -lambda_minus(2) +4.0d0*lambda_minus(1)
- 3.0d0*lambda_minus(0) ) / dx * L(0)
  rhs_L(Nx) = -lambda_minus(Nx) * 0.5d0 * ( L(Nx-2) - 4.0d0*L(Nx-1)
+ 3.0d0*L(Nx) ) / dx &
     - 0.5d0 * ( lambda_minus(Nx-2) - 4.0d0*lambda_minus(Nx-1)
+ 3.0d0*lambda_minus(Nx) ) / dx * L(Nx)
  rhs_phi(0) = alpha(0) *pi(0) + beta(0) *psi(0)
  rhs_phi(Nx) = alpha(Nx)*pi(Nx) + beta(Nx)*psi(Nx)

  if (rk.eq.1) then
    psi = psi_p + rhs_psi * dt
    pi = pi_p + rhs_pi * dt
    phi = phi_p + rhs_phi * dt
    R = R_p + rhs_R * dt
    L = L_p + rhs_L * dt
  else
    psi = 0.5d0 * ( psi_p + psi + rhs_psi * dt)
    pi = 0.5d0 * ( pi_p + pi + rhs_pi * dt)
    phi = 0.5d0 * ( phi_p + phi + rhs_phi * dt)
    R = 0.5d0 * ( R_p + R + rhs_R * dt)
    L = 0.5d0 * ( L_p + L + rhs_L * dt)
  end if

  ! Outgoing wave boundary conditions--calculating pi and psi at the boundaries
  R(0) = 0.0d0
  L(Nx) = 0.0d0
  psi(0) = L(0)
  pi(0) = L(0)
```

```
psi(Nx) = -R(Nx)
pi(Nx)  = R(Nx)

end do
```

! ***** --> Ends Loop Core <--

A final comment is in turn. In this code, we decided to solve all the evolution equations, that is, equations for ϕ, ψ, π, R, and L, although ψ and π are only used for the implementation of the boundary conditions and only the values of π_0, ψ_0, π_{N_x}, and ψ_{N_x} are needed. Thus, one can save allocated memory by avoiding the extra calculation of the evolution of these two variables at the inner points of the domain and only deal with their values at the boundaries.

> **Exercise.** In the examples of this section, we have used time-symmetric initial conditions. Implement initial conditions for a Gaussian pulse with a finite initial velocity to the right.

4.6 1+1 Schrödinger Equation Using the Method of Lines

We again solve the IVP defined for the one-dimensional Schrödinger equation but this time using the method of lines.

In order to solve it differently, instead of using arrays of complex numbers to store the values of the wave function as done in Sect. 3.5, here, we split Schrödinger equation into two equations, one for its real and one for its imaginary parts.

Assuming $\Psi = \Psi_1 + i\Psi_2$, the IVP associated with Schrödinger equation $i\frac{\partial \Psi}{\partial t} = -\frac{1}{2}\frac{\partial^2 \Psi}{\partial x^2} + V(x,t)\Psi$ reads

$$
\begin{aligned}
&\frac{\partial \Psi_1}{\partial t} = -\frac{1}{2}\frac{\partial^2 \Psi_2}{\partial x^2} + V(x,t)\Psi_2 \\
&\frac{\partial \Psi_2}{\partial t} = \frac{1}{2}\frac{\partial^2 \Psi_1}{\partial x^2} - V(x,t)\Psi_1 \qquad \Psi_1 = \Psi_1(x,t),\ \Psi_2 = \Psi_2(x,t) \\
&D = x \in [x_{min}, x_{max}] \times t \in [0, t_f] \qquad Domain \\
&\Psi_1(x,0) = \Psi_{1,0}(x),\ \Psi_2(x,0) = \Psi_{2,0}(x) \qquad Initial\ Conditions \\
&\Psi_1(x_{min}, t),\ \Psi_1(x_{max}, t),\ \Psi_2(x_{min}, t), \\
&\Psi_2(x_{max}, t) \qquad Boundary\ Conditions
\end{aligned}
$$

(4.41)

where $V(x, t)$ is a general potential. In this section, we illustrate the solution for the same two cases explored with the implicit methods in Sect. 3.5, namely, the particle in a box and the particle subject to a harmonic oscillator potential.

The numerical solution is constructed this time using the MoL with the RK3 time integrator from Eq. (4.5) on the same discrete domain as before, $D_d = \{(x_i, t^n) = (x_{min}+i\Delta x, n\Delta t)\}$, where $i = 0, \ldots, N_x$, $\Delta x = (x_{max}-x_{min})/N_x$ and $\Delta t = C\Delta x^2$.

The semi-discrete version of the evolution equations in (4.41), needed for the implementation of the MoL, reads

$$\frac{\partial \Psi_1}{\partial t} = -\frac{1}{2}\frac{\Psi_{2,i+1}^n - 2\Psi_{2,i}^n + \Psi_{2,i-1}^n}{\Delta x^2} + V_i \Psi_{2,i}^n,$$

$$\frac{\partial \Psi_2}{\partial t} = \frac{1}{2}\frac{\Psi_{1,i+1}^n - 2\Psi_{1,i}^n + \Psi_{1,i-1}^n}{\Delta x^2} - V_i \Psi_{1,i}^n, \qquad (4.42)$$

where the second-order accurate formula for the second-order derivative (3.9) has been used and the potential at points of the discrete domain D_d is $V_i^n = V(x_i, t^n)$. These formulas are well defined for $i = 1, \ldots, N_x - 1$, and the missing equations for Ψ_1 and Ψ_2 at the boundaries will depend on the boundary conditions of the problem to solve.

4.6.1 Particle in a 1D Box

For comparison, we proceed with the same parameters as in Sect. 3.5 for this problem. The numerical solution is constructed this time using the MoL with the RK3 time integrator on the same discrete domain as in Sect. 3.5. The numerical domain D_d is defined with $[x_{min}, x_{max}] = [0, 1]$, $t \in [0, 10]$ using $N_x = 100$ or equivalently $\Delta x_1 = 0.01$ as the base resolution, and $\Delta t = C\Delta x^2$ where $C = 0.125$.

The particle is free in the domain $x \in (0, 1)$ and cannot access the rest of $\mathbb{R}$. With this in mind, the continuity conditions on the wave function at the boundaries (x_0, t^n) and (x_{N_x}, t^n) imply Dirichlet conditions $\Psi(0, t) = \Psi(1, t) = 0$. Then Eqs. (4.42) are used only at interior points of the domain, and at boundaries, we set $\Psi_{1,0}^n = \Psi_{2,0}^n = \Psi_{1,N_x}^n = \Psi_{2,N_x}^n = 0$ for all n.

As initial conditions, we use the exact solution (3.66), $\Psi_{n_x}(x) = \sqrt{2}e^{-iE_{n_x}t}\sin(n_x\pi x)$, with $n_x = 2$ at $t = 0$, that is, $\Psi_1(x, 0) = \sqrt{2}\sin(2\pi x)$ and $\Psi_2(x, 0) = 0$, which translated into the numerical domain becomes

$$\Psi_{1,i}^0 = \sqrt{2}\sin(2\pi x_i),$$
$$\Psi_{2,i}^0 = 0,$$

for the array of the real and imaginary parts of the wave function.

4.6 1+1 Schrödinger Equation Using the Method of Lines

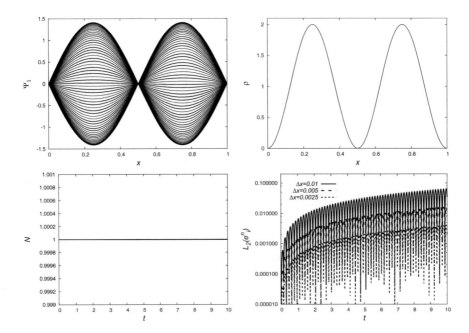

Fig. 4.14 Numerical solution for the particle in a box with $n_x = 2$ using the MoL. At the top, we show snapshots of Ψ_1 and ρ at various times. This shows the wave function oscillates, whereas the density remains stationary. At the bottom, we show that the number of particles $N = \int_0^1 \rho dx$ remains nearly constant in time. Finally, we show the convergence of $L_2(e_i^n)$ to zero

Diagnostics of the Solution. The first physical diagnostics consists in showing the wave function $\Psi = \Psi_1 + i\Psi_2$ oscillates, whereas the density of probability $\rho = \Psi^*\Psi = \Psi_1^2 + \Psi_2^2$ remains time-independent. This is shown in Fig. 4.14. We also show the number of particles $N = \int_0^1 \rho dx$ that has to remain constant in time. The value of N in this solution deviates from one only within 10^{-8} during the evolution, like when using the implicit Crank-Nicolson method in Fig. 3.24.

Finally, the error of the numerical solution is estimated for the real part of the wave function Ψ_1 at every point $(x_i, t^n) \in D_d$:

$$e_i^n = \Psi_{1,i}^n - \Psi_{1,i}^{e,n}. \tag{4.43}$$

where Ψ^e is the exact solution (3.66). In order to show that the numerical solution converges, we calculate $L_2(e_i^n)$ as function of time for $N_x = 100, 200$, and 400 or equivalently resolutions $\Delta x_1 = 0.01$, $\Delta x_2 = 0.005$, and $\Delta x_3 = 0.0025$. The result for $L_2(e_i^n)$ appears in Fig. 4.14. Notice that the errors are of the same order as those obtained using the implicit Crank-Nicolson method of Sect. 3.5. The reason is that the numerical parameters are the same, and the approximations are all second-order accurate as well.

Exercise. In order to understand why C has to be so small using the MoL, it would be interesting to use bigger values of C and find the critical one for which the method is unstable. This should be seen as the runaway increase of the wave function amplitude.

Exercise. Use the Heun RK2 method to integrate this problem and discover a range of values of C for which the solution remains stable.

Exercise. Show that $L_1(e_i^n)$ also converges to zero.

Exercise. At every time-step, calculate the expectation value of the energy

$$E_n = \left\langle \Psi_n \middle| -\frac{1}{2}\frac{\partial^2}{\partial x^2} \middle| \Psi_n \right\rangle = \int_0^1 \left[\Psi_n^* \left(-\frac{1}{2}\frac{\partial^2 \Psi_n}{\partial x^2} \right) \right] dx$$

where n is the quantum number. Verify that it is consistent with the energy values $E_n = (n\pi)^2/2$ in Eq. (3.65). For the example of this section, $n = 2$, where E_2 should have an approximate value of $\sim 2\pi^2$. The integral can be calculated using the formula (3.23).

4.6.2 Particle in a Harmonic Oscillator Potential

In this case, the potential is $V = \frac{1}{2}x^2$ at each point of $(x_i, t^n) \in D_d$ takes the values $V_i^n = \frac{1}{2}x_i^2$, $i = 0, ..., N_x$. This potential is the one to be used in the right-hand sides of the semi-discrete system (4.42).

As initial conditions, we use the exact solution (3.71) at $t = 0$ with $n_x = 3$, which translated to the arrays for real and imaginary parts of the wave function at initial time reads

$$\Psi_{1,i}^0 = \sqrt{\frac{1}{\sqrt{\pi}2^3 3!}} e^{-x_i^2/2} H_3(x_i) = \sqrt{\frac{1}{\sqrt{\pi}2^3 3!}} e^{-x_i^2/2} \left(12x_i - 8x_i^3 \right), \quad (4.44)$$

$$\Psi_{2,i}^0 = 0. \tag{4.45}$$

4.6 1+1 Schrödinger Equation Using the Method of Lines

We use the same numerical parameters used to illustrate the implicit method in Sect. 3.5. The domain is such that $[x_{min}, x_{max}] = [-10, 10]$ and $t \in [0, 10]$. For this problem, the numerical domain D_d uses $N_x = 2000$ or equivalently $\Delta x = 0.01$.

Boundary Conditions. In this case, knowing the spatial size of the domain is sufficient for the wave function to vanish with a Gaussian profile, Neumann boundary conditions are imposed such that the spatial derivative of the wave function is zero at (x_0, t^n) and (x_{N_x}, t^n). In this case, using the forward (3.3) and backward (3.4) discrete formulas of the spatial derivative, these boundary conditions reduce to

$$\Psi_{k,2} - 4\Psi_{k,1} + 3\Psi_{k,0} = 0 \quad \text{and} \quad \Psi_{k,N_x-2} - 4\Psi_{k,N_x-1} + 3\Psi_{k,N_x} = 0,$$

for $k = 1, 2$, from which one obtains values for the wave function at the boundary points

$$\Psi_{k,0} = \frac{1}{3}(4\Psi_{k,1} - \Psi_{k,2}),$$

$$\Psi_{k,N_x} = \frac{1}{3}(4\Psi_{k,N_x-1} - \Psi_{k,N_x-2}),$$

that we implement in the code.

Diagnostics. The wave function Ψ oscillates, whereas the density of probability $\rho = \Psi^*\Psi = \Psi_1^2 + \Psi_2^2$ remains time-independent. We illustrate the oscillations of Ψ_1 and time independence of ρ in Fig. 4.15, so as the number of particles $N = \int_{x_{min}}^{x_{max}} \rho \, dx$, which is nearly constant in time.

The error of the numerical solution is estimated for Ψ_1 at every point $(x_i, t^n) \in D_d$ as in Eq. (4.43) with Ψ^e as in (3.71). Finally, in order to show the numerical solution converges, we calculate $L_2(e_i^n)$ as function of time for $N_x = 2000, 4000$, and 8000 or equivalently resolutions $\Delta x_1 = 0.01$, $\Delta x_2 = 0.005$, and $\Delta x_3 = 0.0025$. The second-order convergence to zero of $L_2(e_i^n)$ is also in Fig. 4.15.

Finally, the code MoLSchroedinger.f90 solves Schrödinger equation for the particle in the box, and the particle in a harmonic oscillator potential is available as Electronic Supplementary Material described in Appendix B.

Exercise. At every time step, calculate the expectation value of the energy

$$E_n = \left\langle \Psi_n \left| -\frac{1}{2}\frac{\partial^2}{\partial x^2} + V \right| \Psi_n \right\rangle = \int_{x_{min}}^{x_{max}} \left[\Psi_n^* \left(-\frac{1}{2}\frac{\partial^2 \Psi_n}{\partial x^2} \right) + V \Psi_n^* \Psi_n \right] dx$$

where n is the number of nodes of the wave function. Verify that it is consistent with the energy values $E_n = n + \frac{1}{2}$ in Eq. (3.71). For the example of this section, $n = 3$, where E_3 should have an approximate value of $\sim 7/2$. The integral can be calculated using the formula (3.23).

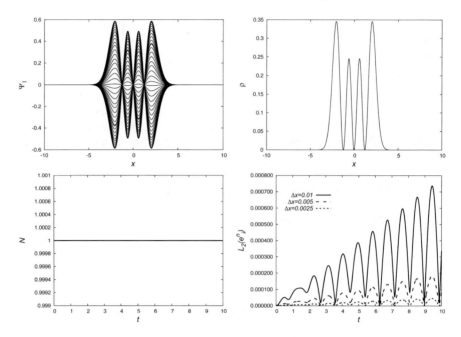

Fig. 4.15 Numerical solution for the particle in a harmonic oscillator potential with $n_x = 3$ using the MoL. At the top, we show snapshots of Ψ_1 and the density of probability ρ at various times. This shows that the wave function oscillates, whereas the density remains nearly stationary. At the bottom, we show that the number of particles $N = \int_{x_{min}}^{x_{max}} \rho dx$ remains nearly constant in time. Finally, we show the second-order convergence to zero of $L_2(e_i^n)$. These results should be compared with those obtained with the implicit method in Fig. 3.26

4.7 Application: Solution of the Wave Equation on Top of a Schwarzschild Black Hole

We saw in Sect. 4.4 that it is relatively easy to solve the wave equation when the gauge functions lapse α and shift β are used to describe the background space-time, and now we present a case of a non-flat space-time.

We first write down the line element in terms of gauge functions as follows: For a spherically symmetric space-time described with spherical coordinates (r, θ, ϕ, t), using the 3 + 1 decomposition, in geometric units $G = c = 1$, the most general line element reads

$$ds^2 = g_{\mu\nu} dx^\mu dx^\nu$$
$$= (-\alpha^2 + \beta_r \beta^r)dt^2 + 2\beta_r dt dr + \gamma_{rr} dr^2 + \gamma_{\theta\theta}(d\theta^2 + \sin^2\theta d\phi^2), \tag{4.46}$$

4.7 Application: Solution of the Wave Equation on Top of a Schwarzschild...

where all the functions α, β^r, γ_{rr}, and $\gamma_{\theta\theta}$ depend on (r, t). It is convenient to write the components of the metric in matrix form and its inverse, because it will be useful to write down the wave equation later on

$$g_{\mu\nu} = \begin{pmatrix} -\alpha^2 + \beta_r \beta^r & \beta_r & 0 & 0 \\ \beta_r & \gamma_{rr} & 0 & 0 \\ 0 & 0 & \gamma_{\theta\theta} & 0 \\ 0 & 0 & 0 & \gamma_{\theta\theta}\sin^2\theta \end{pmatrix},$$

$$g^{\mu\nu} = \begin{pmatrix} -1/\alpha^2 & \beta^r/\alpha^2 & 0 & 0 \\ \beta^r/\alpha^2 & \gamma^{rr} - \frac{\beta^r\beta^r}{\alpha^2} & 0 & 0 \\ 0 & 0 & \gamma^{\theta\theta} & 0 \\ 0 & 0 & 0 & \gamma^{\phi\phi} \end{pmatrix}, \quad (4.47)$$

where γ_{ij} is the 3-metric of the spatial hypersurfaces, with indices running on spatial coordinates only, and satisfies $\gamma^{ik}\gamma_{kj} = \delta^i{}_j$. For metric (4.46), the nonzero entries of γ^{ij} are $\gamma^{rr} = 1/\gamma_{rr}$, $\gamma^{\theta\theta} = 1/\gamma_{\theta\theta}$, $\gamma^{\phi\phi} = 1/(\gamma_{\theta\theta}\sin^2\theta)$.

Considering that g is the determinant of $g_{\mu\nu}$, then $\sqrt{-g} = \alpha\sqrt{\bar\gamma}$, where $\bar\gamma = \gamma_{rr}\gamma_{\theta\theta}^2\sin^2\theta$ is the determinant of the 3-metric γ_{ij} of spatial hypersurfaces of the space-time. Using this, the wave equation using the D'Alambert operator $\Box\phi = \frac{1}{\sqrt{-g}}\partial_\mu[\sqrt{-g}g^{\mu\nu}\partial_\nu\phi]$ for the general metric (4.46) reads

$$\begin{aligned}0 &= \Box\phi \\ &= \frac{1}{\sqrt{-g}}\partial_\mu[\sqrt{-g}g^{\mu\nu}\partial_\nu\phi] \\ &= \frac{1}{\alpha\sqrt{\bar\gamma}}\left(\partial_t[\alpha\sqrt{\bar\gamma}g^{t\nu}\partial_\nu\phi] + \partial_r[\alpha\sqrt{\bar\gamma}g^{r\nu}\partial_\nu\phi]\right) \\ &= \frac{\sin\theta}{\alpha\sqrt{\bar\gamma}\sin\theta}\left(\partial_t[\alpha\sqrt{\bar\gamma}(g^{tt}\partial_t\phi + g^{tr}\partial_r\phi)] + \partial_r[\alpha\sqrt{\bar\gamma}(g^{rt}\partial_t\phi + g^{rr}\partial_r\phi)]\right) \\ &= \frac{1}{\alpha\sqrt{\gamma}}\left(\partial_t\left[\alpha\sqrt{\gamma}\left(-\frac{1}{\alpha^2}\partial_t\phi + \frac{\beta^r}{\alpha^2}\partial_r\phi\right)\right]\right. \\ &\quad\left. + \partial_r\left[\alpha\sqrt{\gamma}\left(\frac{\beta^r}{\alpha^2}\partial_t\phi + \left(\gamma^{rr} - \frac{\beta^r\beta^r}{\alpha^2}\right)\partial_r\phi\right)\right]\right), \end{aligned} \quad (4.48)$$

where we defined $\bar\gamma = \gamma\sin^2\theta$, with $\gamma = \gamma_{rr}\gamma_{\theta\theta}^2$. Defining new variables

$$\pi := \frac{\sqrt{\gamma}}{\alpha}(\partial_t\phi - \beta^r\psi),$$

$$\psi := \partial_r\phi,$$

the above equation transforms into a first-order system of equations:

$$\partial_t \pi = \partial_r(\sqrt{\gamma}\alpha\gamma^{rr}\partial_r\phi + \beta^r\pi),$$

$$\partial_t \psi = \partial_r\left(\frac{\alpha}{\sqrt{\gamma}}\pi + \beta^r\psi\right),$$

$$\partial_t \phi = \frac{\alpha}{\sqrt{\gamma}}\pi + \beta^r\psi, \tag{4.49}$$

where we added the evolution equation for ϕ, obtained from the definition of π.

Now the Black Hole Space-Time. The line element of the Schwarzschild space-time written in Eddington-Finkelstein coordinates reads [4, 5]

$$ds^2 = -\left(1 - \frac{2M}{r}\right)dt^2 + \frac{4M}{r}dtdr + \left(1 + \frac{2M}{r}\right)dr^2 + r^2(d\theta^2 + \sin^2\theta d\phi^2), \tag{4.50}$$

where M is the mass of the black hole and (r, θ, ϕ, t) are the coordinates used to describe the points of the space-time.

Exercise. Show that by equating (4.46) and (4.50), one obtains the explicit expression for metric and gauge functions:

$$\alpha = \frac{1}{\sqrt{1 + \frac{2M}{r}}},$$

$$\beta^r = \frac{2M}{r}\frac{1}{1 + \frac{2M}{r}},$$

$$\gamma_{rr} = 1 + \frac{2M}{r} \Rightarrow,$$

$$\beta_r = \gamma_{rr}\beta^r = \frac{2M}{r},$$

$$\gamma_{\theta\theta} = r^2,$$

$$\gamma_{\phi\phi} = \gamma_{\theta\theta}\sin^2\theta = r^2\sin^2\theta, \tag{4.51}$$

which are needed to specify the system (4.49) for the black hole space-time metric.

Domain. In order to define the domain in which we will solve the wave equation, one needs to keep in mind that at $r = 0$ the Schwarzschild space-time has a gauge

4.7 Application: Solution of the Wave Equation on Top of a Schwarzschild...

invariant singularity to be careful with. On the other hand, it would be helpful to investigate the light cone structure of the space-time in order to know where the domain might offer problems.

Due to the symmetry, the light cones defined in this spherically symmetric space-time are independent of the angular coordinates (θ, ϕ) and are determined by the curves in the tr-plane that satisfy the condition

$$ds^2 = 0 = -\left(1 - \frac{2M}{r}\right) dt^2 + \frac{4M}{r} dt\, dr + \left(1 + \frac{2M}{r}\right) dr^2 \quad \Rightarrow$$

$$0 = -\left(1 - \frac{2M}{r}\right)\left(\frac{dt}{dr}\right)^2 + \frac{4M}{r}\left(\frac{dt}{dr}\right) + \left(1 + \frac{2M}{r}\right) = 0 \quad \Rightarrow$$

$$\frac{dt}{dr} = \begin{cases} \frac{2M-r}{r-2M} = -1 \\ \frac{r+2M}{r-2M} \end{cases} \tag{4.52}$$

whose integration gives an expression for the null trajectories that define light cones:

$$t - t_0 = -r + r_0,$$
$$t - t_0 = 4M \ln\left(\frac{r - 2M}{r_0 - 2M}\right) + r - r_0, \tag{4.53}$$

that describe, respectively, the ingoing and outgoing null geodesics along the radial direction, where t_0 and r_0 are integration constants. This expression allows one to construct the light cone structure shown in Fig. 4.16, described with the ingoing and outgoing null geodesics along the radial direction valid for all (θ, ϕ). Notice first that one of the properties of Eddington-Finkelstein coordinates is that the ingoing null geodesics have slope -1, whereas the outgoing rays pass from having a slope near $+1$ for large r to infinite slope exactly at $r = 2M$ and negative slope for $0 < r < 2M$.

The event horizon is located at $r = 2M$ and notice that light cones are open there. For illustration, there is a pair of bold null rays that cross at the point $(2, 8)M$. There, the outgoing null ray of the cone is vertical, which means that a photon will remain over this surface forever, whereas the ingoing null ray of the cone has slope -1. This means that the world lines of test particles passing through this cone will move inevitably toward $r < 2M$ inside the horizon.

Figure 4.17 illustrates the light cone structure of the space-time when using EF coordinates and how world lines of test particles near the horizon move within light cones and can cross the horizon. For comparison, we also show the cone structure when describing the space-time with Schwarzschild coordinates and illustrate how the light cones are closed near the horizon which prevents particles to cross the horizon. That light cones are open at the horizon is a convenient property if

Fig. 4.16 Null rays that define the light cone structure of the Schwarzschild black hole space-time in Eddington-Finkelstein (EF) coordinates. The outgoing null ray at $r = 2M$ indicates the location of the event horizon, because it is the boundary between rays that can escape toward future null infinity and those that cannot

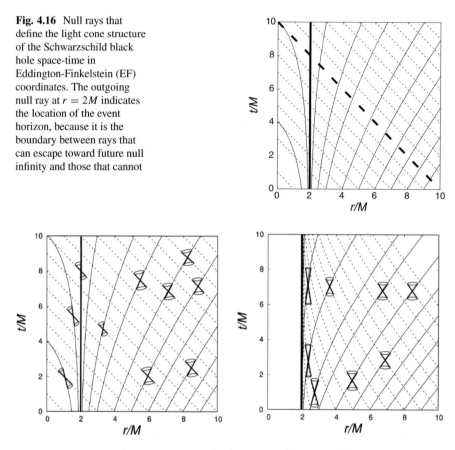

Fig. 4.17 Left: light cone structure of the Schwarzschild space-time in EF coordinates. Right: for comparison, we show the light cone structure when the Schwarzschild space-time is described with Schwarzschild coordinates out of the horizon

one wants to simulate the accretion of matter toward the interior of the hole. Schwarzschild coordinates do not have this property; in that case, both ingoing and outgoing null rays have infinite slope at $r = 2M$, which makes Schwarzschild coordinates inappropriate to describe particles that cross the horizon. Then it is said that Eddington-Finkelstein are *horizon-penetrating* coordinates, whereas Schwarzschild coordinates are not.

The wave equation can be solved in a spatial domain that will allow to **see how a black hole accretes the wave**. Nevertheless, due to the singularity at $r = 0$, it is not possible to use a domain that contains the origin. What is done is to choose a spatial domain $r \in [r_{min}, r_{max}]$ with $0 < r_{min} < 2M$, a method called **excision**, because actually a sphere of radius r_{min} is being excised from the space at all times.

An **important condition** is that the space-time background is fixed and thus the metric and gauge functions α, β^r, γ_{rr}, $\gamma_{\theta\theta}$, and $\sqrt{\gamma}$ are all functions of r only and do not evolve in time.

4.7 Application: Solution of the Wave Equation on Top of a Schwarzschild... 193

It is now time to formulate the IVP for the evolution of a wave on top of a Schwarzschild black hole:

$$
\begin{array}{ll}
\partial_t \pi = \partial_r(\sqrt{\gamma}\alpha\gamma^{rr}\psi + \beta^r \pi) & \\
\partial_t \psi = \partial_r\left(\dfrac{\alpha}{\sqrt{\gamma}}\pi + \beta^r \psi\right) & \\
\partial_t \phi = \dfrac{\alpha}{\sqrt{\gamma}}\pi + \beta^r \psi, & \phi = \phi(r, t), \\
 & \psi = \psi(r, t),\ \pi = \pi(r, t) \\
D = r \in [r_{min}, r_{max}] \times t \in [0, t_f] & \text{Domain} \\
\phi(r, 0) = \phi_0(r),\ \psi(r, 0) = \psi_0(r),\ \pi(r, 0) = \pi_0(r) & \text{Initial Conditions} \\
\phi(r_{min}, t),\ \phi(r_{max}, t) & \\
\psi(r_{min}, t),\ \psi(r_{max}, t) & \\
\pi(r_{min}, t),\ \pi(r_{max}, t) & \text{Boundary Conditions}
\end{array}
$$
(4.54)

Evolution. For the evolution, we need to define the right-hand side of the equations in (4.54) using the gauge and metric functions in (4.51), from which one can explicitly write down the following needed factors:

$$\sqrt{\gamma} = \sqrt{\gamma_{rr}\gamma_{\theta\theta}^2} = r^2\sqrt{1 + \frac{2M}{r}} \quad \Rightarrow$$

$$\frac{\alpha}{\sqrt{\gamma}} = \frac{1}{r^2\left(1 + \frac{2M}{r}\right)},$$

$$\sqrt{\gamma}\alpha\gamma^{rr} = \frac{r^2}{1 + \frac{2M}{r}},$$

which are regular within the numerical domain with excision.

In all the examples, we set $M = 1$, in which, combined with the geometric units $G = c = 1$, time and space are in units of M, the mass of the black hole. Reference [6] contains a detailed conversion from geometric to physical units.

Initial Conditions. We use Dirichlet boundary conditions at the side $(r, 0)$ of the domain:

$$\phi_0(r) = A\frac{\cos(kr)}{r}e^{-(r-r_0)^2/\sigma^2},$$

$$\psi_0(r) = \partial_r\phi_0(r),$$

$$\pi_0(r) = -\frac{\sqrt{\gamma}}{\alpha}\beta^r \psi_0(r), \tag{4.55}$$

corresponding to a spherical shell with Gaussian envelope and wave number k. In the example below, we use $A = 1$, $r_0 = 20$, $\sigma = 2$, and two values of k.

Domain. We set the inner boundary at $r_{min} = 1$ in order to allow the wave to enter the horizon at $r = 2$ and travel within the hole, and the external boundary at $r_{max} = 51$, which is sufficiently far from the horizon. We define the numerical domain as $D_d = \{(r_i, t^n)\}$ with $r_i = r_{min} + i\Delta r$, $i = 0, 1, 2, \ldots, N_r$, $t^n = n\Delta t$, where $\Delta r = (r_{max} - r_{min})/N_r$ and $\Delta t = C \Delta r$ are space and time resolutions. We use the base resolution $N_r = 5000$ or equivalently $\Delta r = 0.01$ and $C = 0.1$.

Boundary Conditions. The inner boundary (r_{min}, t) with $r_{min} = 1 < 2$ lies within the horizon, and according to Fig. 4.16, ingoing and outgoing null rays point toward the origin, and in principle no information will come out from there. We only need to make sure the equations at the boundary $r = r_{min}$ are correctly implemented. The equation for ϕ in (4.54) does not involve any spatial derivatives, whereas the equations for π and ψ involve a radial derivative. Then one needs to use one-sided stencils to calculate such derivatives. Explicitly, using the expression (3.7), the evolution equations at $r_0 = r_{min}$ read

$$\partial_t \pi|_{r=r_0} = -\frac{\left[(\sqrt{\gamma}\alpha\gamma^{rr}\psi + \beta^r\pi)|_{r=r_2} - 4(\sqrt{\gamma}\alpha\gamma^{rr}\psi + \beta^r\pi)|_{r=r_1}\right.}{2\Delta r} + \mathcal{O}(\Delta r^2),$$

$$\partial_t \psi|_{r=r_0} = -\frac{\left[\left(\frac{\alpha}{\sqrt{\gamma}}\pi + \beta^r\psi\right)\bigg|_{r=r_2} - 4\left(\frac{\alpha}{\sqrt{\gamma}}\pi + \beta^r\psi\right)\bigg|_{r=r_1}\right.}{2\Delta r} + \mathcal{O}(\Delta r^2),$$

$$\partial_t \phi|_{r=r_0} = \left(\frac{\alpha}{\sqrt{\gamma}}\pi + \beta^r\psi\right)\bigg|_{r=r_0}. \tag{4.56}$$

At the outer boundary (r_{max}, t), it is possible to implement an outgoing wave boundary condition. This needs the characteristic structure of the system of equations at $r_{max} = r_{N_r}$. If $\mathbf{u} = (\pi, \psi)^T$, the system for π and ψ can be written as

$$\partial_t \mathbf{u} + \partial_r(\mathbf{A}\mathbf{u}) = 0, \quad \text{with} \quad \mathbf{A} = -\begin{pmatrix} \beta^r & \alpha\sqrt{\gamma}\gamma^{rr} \\ \frac{\alpha}{\sqrt{\gamma}} & \beta^r \end{pmatrix}, \tag{4.57}$$

4.7 Application: Solution of the Wave Equation on Top of a Schwarzschild...

where **A** has constant entries as long as the metric functions are evaluated only at $r = r_{N_r}$. The eigenvalues and eigenvectors of **A** are

$$\lambda_\pm = -\beta^r \pm \sqrt{\gamma^{rr}}\alpha, \quad \mathbf{u}_+ = \begin{pmatrix} \sqrt{\gamma\gamma^{rr}} \\ -1 \end{pmatrix} \quad \mathbf{u}_- = \begin{pmatrix} \sqrt{\gamma\gamma^{rr}} \\ 1 \end{pmatrix}. \tag{4.58}$$

The matrix $\mathbf{P} = (\mathbf{u}_+, \mathbf{u}_-)$ diagonalizes **A**, where $\Lambda = \mathbf{P}^{-1}\mathbf{A}\mathbf{P}$ is the equivalent diagonal matrix. Multiplying Eq. (4.57) by $\mathbf{P}^{-1}$ from the left and using that $\mathbf{P}^{-1}\mathbf{A} = \Lambda\mathbf{P}^{-1}$:

$$\partial_t \mathbf{u} + \partial_r(\mathbf{A}\mathbf{u}) = 0|_{r=r_{N_r}}$$

$$\partial_t \mathbf{u} + \mathbf{A}\partial_r \mathbf{u} = 0|_{r=r_{N_r}}$$

$$\mathbf{P}^{-1}(\partial_t \mathbf{u}) + \mathbf{P}^{-1}\mathbf{A}\partial_r \mathbf{u} = 0|_{r=r_{N_r}} \quad \Rightarrow$$

$$\partial_t(\mathbf{P}^{-1}\mathbf{u}) + \mathbf{P}^{-1}\mathbf{A}\partial_r \mathbf{u} = 0|_{r=r_{N_r}} \quad \Rightarrow$$

$$\partial_t \mathbf{w} + \Lambda\partial_r(\mathbf{P}^{-1}\mathbf{u}) = 0|_{r=r_{N_r}} \quad \Rightarrow$$

$$\partial_t \mathbf{w} + \Lambda\partial_r \mathbf{w} = 0|_{r=r_{N_r}}. \tag{4.59}$$

This means that the equations for π and ψ decouple and become a system of advection equations for the entries of the new state vector $\mathbf{w} = \mathbf{P}^{-1}\mathbf{u}$, namely, for the right and left modes:

$$\mathbf{w} = \mathbf{P}^{-1}\mathbf{u} = \mathbf{P}^{-1}\begin{pmatrix} \pi \\ \psi \end{pmatrix} = \frac{1}{2}\begin{pmatrix} \frac{\pi}{\sqrt{\gamma\gamma^{rr}}} - \psi \\ \frac{\pi}{\sqrt{\gamma\gamma^{rr}}} + \psi \end{pmatrix} := \begin{pmatrix} R \\ L \end{pmatrix}. \tag{4.60}$$

For the Eddington-Finkelstein metric, from (4.51), $\sqrt{\gamma\gamma^{rr}} = r^2$, and thus $\sqrt{\gamma\gamma^{rr}}|_{r=r_{N_r}} = r_{N_r}^2$. We are only interested in the values of the characteristic modes at the boundary R_{N_r} and L_{N_r}, not necessarily in the whole domain.

The outgoing wave boundary condition at the outer boundary means that $L = 0$ at $r = r_{N_r}$ because we do not want information reflected back from the boundary. This implies that the values of π and ψ at this boundary satisfy the expressions

$$\begin{aligned}\frac{\pi_{N_r}}{r_{N_r}^2} - \psi_{N_r} &= 2R_{N_r} \\ \frac{\pi_{N_r}}{r_{N_r}^2} + \psi_{N_r} &= 2L_{N_r} = 0\end{aligned} \quad \Rightarrow \quad \begin{aligned}\pi_{N_r} &= r_{N_r}^2 R_{N_r} \\ \psi_{N_r} &= -\frac{\pi_{N_r}}{r_{N_r}^2},\end{aligned} \tag{4.61}$$

and all we need is to know R_{N_r} in order to calculate the values of π_{N_r} and ψ_{N_r}. This can be done with Lagrange extrapolation (for a simple formula, see [7]), of the values of R from the interior neighboring points as follows:

$$R_{N_r} = 3R_{N_r-1} - 3R_{N_r-2} + R_{N_r-3}, \tag{4.62}$$

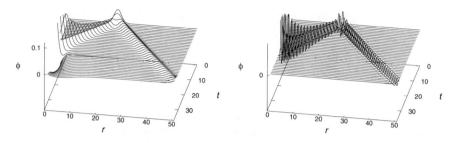

Fig. 4.18 Evolution of ϕ for $k = 0$ and $k = 4$ on D_d

which can be calculated using the definition (4.60) for values of R at points R_{N_r-1}, R_{N_r-2}, and R_{N_r-3}.

The numerical solution is shown in Fig. 4.18 for the wave function ϕ using two wave numbers $k = 0$ and $k = 4$. The initial pulse splits into ingoing and outgoing pulses. The two boundaries absorb the pulse after a finite time. One pulse of the wave function is absorbed by the black hole, and another one escapes toward infinity.

The code MoLEFWave.f90, solves this problem using the third-order Runge-Kutta integrator and is available as Electronic Supplementary Material described in Appendix B.

> **Exercise/Challenge.** It is possible to modify the code above and construct solutions to the Klein-Gordon equation $\Box\phi + m^2\phi = 0$ for a scalar field on top of a black hole. The auxiliary variables and boundary conditions are exactly the same as those constructed here, and it would require only to calculate the appropriate right-hand sides of the evolution equation for the variable π.

4.8 Application: Thermal Diffusion of Earth's Crust

We expand the applicability of diffusion equation in Sect. 3.4 to a more general case with space-dependent diffusion coefficient and a source, the Reaction-Diffusion equation. The application consists in describing the thermal history of Earth's crust. The problem setup is based on the specifications described in reference [8], where a model of the crust temperature from Earth's surface until 130 km under is modeled with the equation:

$$\frac{\partial T}{\partial t} = \frac{\partial}{\partial z}\left(\kappa \frac{\partial T}{\partial z}\right) + \frac{A_{rad} + A_s}{\rho C_P}, \tag{4.63}$$

4.8 Application: Thermal Diffusion of Earth's Crust

where $T = T(z, t)$ is the temperature as function of time and depth z. The thermal diffusivity κ is assumed to be space-dependent, A_{rad} is a heat source due to the radiation projected on the surface, and A_s represents the contribution of strain heating due to shear stress between shells at $z = 35$ km.

The phenomenological functional model for κ as function of T in terms of observations described in [8] is

$$\kappa = \begin{cases} 567.3T^{-1} - 0.062, & T < 846\text{K} \\ 0.732 - 0.000135T, & T \geq 846\text{K} \end{cases} \quad (4.64)$$

where κ has units of $\frac{\text{mm}^2}{\text{s}}$ and T in Kelvin. Also, a phenomenological model of specific heat capacity based on empirical evidence is used in the model, resulting in the following formula [8]:

$$C_P = \begin{cases} 199.5 + 0.0857\, T - 5 \times 10^6\, T^{-2}, & T < 846\text{K} \\ 229.32 + 0.0323\, T - 47.9 \times 10^{-6}\, T^{-2}, & T \geq 846\text{K} \end{cases} \quad (4.65)$$

where C_P has units of $\frac{\text{J}}{\text{mol K}}$, where the molar mass is 221.78 g mol^{-1} and T is in Kelvin.

The model used for A_{rad} is such that $A_{rad} = 2 \times 10^{-6} \text{Wm}^{-3}$ at the Earth's surface and then drops exponentially until 35 km. Between 35 and 70 km of depth, $A_{rad} = 0.2 \times 10^{-6} \text{Wm}^{-3}$ and finally between 70 and 130 km, $A_{rad} = 0.02 \times 10^{-6} \text{Wm}^{-3}$. This model can be summarized as follows:

$$A_{rad} = \begin{cases} 2 \times 10^{-6} e^{-z/15.2}\ \text{Wm}^{-3}, & 0 \leq z < 35\ \text{km} \\ 0.2 \times 10^{-6}\ \text{Wm}^{-3}, & 35 \leq z < 70\ \text{km} \\ 0.02 \times 10^{-6}\ \text{Wm}^{-3}, & 70 \leq z < 130\ \text{km} \end{cases} \quad (4.66)$$

The strain heating source A_s is given by $\bar{A}_s = \tau v/\delta$, where $\tau = 30$ MPa is the shear stress, $v = 3$ cm yr^{-1} is the thrusting velocity, and δ is a characteristic width of the shear zone, modulated by a Gaussian, that is, $A_s = \bar{A}_s e^{-(z-35)^2/\delta^2}$.

Finally, the crust density is assumed to be constant in the whole domain $\rho = 2700$ kg/m^3 and that the average molar mass of the crust is given by 221.78 $\frac{\text{gr}}{\text{mol}}$, which is equivalent to 0.22178 $\frac{\text{kg}}{\text{mol}}$.

Units. Notice that different parameters and quantities use different units. In order to solve the equation, we find appropriate to standardize to units of length in kilometers (km), of time in kiloyears (kyr) and temperature in Kelvin (K), and all parameters are to be transformed accordingly. The terms in Eq. (4.63) in (kyr, km, K) units are

$$\kappa\left[\frac{mm^2}{s}\right] = 3.1536 \times 10^{-2} \kappa\left[\frac{km^2}{kyr}\right] \tag{4.67}$$

$$\frac{A_{rad}}{\rho C_P} = \frac{\bar{A} \times 10^{-6}[\frac{W}{m^3}]}{2.7 \times 10^3[\frac{kg}{m^3}]C_P[\frac{J}{mol\cdot K}]} = \frac{\bar{A} \times 10^{-6}}{2.7 \times 10^3 C_P}\left[\frac{mol\cdot K}{s\cdot kg}\right]$$

$$= \frac{2.2178 \times 3.1536}{2.7} \frac{\bar{A}}{C_P}\left[\frac{K}{kyr}\right] \tag{4.68}$$

$$A_s = \bar{A}_s e^{-\frac{(z-35)^2}{\delta^2}}$$

$$\bar{A}_s = \frac{\tau \nu}{\delta} = \frac{3 \times 10^7[\frac{N}{m^2}]3[\frac{cm}{yr}]}{\delta[km]} = \frac{9}{3.1536\cdot\delta} \times 10^{-5}\left[\frac{kg}{m\cdot s^3}\right] \tag{4.69}$$

$$\frac{A_s}{\rho C_P} = \frac{\frac{9}{3.1536\cdot\delta} \times 10^{-5}[\frac{kg}{m\cdot s^3}]}{2.7 \times 10^3[\frac{kg}{m^3}]C_P[\frac{J}{mol\cdot K}]} e^{-\frac{(z-35)^2}{2\delta^2}}$$

$$= 73.9266666666667 \frac{1}{\delta\, C_P} e^{-\frac{(z-35)^2}{2\delta^2}}\left[\frac{K}{kyr}\right] \tag{4.70}$$

where we used the conversion factor for units of time 1kyr $= 3.1536 \times 10^{10}$s.

The problem is solved in the domain $z \in [0, 135]$km during a time window of 40 million years (Myr). We describe the domain $D = [z_{min}, z_{max}] \times [0, t_f]$ with $z_{min} = 0$, $z_{max} = 135$ km, and $t_f = 40$Myr. The discrete domain D_d is the set of points (z_i, t^n) where $z_i = i\Delta z$ with $\Delta z = 135/N_z$ where $N_z + 1$ is the number of points along the spatial direction and $t^n = n\Delta t$ with $\Delta t = CFL\Delta z^2$.

The value of the various functions (4.67)–(4.70) involved in the equation at point (z_i, t^n) are κ_i, $C_{P,i}$, $A_{rad,i}$, and $A_{s,i}$, since κ and C_P are functions of z implicitly through the functional dependence on T in (4.64) and (4.65), whereas the sources A_{rad} and A_s are explicitly space-dependent. Ready for integration, the semi-discrete version of the evolution equation for T (4.63) at point (z_i, t^n), suitable for the method of lines, is:

$$\left.\frac{\partial T}{\partial t}\right|_{(z_i,t^n)} = \kappa_i \frac{T_{i+1}^n - 2T_i^n + T_{i-1}^n}{\Delta z^2} + \frac{1}{4}\frac{\kappa_{i+1} - \kappa_{i-1}}{\Delta z}\frac{T_{i+1}^n - T_{i-1}^n}{\Delta z} + \frac{A_{rad,i} + A_{s,i}}{\rho_i C_{P,i}}. \tag{4.71}$$

At boundary points $i = 0$ and $i = N_z$, time-independent Dirichlet boundary conditions are set to $T(z_0 = 0\,km) = 290K$ and at $T(z_{N_z} = 130\,km) = 1300C = 1573.15K$. The one at $z = 0$ is motivated by a time-independent average surface temperature, whereas the one at $z = 130$ km is motivated by the heat transfer by convection that keeps the temperature stationary at that depth.

Initial Conditions. There is no prescription for initial conditions in [8], although diffusion and source terms are expected to drive the evolution of T. Then, as initial condition, we use the line

4.9 How to Improve the Methods in This Chapter

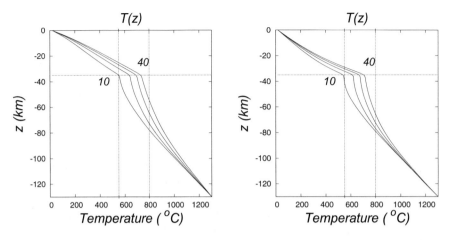

Fig. 4.19 Temperature of the crust as function of depth z at 10, 20, 30, and 40 Myr, although we only indicate the 10- and the 40-Myr geotherms. At the left, the case of constant $\kappa = 1 \left[\frac{mm^2}{s}\right]$ and at the right κ as given by the empirical expression (4.64). This example uses $\delta = 1.5$ km

$$T(z,0) = T(0,0) + \frac{T(130,0) - T(0,0)}{130} z = 290 + \frac{1573.15 - 290}{130} z$$

which is used to start the evolution.

The discrete domain is defined with $N_z = 1300$ cells or equivalently spatial resolution $\Delta z = 0.1$ km, whereas time resolution is $\Delta t = 0.25 \Delta z^2 = 2.5$ yr. The evolution of temperature with these parameters for $\delta = 1.5$ km is shown in Fig. 4.19, with plots every 10 million years. The strain heating at 35 km distorts the profile, whereas diffusion pushes geotherms to the right. These results illustrate how temperature grows in time in the scale of millions of years. Finally, this plot can be compared with Figure 3 of reference [8].

The code `EarthCrust.f90` that serves to solve this problem is available as Electronic Supplementary Material described in Appendix B.

4.9 How to Improve the Methods in This Chapter

The methods in the chapter can be improved in at least two ways: The first one is the time integration, for which a number of ODE solvers can be used, both explicit, implicit, or mixed implicit-explicit, each with different dissipation and stability properties that may depend on the equations to be solved. The second one is the accuracy of the discretization of spatial derivative operators, which can easily improve from second to fourth or sixth order, operators that can be constructed with the ideas of Sect. 3.1.

4.10 Projects

With the methods described in this chapter, it is possible to carry out the following projects:

4.10.1 Solution of the Wave Equation in the Whole Space-Time

It is possible to construct solutions of the $1 + 1$ wave equation in the whole space-time. For this, one has to compactify the space-time into a finite numerical domain. There are various ways to compactify the space-time, for example, by redefining the spatial coordinate as $\tilde{x} = \frac{x}{1-x^2}$ that brings the spatial infinity $\pm\infty$ to ± 1.

In the case of the wave equation, which travels at the speed of light, the boundary is at future null infinity, or $\mathcal{J}^+$, and a foliation using constant mean curvature spatial slices compactifies $\mathcal{J}^+$. In order to know what this means, in Fig. 4.20, we show the Penrose diagrams of some particular examples of this compactification, and in Fig. 4.21, we show the resulting light cone structure of the space-time in these compactified coordinates.

The project consists in solving the wave equation in coordinates of this type following the detailed description in [9]. Further projects can include the solution of the spherically symmetric wave equation in the $3 + 1$ Minkowski space-time that reduces to a $1 + 1$ IVP and in fact the solution of the wave equation on top of a Schwarzschild black hole using compactified versions of the space-time. The technical information and the required details can be found in the academic paper [9].

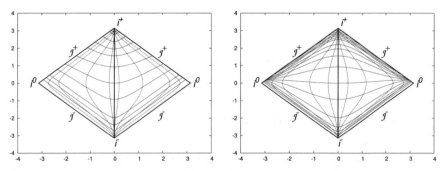

Fig. 4.20 Penrose diagrams of two different foliations that compactify future null infinity of the $1 + 1$ Minkowski space-time

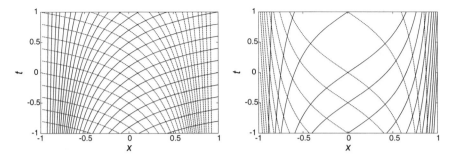

Fig. 4.21 Light cone structure of the two foliations in the previous figure, which compactify future null infinity of the 1 + 1 Minkowski space-time

4.10.2 Nonlinear Absorption of a Scalar Field

In Sect. 4.7, we described the methods to solve the wave equation with spherical symmetry, on top of a Schwarzschild black hole. This problem can be generalized and consider also the evolution of the space-time coupled to the evolution of the Klein-Gordon field from the equation in the **challenge** above $\Box\phi + m^2\phi = 0$.

Scalar field and Einstein field equations can be defined as an IVP for the field and the geometry. A simple method to solve Klein-Gordon equation coupled to a spherically symmetric space-time is found in [10]. The additional work consists in the implementation of the solution of Einstein equations, which are written in terms of the 3+1 decomposition of space-time, using the ADM formulation [4, 5]. Then, evolution equations for the components of the 3-metric of space-like hypersurfaces used to foliate the space-time and the components of their extrinsic curvature are coupled to the evolution of the scalar field [10].

Interesting diagnostics appears in this case, because it is possible to track the growth of the black hole horizon caused by the accretion of ϕ. Methods for the evolution and diagnostics are detailed in [10].

4.10.3 Black Hole Shrinking

The evolution of the space-time in the previous project can be improved by using the GBSSN formulation of Einstein equations instead of the ADM formulation [4, 5].

The idea here is to, additional to the change of formulation, evolve a scalar field with negative energy density and show that the black hole horizon shrinks. The equations to be solved are described in reference [11], and the methods for the evolution of the scalar field and space-time geometry are those described in this chapter.

4.10.4 Bose Condensates in Optical Lattices

The dynamics of a dilute Bose Gas at very low temperatures can be described by the Gross-Pitaevskii equation, which mathematically is Schrödinger equation with the nonlinear contribution $|\Psi|^2\Psi$, and has a macroscopic, collective interpretation. The nonlinear term models the interaction between atoms and is related to the number of condensed particles.

Considering that we have solved problems with periodic boundary conditions, an application of Schrödinger equation consists in the simulation of the evolution of a Bose Condensate on a lattice, modeled with a repulsive, periodic potential [12]. The equation to be solved is

$$i\frac{\partial \Psi}{\partial t} = -\frac{1}{2}\frac{\partial^2 \Psi}{\partial x^2} + V_0 \cos(x)\Psi + C|\Psi|^2\Psi.$$

The project consists in calculating the current of the condensate as function of time in terms of the number of condensed atoms, a number encrypted in the constant C. Other phenomena like Bloch oscillations or transport blocking can also be studied [13]. The details needed to follow this analysis can be found in [14].

References

1. C.W. Shu, S.J. Osher, Efficient implementation of essentially non-oscillatory shock-capturing schemes, II. J. Comput. Phys. **83**, 32 (1989). https://doi.org/10.1016/0021-9991(89)90222-2
2. B. Gustafsson, H.-O. Kreiss, J. Oliger, *Time Dependent Problems and Difference Methods*. (Wiley Interscience Series of Texts, Monographs, and Tracts, New York, 1995)
3. R.J. LeVeque, *Numerical Methods for Conservation Laws*, 2nd edn. (Birkhäuser, Basel, 1992)
4. M. Alcubierre, *Introduction to 3+1 Numerical Relativity* (Oxford Science Publications, Oxford, 2008)
5. T.W. Baumgarte, S.L. Shapiro, *Numerical Relativity: Solving Einstein's Equations on the Computer* (Cambridge University, Cambridge, 2010)
6. L. Rezzolla, O. Zanotti, *Relativistic Hydrodynamics* (Oxford University, Oxford, 2013)
7. W.H. Press, S.A Teukolsky, W.T. Vettering, B.P. Flannery, *Numerical Recipes in Fortran 77: The Art of Scientific Computing*, 2nd edn. (Cambridge University, Cambridge, 2003)
8. A.G. Whittington, A.M. Hofmeister, P.I. Nabelek, Temperature-dependent thermal diffusivity of the Earth's crust and implications for magmatism. Nature **458**, 319–321 (2009). https://www.nature.com/articles/nature07818. Ibid. Erratum: Temperature-dependent thermal diffusivity of the Earth's crust and implications for magmatism. Nature **458**, 319–321 (2009). https://www.nature.com/articles/nature08037
9. A. Cruz-Osorio, A. González-Juárez, F.S. Guzmán, F.D. Lora-Clavijo. Numerical solution of the wave equation on particular space-times using CMC slices and scri-fixing conformal compactification. Rev. Mex. Fis. **56**, 456–468 (2010). arXiv:1007.3776 [gr-qc] https://doi.org/10.48550/arXiv.1007.3776
10. F.S. Guzmán, F.D. Lora-Clavijo, Spherical nonlinear absorption of cosmological scalar fields onto a black hole. Phys. Rev. **D 85**, 024036 (2012). arXiv:1201.3598 [astro-ph.CO] https://doi.org/10.1103/PhysRevD.85.024036

11. J.A. González, F.S. Guzmán, Accretion of phantom scalar field into a black hole. Phys. Rev. **D 79**, 121501(R) (2009). arXiv:0903.0881 [gr-qc] https://doi.org/10.1103/PhysRevD.79.121501
12. D.-Il Choi, Q. Niu, Bose-Einstein condensates in an optical lattice. Phys. Rev. Lett. **82**, 2022 (1999). https://doi.org/10.1103/PhysRevLett.82.2022
13. D.-Il Choi, B. Wu, To detect the looped Bloch bands of Bose–Einstein condensates in optical lattices. Phys. Lett. **A 318**, 558–563 (2003). https://doi.org/10.1016/j.physleta.2003.09.066
14. D.-Il Choi, PhD Thesis *Numerical studies of nonlinear Schrödinger and Klein-Gordon systems: Techniques and applications* (The University of Texas, Austin, 1998). http://laplace.physics.ubc.ca/People/matt/Doc/Theses/Phd/choi.pdf

Chapter 5
Finite Volume Methods

Abstract We use Burgers equation to illustrate how the methods based on finite difference approximations fail to resolve IVP involving quasilinear PDEs. We then move straightforwardly toward the solution of Euler equations in $1 + 1$ dimensions, using a simple combination of the elements composing the finite volume approach, namely, simple variable reconstructors and the HLLE flux formula. We invest efforts in the construction of the HLLE formula and the characteristic structure of systems of equations first seen when solving the generalized wave equation. We illustrate the use of the method with the simulation of the formation of the stationary solar wind assuming spherical symmetry. We also show the power and need of the method in relativistic physics by solving the relativistic Euler equations in $1 + 1$ dimensions and with the problem of the accretion of a perfect fluid onto a Schwarzschild black hole.

Keywords Partial Differential Equations · Finite volumes · Hydrodynamics

We use Burgers equation to illustrate how the methods based on finite difference approximations fail to resolve IVP involving quasilinear PDEs. We then move straightforwardly toward the solution of Euler equations in $1+1$ dimensions, using a simple combination of the elements composing the finite volume approach, namely, simple variable reconstructors and the HLLE flux formula. We invest efforts in the construction of the HLLE formula and the characteristic structure of systems of equations first seen when solving the generalized wave equation. We illustrate the use of the method with the simulation of the formation of the stationary solar wind assuming spherical symmetry. We also show the power and need of the method in relativistic physics by solving the relativistic Euler equations in $1 + 1$ dimensions and with the problem of the accretion of a perfect fluid onto a Schwarzschild black hole.

F. S. Guzmán, *Numerical Methods for Initial Value Problems in Physics*,
https://doi.org/10.1007/978-3-031-33556-3_5

5.1 Characteristics in Systems of Equations

Consider the IVP associated with the advection equation:

$$\begin{array}{ll} \dfrac{\partial u}{\partial t} + a \dfrac{\partial u}{\partial x} = 0, & u = u(x,t) \\ D = x \in (-\infty, \infty) \times t \geq 0 & Domain \\ u(x, 0) = u_0(x) & Initial\ Conditions \\ u(x_{min}, t),\ u(x_{max}, t) & Boundary\ Conditions \end{array} \quad (5.1)$$

We know from Sect. 4.3 that a is the velocity of propagation of the initial data $u_0(x)$. This example is used to construct the **characteristic curves**, that is, curves along which the initial data propagate without change. Notice that the domain of the IVP is such that boundaries will not affect the development of initial data.

Calculation of the total derivative of u with respect to time gives

$$\frac{du}{dt} = \frac{\partial u}{\partial t} + \frac{\partial u}{\partial x}\frac{dx}{dt},$$

that equals zero for $dx/dt = a$ according to (5.1). This restriction implies that $x = x_0 + at$ and that the initial conditions at point x_0 will remain constant along the curve $x = x_0 + at$. The value of u along this line will be the same within the whole domain D shifted along the spatial coordinate. Thus, for every x_0 at initial time, the solution to the equation in (5.1) will be given in terms of the initial conditions $u_0(x)$ as

$$u(x, t) = u_0(x - at), \quad (5.2)$$

which can be verified by direct substitution into the advection Eq. (5.1).

Now, let us consider an IVP associated with the following system of equations similar but more general than the advection equation:

$$\frac{\partial u_i}{\partial t} + \sum_{j=1}^{m} a_{ij}(x, t, u_1, \ldots, u_m) \frac{\partial u_j}{\partial x} + b_i(x, t, u_1, \ldots, u_m) = 0, \quad (5.3)$$

for $i = 1, \ldots, n$. This is a system with n equations and n unknowns that depend on x and t, which can be written as

$$\frac{\partial \mathbf{u}}{\partial t} + \mathbf{A}\left(\frac{\partial \mathbf{u}}{\partial x}\right) + \mathbf{b} = 0 \quad (5.4)$$

5.1 Characteristics in Systems of Equations

where

$$\mathbf{u} = \begin{pmatrix} u_1 \\ u_2 \\ \vdots \\ u_n \end{pmatrix}, \quad \mathbf{A} = \begin{pmatrix} a_{11} & a_{12} & \cdots & a_{1n} \\ a_{21} & a_{22} & \cdots & a_{2n} \\ \vdots & & & \vdots \\ a_{n1} & a_{n2} & \cdots & a_{nn} \end{pmatrix}, \quad \mathbf{b} = \begin{pmatrix} b_1 \\ b_2 \\ \vdots \\ b_n \end{pmatrix}. \tag{5.5}$$

In the particular case of constant entries a_{ij} and b_i, this system is **linear with constant coefficients**.

If $a_{ij} = a_{ij}(x,t)$ and $b_j = b_j(x,t)$, the system is **linear with variable coefficients**.

If the entries of $\mathbf{A}$ depend on $\mathbf{u}$, that is, $\mathbf{A} = \mathbf{A}(\mathbf{u})$, the system is **quasilinear** because it may contain products between the variables u_i and their first-order spatial derivatives $\partial u_j / \partial x$.

There are various degrees of hyperbolicity of a system, which are important from the mathematical point of view. The system is called **weakly hyperbolic** if the eigenvalues of $\mathbf{A}$ are real, **strongly hyperbolic** if the eigenvalues of $\mathbf{A}$ are real and it has a complete set of eigenvectors, **strictly hyperbolic** is the eigenvalues are real and distinct, and **symmetric hyperbolic** if $\mathbf{A}$ is Hermitian. So far, what we need for now is to know that eigenvalues represent speeds of information at each point of the domain and that they will be helpful to describe the Finite Volume method. On more formal matters, the degree of hyperbolicity of the system of equations is of great importance to determine whether an IVP is well posed. Recommended reading on this subject is [1, 2].

Exercise. Determine the type of hyperbolicity of the first order system (4.25)–(4.27), developed for the wave equation.

5.1.1 Constant Coefficient Case

Let us consider the homogenous case $\mathbf{b} = \mathbf{0}$ and that the system is strongly hyperbolic. Assume the entries of the matrix $\mathbf{A}$ are all constant, which has eigenvalues $\lambda_1, \ldots, \lambda_n$, and a set of linearly independent eigenvectors $\mathbf{r}_1, \ldots, \mathbf{r}_n$. Then the system can be split into a set of n advection equations as follows, provided $\mathbf{A}$ can be diagonalized. Using the matrix whose columns are the eigenvectors $\mathbf{P} = (\mathbf{r}_1, \ldots, \mathbf{r}_n)$, then $\mathbf{A} = \mathbf{P}\mathbf{\Lambda}\mathbf{P}^{-1}$ where

$$\mathbf{\Lambda} = \begin{pmatrix} \lambda_1 & 0 & \cdots & 0 \\ 0 & \lambda_2 & \cdots & 0 \\ \vdots & & & \vdots \\ 0 & 0 & \cdots & \lambda_n \end{pmatrix}, \tag{5.6}$$

where $\mathbf{r}_i$ is the eigenvector associated with the eigenvalue λ_i. Defining the vector $\mathbf{w} := \mathbf{P}^{-1}\mathbf{u}$ and multiplying (5.4) by $\mathbf{P}^{-1}$ from the left, we obtain

$$\mathbf{P}^{-1}\frac{\partial \mathbf{u}}{\partial t} + \mathbf{P}^{-1}\mathbf{A}\frac{\partial \mathbf{u}}{\partial x} = 0 \quad \Rightarrow$$

$$\mathbf{P}^{-1}\frac{\partial \mathbf{u}}{\partial t} + \mathbf{\Lambda}\mathbf{P}^{-1}\frac{\partial \mathbf{u}}{\partial x} = 0 \quad \Rightarrow$$

$$\frac{\partial \mathbf{w}}{\partial t} + \mathbf{\Lambda}\frac{\partial \mathbf{w}}{\partial x} = 0,$$

where we have used the fact that a_{ij} are constant. The last equation **indicates** that the system decouples into n **advection equations** for the variables $w_i = (\mathbf{P}^{-1})_{ij} w_j$:

$$\frac{\partial w_i}{\partial t} + \lambda_i \frac{\partial w_i}{\partial x} = 0, \quad i = 1, \ldots, n. \tag{5.7}$$

Notice the *key result here*; the eigenvalue λ_i of the matrix $\mathbf{A}$ is the *velocity* associated with the advection equation for the variable w_i—that the eigenvalues are interpreted as velocities is *key*, as seen for the wave equation in Sects. 4.4 and 4.5, where (R, L) are the characteristic variables w_1 and w_2 for the wave equation. Its use will also be seen in flux balance laws below.

The solution of these equations is exactly the same as for the advection equation constructed above, that is, for given initial conditions $w_{i,0}(x)$ of the variable w_i, its solution in the whole domain is

$$w_i(x, t) = w_{i,0}(x - \lambda_i t),$$

and likewise for all $i = 1, \ldots, n$. This result corresponds to the homogeneous system; however, the characteristic velocities are the same for nonhomogeneous cases where $\mathbf{b} \neq \mathbf{0}$, because only the principal part of the equations is considered, although the exact solution is not the same due to the source.

5.1.2 Variable Coefficient Case

In general, the entries of the matrix $\mathbf{A}$ in (5.4)–(5.5) are not constant, because they depend either on the independent variables (x, t) or on the unknown functions u_i themselves.

5.1 Characteristics in Systems of Equations

Let us consider the following IVP, a slight generalization of the advection equation problem:

$$\begin{aligned}
&\frac{\partial u}{\partial t} + \lambda(u)\frac{\partial u}{\partial x} = 0, & & u = u(x,t) \\
&D = x \in [x_{min}, x_{max}] \times t \in [0, t_f] & & \text{Domain} \\
&u(x,0) = u_0(x) & & \text{Initial Conditions} \\
&u(x_{min}, t),\ u(x_{max}, t) & & \text{Boundary Conditions}
\end{aligned} \quad (5.8)$$

Following (5.2), the solution in this case would be $u(x,t) = u_0(x - \lambda(u)t)$ for initial conditions $u_0(x) = u(x,0)$. Notice however that λ depends on the unknown itself. The paradigm of this problem is the **Burgers equation**. In this case, $\lambda(u) = u$, and the corresponding IVP reads

$$\begin{aligned}
&\frac{\partial u}{\partial t} + u\frac{\partial u}{\partial x} = 0, & & u = u(x,t) \\
&D = x \in [x_{min}, x_{max}] \times t \in [0, t_f] & & \text{Domain} \\
&u(x,0) = u_0(x) & & \text{Initial Conditions} \\
&u(x_{min}, t),\ u(x_{max}, t) & & \text{Boundary Conditions}
\end{aligned} \quad (5.9)$$

whose solution is $u(x,t) = u_0(x - u_0(x_0)t)$ for all x_0 in the spatial domain. Notice the following: Assume there are two nearby space points $x_1 < x_2$, where initial conditions are such that $u_0(x_1) > u_0(x_2) > 0$. The characteristic line passing through x_1 moves faster to the right than the one passing through x_2. The crossing of characteristic lines produces the formation of a discontinuity. **Then the finite differences method used in previous chapters fails due to the formation of a discontinuity on the unknown function**, and consequently the formulas for the calculation of derivatives from Sect. 3.1 are not useful anymore, because discontinuous functions do not admit a Taylor series expansion.

Let us see what happens when solving Burgers equation using the method of lines with a finite difference discretization. Consider the finite domain $[x_{min}, x_{max}] = [-0.5, 0.5]$, with $t_f = 2$, and the numerical domain $D_d = \{(x_i, t^n)\}$, with $x_i = x_{min} + i\Delta x$, $i = 0, \ldots, N_x$,, $t^n = n\Delta t$, using $N_x = 1000$ or equivalently $\Delta x = 0.001$, and $\Delta t = C\Delta x$ with $C = 0.1$. Assume the initial condition is the Gaussian $u_0(x) = e^{-x^2/0.01}$. The semi-discrete version at the inner points of the numerical domain D_d will be

$$\frac{\partial u}{\partial t} = -u_i \frac{u_{i+1} - u_{i-1}}{2\Delta x} + \mathcal{O}(\Delta x^2), \quad (5.10)$$

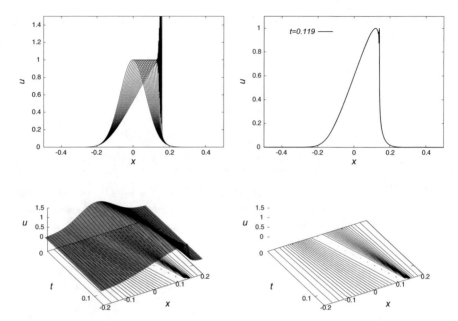

Fig. 5.1 At the top left, we show the numerical solution of Burgers equation using the method of lines at various times, using the semi-discrete version (5.10). At the top right, the solution at $t = 0.119$, just after the discontinuity forms. In the bottom, we show the solution in the space-time domain and its isocontours, which are an approximation of characteristic lines and illustrate how these behave before the discontinuity forms

with second-order accuracy in the right-hand side. The result using the RK2 time integrator is shown in Fig. 5.1. With these numerical parameters, by time $t \sim 0.119$, the solution becomes discontinuous and explodes later on.

The **lessons** here are that when the equations are quasilinear, the solutions, **even if initial data are smooth**, can develop discontinuities and that the method of lines with a semi-discrete version that uses a finite difference approximation will not be able to solve the problem after a discontinuity forms.

Now, let us return to the case of a system of various equations. In this case, the system of equations can be written as

$$\frac{\partial \mathbf{u}}{\partial t} + \mathbf{A}(\mathbf{u})\frac{\partial \mathbf{u}}{\partial x} = \mathbf{S}(\mathbf{u}). \tag{5.11}$$

Likewise for Burgers equation, the characteristic velocities, namely, the eigenvalues of the matrix $\mathbf{A}$, depend on (x, t) through $\mathbf{u}$, that is, they are not constant like in the constant coefficient case and can develop discontinuities. This is the reason why we need to know how to handle discontinuities, for which the **finite volume** is a useful method that will help.

5.1.3 Finite Volume Method in 1+1 Dimensions

We have seen that the nonlinearity of the equations of an IVP can lead to the formation of discontinuities. For this reason, numerical methods based on the continuity and smoothness of the involved functions, like the finite difference approximation of derivatives, are not suitable to solve this kind of equations because they are based on Taylor series expansions that need functions to be smooth.

One of the most widely used approaches to handle discontinuous functions is the finite volume method, which considers a discrete domain with a cell-centered view in the space-time domain. Explicitly, the discrete domain D_d is similar to that we have defined so far. Consider the domain $D = [x_{min}, x_{max}] \times [0, t_f]$, and then assume as we have done that the discrete domain is $D_d = \{(x_i, t^n) \mid x_i = x_{min} + i\Delta x, t^n = n\Delta t\}$, where $\Delta x = (x_{max} - x_{min})/N_x$, $\Delta t = C\Delta x$, with $i = 0, \ldots, N_x$ and the integer label n, **except** that points of D_d are the corners of **cells**, the bricks of the finite volume discretization.

In view of the formation of discontinuities, one cannot consider the state variables u_i to be smooth. Therefore, the general scenario has to be that of discontinuous function that can be defined in general as a **Riemann Problem**, which is an IVP with evolution equations for a set of variables **u**, with initial conditions

$$\mathbf{u}_0(x) = \begin{cases} \mathbf{u}_L, & \text{for } x < x_0 \\ \mathbf{u}_R, & \text{for } x > x_0 \end{cases} \quad (5.12)$$

where $\mathbf{u}_L$ and $\mathbf{u}_R$ are constant values, different in general, of the variables at the left and right from the point x_0. In summary, initial conditions for **u** are discontinuous at x_0. The evolution of $\mathbf{u}_0$ will depend on the evolution equations of the IVP.

One therefore has to assume that at each point of the numerical domain there may be a discontinuity of the variables that cannot be solved using finite differences, that is, a Riemann problem. The finite volume method is designed to approximately solve Riemann problems at each point of the domain, constructed using the integral version of the system of equations of the IVP. The integral version is assumed to hold at cells of the domain, unlike at points of D_d. In Fig. 5.2, we show the relation between points of D_d and the cells of the domain.

Let us illustrate the finite volume method considering the **system of flux balance law** equations:

$$\frac{\partial \mathbf{u}}{\partial t} + \frac{\partial \mathbf{F}(\mathbf{u})}{\partial x} = \mathbf{S}, \quad (5.13)$$

where **u** is the vector of **conserved variables**, **F(u)** is a **vector of fluxes** which depends on the variables **u**, and **S** is a **vector of sources**. First, the integral version of (5.13) considers the averaged integration of the equation over the volume of a cell $C_i^{n+1/2}$ as follows:

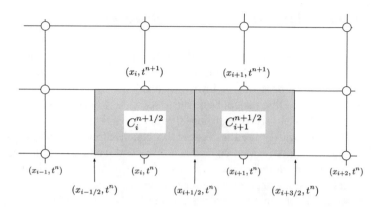

Fig. 5.2 In this figure, we present the discretization and cell structure of the space-time domain. White circles indicate the location of the elements of D_d. The cell $C_i^{n+1/2}$ is centered at $(x_i, t^{n+1/2})$, and its space-time volume is $\Delta t \Delta x$

$$\frac{1}{\Delta t \Delta x} \int_{x_{i-1/2}}^{x_{i+1/2}} \int_{t^n}^{t^{n+1}} \frac{\partial \mathbf{u}}{\partial t} \, dx dt + \frac{1}{\Delta t \Delta x} \int_{x_{i-1/2}}^{x_{i+1/2}} \int_{t^n}^{t^{n+1}} \frac{\partial \mathbf{F(u)}}{\partial x} \, dx dt =$$

$$= \frac{1}{\Delta t \Delta x} \int_{x_{i-1/2}}^{x_{i+1/2}} \int_{t^n}^{t^{n+1}} \mathbf{S} \, dx dt, \quad (5.14)$$

where the volume of the cell is $\Delta x \Delta t$. Second, using Gauss theorem, this last equation can be integrated to obtain a discrete version of the integral form of the system of Eqs. (5.13):

$$\frac{\bar{\mathbf{u}}_i^{n+1} - \bar{\mathbf{u}}_i^n}{\Delta t} + \frac{\left(\bar{\mathbf{F}}_{i+1/2}^{n+1/2} - \bar{\mathbf{F}}_{i-1/2}^{n+1/2} \right)}{\Delta x} = \bar{\mathbf{S}}_i^{n+1/2}, \quad (5.15)$$

where $\bar{\mathbf{u}}_i^n$ are the spatial averages of the conservative variables at time t^n:

$$\bar{\mathbf{u}}_i^n = \frac{1}{\Delta x} \int_{x_{i-1/2}}^{x_{i+1/2}} u(x, t^n) dx,$$

and $\bar{\mathbf{F}}_{i+1/2}^{n+1/2}$ are the temporal averages of the fluxes at space position $x_{i+1/2}$:

$$\bar{\mathbf{F}}_{i+1/2}^{n+1/2} = \frac{1}{\Delta t} \int_{t^n}^{t^{n+1}} \mathbf{F}(\mathbf{u}(x_{i+1/2}, t)) dt.$$

5.2 1+1 Euler Equations

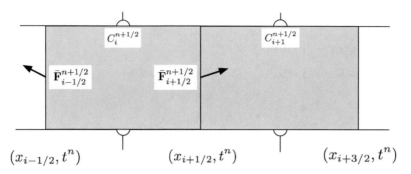

Fig. 5.3 Illustration of the fluxes at the boundaries of the cell $C_i^{n+1/2}$

Finally, $\bar{\mathbf{S}}_i^{n+1/2}$ is the space-time average of the sources

$$\bar{\mathbf{S}}_i^{n+1/2} = \frac{1}{\Delta x \Delta t} \int_{x_{i-1/2}}^{x_{i+1/2}} \int_{t^n}^{t^{n+1}} \mathbf{S}(t,x) dx dt.$$

Equation (5.15) can be seen as an evolution equation for the state averaged vector of conservative variables $\bar{\mathbf{u}}$, whose a semi-discrete version developed around $(x_i, t^{n+1/2})$ is

$$\frac{\partial \bar{\mathbf{u}}}{\partial t} = -\frac{\left(\bar{\mathbf{F}}_{i+1/2}^{n+1/2} - \bar{\mathbf{F}}_{i-1/2}^{n+1/2}\right)}{\Delta x} + \bar{\mathbf{S}}_i^{n+1/2}, \qquad (5.16)$$

where the fluxes are evaluated at the boundaries between cells as illustrated in Fig. 5.3. In this view, Eq. (5.16) is a semi-discrete version of a system of equations that can be integrated, provided the fluxes at the intercell boundaries are known. The construction of these fluxes is essential to the method, and we exemplify its construction and implementation with the solution of Euler equations.

5.2 1+1 Euler Equations

Euler equations describe the evolution of a compressible, inviscid fluid, characterized by its density, pressure, velocity field, and energy. Here, we focus on the one spatial dimension case, where all variables depend on a spatial coordinate x and time t. In these terms, the IVP involving Euler equations reads (see, for instance, [3])

$$\begin{aligned}
&\partial_t \rho + \partial_x(\rho v) = 0 \\
&\partial_t(\rho v) + \partial_x(\rho v^2 + p) = 0 \\
&\partial_t E + \partial_x[v(E+p)] = 0 \qquad &&\rho = \rho(x,t),\ p = p(x,t),\ v = v(x,t), \\
& &&E = E(x,t) \\
&D = x \in [x_{min}, x_{max}] \times t \in [0, t_f] \qquad &&Domain \\
&\rho(x,0) = \rho_0(x),\ p(x,0) = p_0(x) \qquad &&Initial\ Conditions \\
&v(x,0) = v_0(x),\ E(x,0) = E_0(x) \\
&\rho(x_{min},t),\ \rho(x_{max},t) \qquad &&Boundary\ Conditions \\
&p(x_{min},t),\ p(x_{max},t) \\
&v(x_{min},t),\ v(x_{max},t) \\
&E(x_{min},t),\ E(x_{max},t)
\end{aligned} \qquad (5.17)$$

where each volume element of the fluid has density ρ, velocity v along x, total energy $E = \rho(\frac{1}{2}v^2 + e)$, internal energy e, and pressure p. The system is underdetermined, with three equations and four unknowns (ρ, v, E, p), which is closed using an equation of state (EoS) relating three variables, for example, $p = p(\rho, e)$.

It is usual to consider the ideal gas EoS, which is given by $p = \rho e(\gamma - 1)$, where $\gamma = c_p/c_v$ is the adiabatic index, the ratio between the specific heats of the fluid.

Like Burgers equation in Sect. 5.1, these equations are quasilinear, and the variables may develop discontinuities even if they are smooth. This makes the IVP illustrative of how the finite volume method captures the typical discontinuities developed in fluid dynamics.

We write Euler equations as a system of flux balance laws:

$$\frac{\partial \mathbf{u}}{\partial t} + \frac{\partial \mathbf{F}(\mathbf{u})}{\partial x} = \mathbf{S}(\mathbf{u}), \qquad (5.18)$$

where $\mathbf{u}$ is the state vector, $\mathbf{F}$ a vector of fluxes, and $\mathbf{S}$ a vector with sources. It is important to name $\rho, v, E,$ and p as **primitive variables**, which contain the physical information of the fluid. As a counterpart, one defines **conservative variables** $u_1 = \rho$, $u_2 = \rho v$, $u_3 = E$, which will be the entries of $\mathbf{u}$. Euler equations in (5.17) can be written as a set of flux balance laws like (5.18) with

$$\mathbf{u} = \begin{bmatrix} \rho \\ \rho v \\ E \end{bmatrix} := \begin{bmatrix} u_1 \\ u_2 \\ u_3 \end{bmatrix}, \quad \mathbf{F} = \begin{bmatrix} \rho u \\ \rho v^2 + p \\ v(E+p) \end{bmatrix} = \begin{bmatrix} u_2 \\ u_2^2/u_1 + p \\ u_2(u_3+p)/u_1 \end{bmatrix} \qquad (5.19)$$

5.2 1+1 Euler Equations

and $\mathbf{S} = \mathbf{0}$. With the equations written in this fashion, *one can use the MoL* given an appropriate discretization of the divergence of the flux vector and the sources as suggested in (5.16).

The **finite volume** discretization considers the points of the discrete domain $D_d = \{(x_i, t^n)\}$, with $x_i = x_{min} + i\Delta x$, $t^n = n\Delta t$, as the corners of space-time cells C_i^n. The cell-centered domain is illustrated in Fig. 5.2, where two neighboring cells $C_i^{n+1/2}$ and $C_{i+1}^{n+1/2}$ and some of the points of D_d are shown.

The discretization of the divergence of fluxes requires the characteristic analysis that determines the characteristic variables of the system and their velocities of propagation, as seen in Sects. 5.1.1 and 5.1.2. An approximate version of (5.18) uses the chain rule to write down the system

$$\frac{\partial \mathbf{u}}{\partial t} + \frac{d\mathbf{F}}{d\mathbf{u}} \frac{\partial \mathbf{u}}{\partial x} = \mathbf{S},$$

$$\frac{\partial \mathbf{u}}{\partial t} + \mathbf{A} \frac{\partial \mathbf{u}}{\partial x} = \mathbf{S}, \qquad (5.20)$$

where $\mathbf{A}$ is the Jacobian matrix

$$\mathbf{A} = \begin{bmatrix} \partial F_1/\partial u_1 & \partial F_1/\partial u_2 & \partial F_1/\partial u_3 \\ \partial F_2/\partial u_1 & \partial F_2/\partial u_2 & \partial F_2/\partial u_3 \\ \partial F_3/\partial u_1 & \partial F_3/\partial u_2 & \partial F_3/\partial u_3 \end{bmatrix}. \qquad (5.21)$$

At this point, it is useful to apply the characteristic analysis from the previous section.

Let us construct the characteristic structure of $\mathbf{A}$ for the EoS $p = \rho e(\gamma - 1)$ used here. For this particular case in terms of the conservative variables, the vector of fluxes from (5.19) is

$$\mathbf{F}(\mathbf{u}) = \begin{bmatrix} u_2 \\ \frac{1}{2}(3-\gamma)\frac{u_2^2}{u_1} + (\gamma - 1)u_3 \\ \gamma \frac{u_2}{u_1} u_3 - \frac{1}{2}(\gamma - 1)\frac{u_2^3}{u_1^2} \end{bmatrix}, \qquad (5.22)$$

and the matrix $\mathbf{A} = \frac{d\mathbf{F}}{d\mathbf{u}}$ in (5.21) reads

$$\mathbf{A} = \begin{bmatrix} 0 & 1 & 0 \\ -\frac{1}{2}(\gamma - 3)(\frac{u_2}{u_1})^2 & (3-\gamma)(\frac{u_2}{u_1}) & \gamma - 1 \\ -\frac{\gamma u_2 u_3}{u_1^2} + (\gamma - 1)(\frac{u_2}{u_1})^3 & \frac{\gamma u_3}{u_1} - \frac{3}{2}(\gamma - 1)(\frac{u_2}{u_1})^2 & \gamma(\frac{u_2}{u_1}) \end{bmatrix}. \qquad (5.23)$$

An important velocity scale is the **speed of sound**. For this EoS, the eigenvalues and eigenvectors of the Jacobian matrix (5.23) read (see, for example, [3]):

$$\lambda_1 = v - a, \quad \lambda_2 = v, \quad \lambda_3 = v + a, \tag{5.24}$$

$$\mathbf{r}_1 = \begin{bmatrix} 1 \\ v - a \\ \frac{1}{2}v^2 + e + p/\rho - va \end{bmatrix}, \quad \mathbf{r}_2 = \begin{bmatrix} 1 \\ v \\ \frac{1}{2}v^2 \end{bmatrix}, \quad \mathbf{r}_3 = \begin{bmatrix} 1 \\ v + a \\ \frac{1}{2}v^2 + e + p/\rho + va \end{bmatrix}. \tag{5.25}$$

where $a = \sqrt{\gamma \frac{p}{\rho}}$ is the speed of sound of the fluid. These elements are used to construct the solution to the IVP (5.17).

Keep in mind that in order to implement the MoL for (5.16), one needs an appropriate discrete version of the second term in (5.18).

In this book, we will use only the simplest of approximate flux formulas, the HLLE [3], which requires only the eigenvalues of the Jacobian matrix. Other formulas can be programmed on top of the codes provided later on.

The essence of approximate formulas like HLLE (Harten, Lax, van Leer, Einfeldt, [4, 5]) is that they provide approximate values for the fluxes at intercell boundaries. In Fig. 5.4, we show a zoomed version of Fig. 5.2, illustrating the intercell boundary at $x_{i+1/2}$. Variables to the left and to the right from this boundary will use L and R labels, specifically variables $\rho^L, \rho^R, v^L, v^R, E^L, E^R, p^L, p^R$ and conservative variables $u_1^L, u_1^R, u_2^L, u_2^R, u_3^L, u_3^R$, to the left and to the right from the intercell boundary.

Notice that in terms of L variables there will be a flux $\mathbf{F}(\mathbf{u}_{i+1/2}^L)$ entering the intercell boundary at $x_{i+1/2}$ *from the left* and in terms of R variables there will be a flux $\mathbf{F}(\mathbf{u}_{i+1/2}^R)$ entering *from the right*.

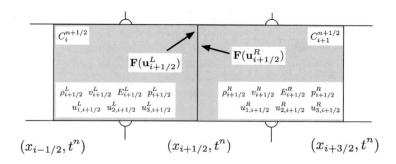

Fig. 5.4 Intercell boundary located between x_i and x_{i+1}. State variables, both primitive and conservative, at the left and at the right from this boundary are labeled with L and R, respectively. The arrows indicate the fluxes approaching from the left and from the right of the intercell boundary

5.2 1+1 Euler Equations

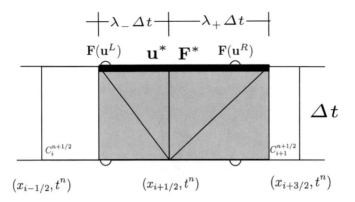

Fig. 5.5 Illustration of the intercell boundary with fluxes at the left $\mathbf{F}(\mathbf{u}^L)$, fluxes at the right $\mathbf{F}(\mathbf{u}^R)$, and variables $\mathbf{u}^*$ and fluxes $\mathbf{F}^*$ at the star region, indicated with a very bold line. The distance travelled by the fastest mode to the left in the lapse Δt is $\lambda_- \Delta t$, whereas the distance travelled by the fastest mode to the right in the lapse Δt is $\lambda_+ \Delta t$. The shaded rectangle is auxiliary in the construction of the HLLE formula (5.31)

The HLLE approximate formula **assumes** that the interesting dynamics happens between the fastest mode moving to the left with velocity λ_- and the fastest mode moving to the right with velocity λ_+, where the velocity of each mode is an eigenvalue of the Jacobian matrix $\mathbf{A}$. Figure 5.5 illustrates the region of causal influence of the information at the point $x_{i+1/2}$ that will help to construct the HLLE formula.

The formula involves only the characteristic structure of the conservation law (5.18), related to the principal part of the equation, and the possible source vector is ignored:

$$\frac{\partial \mathbf{u}}{\partial t} + \frac{\partial \mathbf{F}(\mathbf{u})}{\partial x} = 0. \qquad (5.26)$$

The integration of (5.26) in the volume $[\lambda_- \Delta t, , \lambda_+ \Delta t] \times [0, \Delta t]$ gives

$$\int_{\lambda_- \Delta t}^{\lambda_+ \Delta t} \int_0^{\Delta t} \frac{\partial \mathbf{u}}{\partial t} dt dx + \int_{\lambda_- \Delta t}^{\lambda_+ \Delta t} \int_0^{\Delta t} \frac{\partial \mathbf{F}}{\partial t} dt dx = 0 \Rightarrow$$

$$\int_{\lambda_- \Delta t}^{\lambda_+ \Delta t} u(x, \Delta t) dx - \int_{\lambda_- \Delta t}^{\lambda_+ \Delta t} u(x, 0) dx + \int_0^{\Delta t} \mathbf{F}(\lambda_+ \Delta t, t) dt$$

$$- \int_0^{\Delta t} \mathbf{F}(\lambda_- \Delta t, t) dt = 0.$$

The first integral is rewritten in terms of $\mathbf{u}^*$, the state vector in the *star zone*, whereas the second integral is split into two integration intervals $[\lambda_- \Delta t, 0]$ and $[0, \lambda_+ \Delta t]$ as follows:

$$\lambda_+ \Delta t \mathbf{u}^* - \lambda_- \Delta t \mathbf{u}^* - \int_{\lambda_- \Delta t}^{0} \mathbf{u}(x,0) dx - \int_{0}^{\lambda_+ \Delta t} \mathbf{u}(x,0) dx + \Delta t \mathbf{F}(\mathbf{u}^R)$$
$$- \Delta t \mathbf{F}(\mathbf{u}^L) = 0, \Rightarrow \quad (5.27)$$
$$\lambda_+ \Delta t \mathbf{u}^* - \lambda_- \Delta t \mathbf{u}^* + \lambda_- \Delta t \mathbf{u}^L - \lambda_+ \Delta t \mathbf{u}^R + \Delta t \mathbf{F}(\mathbf{u}^R) - \Delta t \mathbf{F}(\mathbf{u}^L) = 0. \quad (5.28)$$

From this equation, one can solve for $\mathbf{u}^*$

$$\mathbf{u}^* = \frac{\lambda_+ \mathbf{u}^R - \lambda_- \mathbf{u}^L - \mathbf{F}(\mathbf{u}^R) + \mathbf{F}(\mathbf{u}^L)}{\lambda_+ - \lambda_-}. \quad (5.29)$$

In order to calculate the numerical flux, namely, $\mathbf{F}^*$, one has to add an extra equation, the integration of (5.26) either in the volume $[\lambda_- \Delta t, 0] \times [0, \Delta t]$ or in the volume $[0, \lambda_+ \Delta t] \times [0, \Delta t]$ illustrated in Fig. 5.5. Here, we develop the integral using the second option:

$$\int_{0}^{\lambda_+ \Delta t} \mathbf{u}(x, \Delta t) dx - \int_{0}^{\lambda_+ \Delta t} \mathbf{u}(x, 0) dx + \Delta t (\mathbf{F}(\mathbf{u}^R) - \mathbf{F}^*) = 0 \Rightarrow$$
$$\lambda_+ \Delta t \mathbf{u}^* - \lambda_+ \Delta t \mathbf{u}^R + \Delta t (\mathbf{F}(\mathbf{u}^R) - \mathbf{F}^*) = 0 \Rightarrow$$
$$\mathbf{F}^* = \lambda_+ \mathbf{u}^* - \lambda_+ \mathbf{u}^R - \mathbf{F}(\mathbf{u}^R). \quad (5.30)$$

Substitution of $\mathbf{u}^*$ from (5.29) into (5.30) gives

$$\mathbf{F}^* = \frac{\lambda_- \lambda_+ (\mathbf{u}^R - \mathbf{u}^L) + \lambda_+ \mathbf{F}(\mathbf{u}^L) - \lambda_- \mathbf{F}(\mathbf{u}^R)}{\lambda_+ - \lambda_-}. \quad (5.31)$$

This is a formula for the fluxes at the intercell boundary $x_{i+1/2}$ in terms of the fluxes at the left $\mathbf{F}(\mathbf{u}^L)$ and at the right $\mathbf{F}(\mathbf{u}^R)$ from $x_{i+1/2}$. This flux is directly associated with the point $x_{i+1/2}$ boundary and is finally written as [3, 6]

$$\bar{\mathbf{F}}_{i+1/2}^{HLLE} = \frac{\lambda_+ \mathbf{F}\left(\mathbf{u}_{i+1/2}^L\right) - \lambda_- \mathbf{F}\left(\mathbf{u}_{i+1/2}^R\right) + \lambda_+ \lambda_- (\mathbf{u}_{i+1/2}^R - \mathbf{u}_{i+1/2}^L)}{\lambda_+ - \lambda_-} \quad (5.32)$$

where, as said before, λ_+ and λ_- are defined as the fastest velocity to the right and to the left, respectively:

$$\lambda_+ = \max\left(0, \lambda_1^R, \lambda_2^R, \lambda_3^R, \lambda_1^L, \lambda_2^L, \lambda_3^L\right),$$
$$\lambda_- = \min\left(0, \lambda_1^R, \lambda_2^R, \lambda_3^R, \lambda_1^L, \lambda_2^L, \lambda_3^L\right), \quad (5.33)$$

5.2 1+1 Euler Equations

where the eigenvalues L and R of the Jacobian matrix are defined following (5.24) in terms of the primitive variables:

$$\lambda_1^L = v_L - a_L,$$
$$\lambda_2^L = v_L,$$
$$\lambda_3^L = v_L + a_L,$$
$$\lambda_1^R = v_R - a_R,$$
$$\lambda_2^R = v_R,$$
$$\lambda_3^R = v_R + a_R, \tag{5.34}$$

where

$$a_L = \sqrt{p_L \gamma / \rho^L}$$
$$a_R = \sqrt{p_R \gamma / \rho^R}$$

is the speed of sound for an ideal gas.

Let us collect some results. Fluxes depend on conservative variables, and the eigenvalues depend on primitive variables, both L and R. This means that we still have to provide the values for all these variables at L and R cells from the intercell boundary $x_{i+1/2}$. How these variables are constructed defines **different flavors** of the method. This step is called **reconstruction of variables**.

A simple reconstruction of variables assumes these are constant at each cell. This is the **Godunov reconstruction** illustrated in Fig. 5.6. Formally, this reconstructor assigns values to either primitive or conservative variables w as follows:

$$w_{i+1/2}^L = w_i,$$
$$w_{i+1/2}^R = w_{i+1}, \tag{5.35}$$

see Fig. 5.4. This method is *first-order accurate*. A second-order accurate method is the **linear reconstruction** and assumes the variables to be a line at each cell. One of the most popular linear reconstructor is the minmod, defined as follows:

$$w_{i+1/2}^L = w_i + \sigma_i (x_{i+1/2} - x_i),$$
$$w_{i+1/2}^R = w_{i+1} + \sigma_{i+1} (x_{i+1/2} - x_{i+1}), \tag{5.36}$$
$$\sigma_i = \text{minmod}(m_{i-1/2}, m_{i+1/2})$$

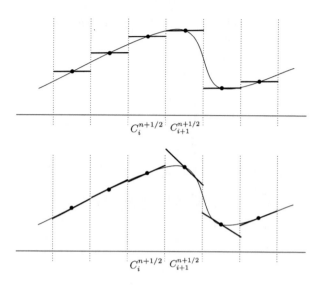

Fig. 5.6 Illustration along the spatial direction of the domain, of Godunov (top) and minmod (bottom) reconstruction of variables within cells. The curve represents a function, the dotted lines are the intercell boundaries, and the thick straight segments approximate the function assuming constant piecewise and linear piecewise approximations

$$\text{minmod}(a, b) = \begin{cases} a & \text{if } |a| < |b| \text{ and } ab > 0 \\ b & \text{if } |a| > |b| \text{ and } ab > 0 \\ 0 & \text{if } ab < 0. \end{cases}$$

The arguments $m_{i-1/2}$ and $m_{i+1/2}$ are the slopes of the variable w at intercell boundaries:

$$m_{i+1/2} = \frac{w_{i+1} - w_i}{x_{i+1} - x_i},$$

$$m_{i-1/2} = \frac{w_i - w_{i-1}}{x_i - x_{i-1}}.$$

Thus, reconstruction (5.36) assumes the variables are linear piecewise with slopes σ_i and σ_{i+1}. This is a better approximation than constant piecewise, and is second-order accurate.

Let us wrap up everything together into a code that solves the IVP in (5.17).

Workhorse Example: The Shock-Tube Problem
This is literally the Riemann problem in Eq. (5.12) for two states of the same fluid, that is, density, velocity, and pressure are initially in two constant states, separated by a discontinuity at the center of the spatial domain. Euler equations drive the further evolution of the fluid variables.

5.2 1+1 Euler Equations

We solve the IVP in the domain $D = [x_{min}, x_{max}] \times [0, t_f] = [0, 1] \times [0, 1]$ and define the discrete domain as $D_d = \{(x_i, t^n)\}$, $x_i = x_{min} + i\Delta x$, $i = 0, \ldots, N_x$, $t^n = n\Delta t$ with $\Delta x = (x_{max} - x_{min})/N_x$ and $\Delta t = 0.25\Delta x$. The steps needed for a sufficiently ordered implementation are listed next:

1. *Initial Conditions.* The initial conditions correspond to a fluid in two states separated by a wall at $x = 1/2$ and are

$$\rho = \begin{cases} \rho_1, & x \leq 1/2 \\ \rho_2, & x > 1/2 \end{cases}, \quad p = \begin{cases} p_1, & x \leq 1/2 \\ p_2, & x > 1/2 \end{cases}, \quad v = \begin{cases} v_1, & x \leq 1/2 \\ v_2, & x > 1/2 \end{cases},$$

where $\rho_1, \rho_2, p_1, p_2, v_1,$ and v_2 have constant values that define two different states.

With these initial conditions on primitive state variables, it is possible to define the initial value of conservative variables for the ideal gas EoS $p = \rho e(\gamma - 1)$, according to definitions in (5.19) as follows:

$$u_1 = \rho,$$
$$u_2 = \rho v,$$
$$u_3 = E = \rho\left(\frac{1}{2}v^2 + e\right) = \frac{1}{2}\rho v^2 + \frac{p}{\gamma - 1}. \tag{5.37}$$

2. *Calculation of HLLE Fluxes.* From the first step on, we need to calculate the numerical fluxes in order to calculate the right-hand side in (5.16). These fluxes are given by formula (5.32) in terms of conservative variables, evaluated at the left and right from intercell boundaries, whereas eigenvalues of the Jacobian matrix (5.24) are in terms of the primitive variables evaluated also at left and right from intercell boundaries.

An appropriate order of instructions is as follows:

(a) Reconstruct L and R values of primitive and conservative variables using Godunov (5.35) or minmod (5.36) methods.
(b) Calculate the eigenvalues of the Jacobian matrix (5.21) using (5.34).
(c) Calculate λ_+ and λ_- using (5.33).
(d) Calculate the HLLE fluxes $\mathbf{F}(\mathbf{u}^L_{i+1/2})$ and $\mathbf{F}(\mathbf{u}^R_{i+1/2})$ needed in (5.32), using the results from (a) and expressions for $\mathbf{F}$ in formula (5.19).
(e) Implement formula (5.32) for the fluxes $\mathbf{F}^{HLLE}$ at $x_{i+1/2}$.
(f) Use the MoL with the expression (5.16) to integrate in time.
(g) Implement boundary conditions on the conservative variables.
(h) The evolution updates the conservative variables u_1, u_2, and u_3, and primitive variables are needed for the calculation of the eigenvalues of $\mathbf{A}$ and for output. In the case of Euler equations, it is easy to recover the primitive variables by inversion of (5.37):

$$\rho = u_1,$$

$$v = \frac{u_2}{u_1},$$

$$p = (\gamma - 1)\left[u_3 - \frac{1}{2}\frac{u_2^2}{u_1}\right]. \tag{5.38}$$

We call the attention to item (g) of boundary conditions we have to describe now.

Outflow Boundary Conditions. This type of conditions are useful to study isolated systems that expel material, and the numerical boundaries are expected to be transparent. A basic construction of these conditions in a 1+1 problem copies the values of conservative variables at boundary points x_0 and x_{N_x} from the nearest inner neighbors:

$$\mathbf{u}_0 = \mathbf{u}_1$$

$$\mathbf{u}_{N_x} = \mathbf{u}_{N_x-1} \tag{5.39}$$

for all entries of $\mathbf{u}$. The effect is that the Riemann problems at intercell boundaries $x_{1/2}$ and $x_{N_x-1/2}$ are trivial, because the states at the left and right are the same. Therefore, there will be no dynamics produced at these intercell boundaries that could interfere with the dynamics at interior cells of the numerical domain. This explanation suffices for the reminder of the book, although we recommend the interested reader to follow higher-order and more elaborate approaches to these conditions in [3, 6].

3. *Repeat All the Steps in 2.* for the evolution of the system from t^n to t^{n+1} for all n in the numerical domain.

All these steps are specified in the code MoLEuler.f90, that we use to solve the shock-tube problem, and are listed in Appendix B.

There are four scenarios that result from the shock-tube problem that we illustrate using the parameters in Table 5.1. These four scenarios are characterized by the waves developed during the evolution, explained in a pedagogical fashion in the educative paper [7]. For illustration, we use D_d such that the spatial domain is $[x_{min}, x_{max}] = [0, 1]$, discretized with $N_x = 1000$ cells of resolution $\Delta x = 0.001$.

Table 5.1 Initial conditions for the shock-tube problem in the four possible scenarios. In all cases, we use $\gamma = 1.4$

Case	ρ_L	ρ_R	v_L	v_R	p_L	p_R
Rarefaction-shock	1.0	0.125	0.0	0.0	1.0	0.1
Shock-rarefaction	0.125	1.0	0.0	0.0	0.1	1.0
Rarefaction-rarefaction	1.0	1.0	−1.0	1.0	0.4	0.4
Shock-shock	1.0	1.0	1.0	−1.0	0.4	0.4

5.2 1+1 Euler Equations

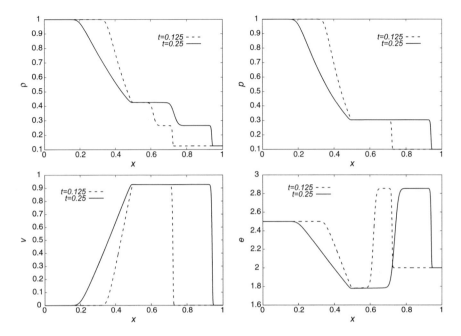

Fig. 5.7 Two snapshots of the solution for the shock-tube problem in the rarefaction-shock case

For the evolution, we set $\Delta t = C \Delta x$ with $C = 0.25$ and use the RK2 integrator. We show snapshots corresponding to these four cases of density ρ, pressure p, velocity v, and internal energy $e = \frac{p}{\rho(\gamma-1)}$ of the fluid. The four cases are solved using the Godunov reconstruction, and later on, we compare the results with those using minmod.

Rarefaction-shock case. The results obtained using the code above, at two values of time, are shown in Fig. 5.7. The density starts decreasing at the left which is a sign of rarefaction, whereas a shock wave is being propagated to the right. At time $t = 0.25$, in the plot of ρ, one observes the shock wave is located at around $x \sim 0.94$, whereas a contact discontinuity appears at around $x \sim 0.72$. Notice that the pressure is continuous at the contact discontinuity. The velocity, initially zero, becomes positive, which indicates that the fluid moves to the right.

Shock-rarefaction case. This is the antisymmetric to the previous case whose solution is shown in Fig. 5.8. The higher density and pressure correspond to the state on the right and the fluid moves to the left. Even though the case is very similar to the previous one, it is important, because the fluxes have opposite direction and it serves to verify the methods and code work in the two directions.

Rarefaction-rarefaction case. Density and pressure are the same in the two states at initial time; however, the initial velocities point outward. This produces a depletion of the central part of the domain, and two rarefaction waves are produced during the evolution, both moving outward. The results appear in Fig. 5.9.

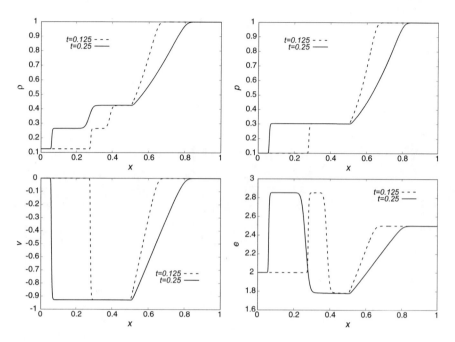

Fig. 5.8 Two snapshots of the solution for the shock-tube problem in the shock-rarefaction case

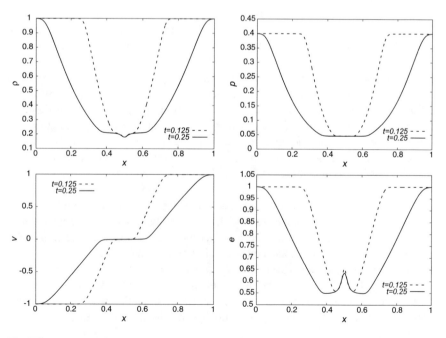

Fig. 5.9 Two snapshots of the solution for the shock-tube problem in the rarefaction-rarefaction case

5.2 1+1 Euler Equations

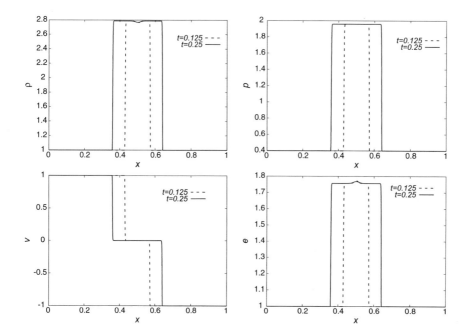

Fig. 5.10 Two snapshots of the solution for the shock-tube problem in the shock-shock case

Shock-shock scenario. In this case also, density and pressure are the same in the two states, and now the initial velocity of the fluid points toward the center of the domain. Two shock waves are formed in the process as shown in Fig. 5.10.

Effects of using different reconstructors. For comparison, we use the rarefaction-shock case using the same numerical parameters and initial conditions. In one case, the variables are reconstructed using the Godunov method (5.35), and in a second case, we use the minmod reconstruction (5.36). The results at $t = 0.25$ appear in Fig. 5.11. Notice that the discontinuities are better captured with the minmod reconstructor. It would be ideal to compare with the exact solution; however, for this, we let the following exercise:

Exercise. Using the recipe in [7], add the exact solution to the program `MolEuler.f90`, and calculate the error of the density using the two reconstructors. Also, attempt a convergence test and show that when using the minmod reconstructor the convergence is first order. What is the convergence order when using Godunov?

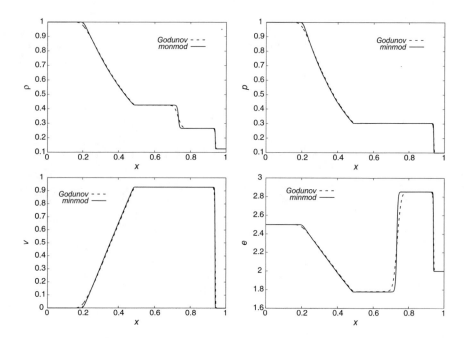

Fig. 5.11 Comparison of the solution using Godunov and minmod reconstructors for the rarefaction-shock case at time $t = 0.25$. The differences consist in that the contact discontinuity and the shock discontinuity are better captured by the minmod reconstructor

Exercise. Program a code that solves Burgers equation (5.9) using the finite volume approach, and compare the results with those in Fig. 5.1.

5.3 Application: Spherically Symmetric Hydrodynamical Solar Wind

We will use a generalization of the previous code to simulate a spherically symmetric hydrodynamic solar wind. A hydrodynamic wind is a relatively basic model of wind that does not react to magnetic fields, in which case one would need to solve the equations of magnetohydrodynamics.

Stationary Wind. The heliosphere is characterized by a nearly stationary wind that permeates the whole interplanetary space and is permanently being pumped from the solar surface. Here, we simulate the formation of such stationary solar wind.

5.3 Application: Spherically Symmetric Hydrodynamical Solar Wind

Let us rewrite down Euler equations in general:

$$\partial_t \rho + \nabla \cdot (\rho \mathbf{v}) = 0,$$
$$\partial_t (\rho \mathbf{v}) + \nabla \cdot (\mathbf{v} \otimes \rho \mathbf{v}) + \nabla p = -\rho \nabla \phi,$$
$$\partial_t E + \nabla \cdot [(E + p)\mathbf{v}] = -\rho \mathbf{v} \cdot \nabla \phi, \tag{5.40}$$

where ρ, $\mathbf{v}$, and p are the density, velocity field, and pressure of the gas volume elements; $E = \rho(\frac{1}{2}v^2 + e)$ is its total energy, with e the internal energy related to ρ and p through the ideal gas equation of state $p = \rho e(\gamma - 1)$; and finally ϕ is the potential responsible for an external force acting on the fluid, which in our case will be the gravitational potential due to the Sun. Notice that in general, all variables involved depend on time and space.

The system (5.40) is coordinate-independent and can be simplified under certain conditions. Assuming the wind is spherically symmetric means that the flux is radial, and the system simplifies when using spherical coordinates (r, θ, φ). In this case, $\mathbf{v} = (v^r, 0, 0)$ and ρ, p, v^r, e, E are all functions of (r, t), whereas ϕ, we suppose, depends only on r considering the Sun produces a stationary, spherically symmetric, gravitational field.

Euler equations for a spherically symmetric flow reduce to three equations:

$$\frac{\partial \rho}{\partial t} + \frac{1}{r^2}\frac{\partial (r^2 \rho v^r)}{\partial r} = 0,$$

$$\frac{\partial (\rho v^r)}{\partial t} + \frac{1}{r^2}\frac{\partial}{\partial r}[r^2(\rho v^{r2} + p)] = \frac{2p}{r} - \rho G \frac{M_\odot}{r^2},$$

$$\frac{\partial E}{\partial t} + \frac{1}{r^2}\frac{\partial}{\partial r}\left[r^2 \rho v^r \left(\frac{\gamma p}{\rho(\gamma - 1)} + \frac{1}{2}v^{r2}\right)\right] = -\rho G v^r \frac{M_\odot}{r^2}, \tag{5.41}$$

to be solved in the domain $(r, t) \in [r_{min}, r_{max}] \times [0, t_f]$, which means that for spherically symmetric scenarios, described in spherical coordinates, the problem in three spatial dimensions can be reduced to a 1+1 problem in the (r, t) domain.

There are various differences between this 1+1 problem and that in the previous section, where the spatial coordinate was Cartesian, not spherical. First is the factor $1/r^2$ in front of the divergence term, together with the factor r^2 within the flux components, which are consequence of the 3D spherical coordinates; second, the sources are nonzero. However, there is a very useful similarity, namely, the fluxes have the same expressions as for the one-dimensional Cartesian coordinate, and the system (5.41) can be written as the flux balance law:

$$\frac{\partial \mathbf{u}}{\partial t} + \frac{1}{r^2}\frac{\partial (r^2 \mathbf{F})}{\partial r} = \mathbf{S}, \tag{5.42}$$

where

$$\mathbf{u} = \begin{bmatrix} \rho \\ \rho v^r \\ E \end{bmatrix} := \begin{bmatrix} u_1 \\ u_2 \\ u_3 \end{bmatrix}, \quad \mathbf{F} = \begin{bmatrix} u_2 \\ \frac{1}{2}(3-\gamma)\frac{u_2^2}{u_1} + (\gamma-1)u_3 \\ \gamma \frac{u_2}{u_1} u_3 - \frac{1}{2}(\gamma-1)\frac{u_2^3}{u_1^2} \end{bmatrix}, \quad \mathbf{S} = \begin{bmatrix} 0 \\ \frac{2p}{r} - \rho G \frac{M_\odot}{r^2} \\ -\rho G v^r \frac{M_\odot}{r^2} \end{bmatrix},$$
(5.43)

and the flux entries as exactly like those in (5.22). Then the eigenvalues of the Jacobian matrix are (5.24) that we rewrite here:

$$\lambda_1 = v^r - a, \quad \lambda_2 = v^r, \quad \lambda_3 = v^r + a,$$
(5.44)

where $a = \sqrt{\gamma \frac{p}{\rho}}$ is the speed of sound in the gas.

The equations are to be solved in the spatial domain assuming the Sun is at the coordinate origin. The spatial domain will be the volume within the spheres of radius r_{min} and r_{max}. The inner radius r_{min} is chosen such that the characteristic velocities of the system (5.41) point all outward from the Sun, typically $r_{min} \sim 20 R_\odot$, whereas the outer r_{max} is chosen such that the domain contains at least Earth's position.

The use of this domain has two important advantages: first is that at r_{min} the information enters the domain, and, second, maintaining the state variables constant at r_{min} guarantees that the flow will enter the domain and point outward from the Sun until the flow becomes stationary; another implication is that the domain does not contain the origin $r = 0$ and it is not necessary to deal with the regularization of the equations there.

From Physical to Code Units. Solar wind parameters are usually the density, velocity, and temperature. Physical and code units are related through a scaling relation for each variable, $\rho_{phys} = \rho_0 \rho$, pressure $p_{phys} = p_0 p$, length $l_{phys} = l_0 l$, time $t_{phys} = t_0 t$, velocity $v_{phys} = v_0 v$, and temperature $T_{phys} = T_0 T$, where the naught index indicates the scale factors of each variable, the index *phys* indicates the quantity in physical units, and the quantities without index are in code units. We choose length units to be in solar radius and time in hours, that is, $l_0 = R_\odot$ and $t_0 = 3600$ s, with which the scale of velocity is $v_0 = l_0/t_0$. Temperature scale is given by $T_0 = m_H v_0^2 / k_B$ where m_H is the hydrogen molecular mass and k_B Boltzmann's constant. Pressure and energy scale with constants given by $p_0 = \rho_0 v_0^2$ and $E_0 = \rho_0 v_0^2$, respectively.

When the equations of the system (5.41) are scaled according to these rules, the second equation has an overall factor on the left $\frac{\rho_0 v_0^2}{l_0}$ and on the gravitational source a factor $\frac{\rho_0}{l_0^2}$; then for the equation to hold in code units, one has to multiply the source term by $\frac{1}{l_0 v_0^2}$. Analogously, an overall factor in the left of the third equation is $\frac{\rho_0 v_0^3}{l_0}$

5.3 Application: Spherically Symmetric Hydrodynamical Solar Wind

and on the right a factor $\frac{\rho_0 v_0}{l_0^2}$, and then, in order for the third equation to hold, one has to multiply the source term by the factor $\frac{1}{l_0 v_0^2}$.

Physical and Numerical Domain. The mean Sun-Earth distance is $\sim 214.83452 R_\odot$; then we can set $r_{min} = 20 R_\odot$ and $r_{max} = 220 R_\odot$ so that the domain contains the Earth. Since $l_0 = R_\odot$, the domain in code units is simply $r_{min} = 20$ and $r_{max} = 220$. Thus, we define the numerical domain $D_d = \{(r_i, t^n)\}$ with $r_i = r_{min} + i \Delta r$, $i = 0, \ldots, N_r$, $t^n = n \Delta t$ and resolutions $\Delta r = (r_{max} - r_{min})/N_r$, $\Delta t = C \Delta r$. The base resolution uses $N_r = 1000$ and $C = 0.25$.

Initial Conditions. Traditionally, the properties of the solar wind are given for density ρ_{wind}, velocity v_{wind}^r, and temperature T_{wind}, and pressure has to be set in terms of density and temperature by $p = \rho T$. We define the initial conditions for a spherical Riemann problem, namely, a sphere of high density and pressure, whereas outside we assume an environment at rest with low density and pressure. We set the interface between the two states at the first intercell boundary of the domain D_d at $r_1 = r_{min} + \Delta r$ as follows:

$$\rho_0(r) = \begin{cases} \rho_{wind}, & \text{if } r \leq r_1 \\ 0.1 \rho_{wind}, & \text{if } r > r_1 \end{cases}$$

$$v_0^r(r) = \begin{cases} v_{wind}^r, & \text{if } r \leq r_1 \\ 0, & \text{if } r > r_1 \end{cases}$$

$$p_0(r) = \begin{cases} \rho_{wind} T_{wind}, & \text{if } r \leq r_1 \\ 0.1 \rho_{wind} T_{wind}, & \text{if } r > r_1 \end{cases} \quad (5.45)$$

where the factor 0.1 in the outside region is defined only to define a small density and pressure that will eventually be pushed away by the injected wind state variables. An important difference with the shock-tube tests showed earlier is that we do not simply release the fluid and let it evolve. In this case, we want to **permanently inject** the state variables of the wind through the inner boundary at r_{min}, in order to simulate the Sun permanently pumping plasma into the heliosphere.

Evolution. We solve the evolution of state variables in (5.41) using a slightly modified version of the code used for the shock-tube problem from the previous section, with the RK2 Heun time integrator and the same HLLE flux formula. The two main tweaks are the addition of the sources in the equations and the permanent injection of the wind variables. This is done by resetting the variables with the values in (5.45) on the first cell of the domain at all times.

Formation of the stationary wind. Following [8], as an example, we use the wind variables $\rho_{wind} = 2100$, $v_{wind}^r = 2.5 \times 10^5$ m/s, and $T_{wind} = 5 \times 10^5$ K. In

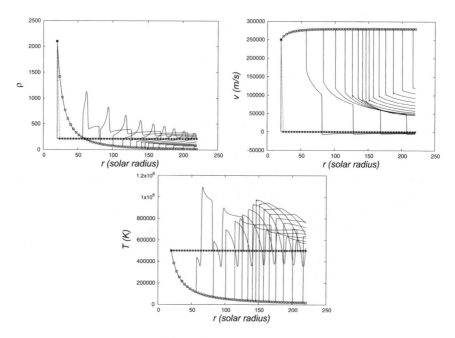

Fig. 5.12 Snapshots of density, velocity, and temperature every 50 time units, from initial time until the solar wind flow becomes stationary. The pulse of each variable propagates to the right and fills in the space with the solar wind fluid, between the spheres of radius r_{min} and r_{max}. The stars indicate the profile at initial time, and squares correspond to the final stationary profile

Fig. 5.12, we show snapshots of the state variables that illustrate how the initial shock propagates along the radial coordinate, until the flow stabilizes and each variable becomes stationary by $t \sim 775$.

Injection of a Coronal Mass Ejection. A coronal mass ejection (CME) is a burst of plasma ejected from the solar surface, due to the highly intense magnetic field activity and its topological changes that trigger the formation of small and major ejections of plasma. CMEs are the most important concept in space weather, because they transport plasma across the Solar System, with a certain front shock speed, and leave a trace of high temperature behind, and eventually interact with planetary atmospheres, producing magnetic storms. Prediction of the properties of the plasma carried by CMEs is of major importance, because it allows to predict effects on Earth and the satellite technology.

Once we have a code that produces a stationary wind, we want to present an oversimplified version of a CME, yet illustrative of how it could be simulated in a more general scenario. It is simple, because again it does not carry a magnetic field and because in spherical symmetry it has to be a spherically symmetric blast, very different from CMEs, which are localized bursts, like those we show later in Fig. 6.18.

5.3 Application: Spherically Symmetric Hydrodynamical Solar Wind

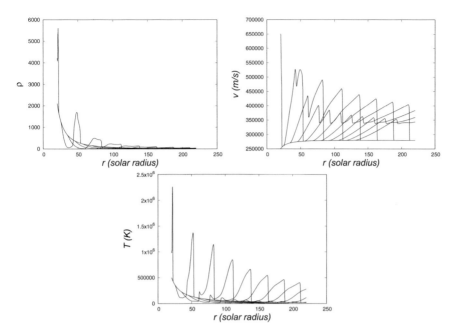

Fig. 5.13 Snapshots every 10 hours, starting from $t = 800$h, of density, velocity, and temperature during the propagation of the CME. The pulses of the three variables propagate to the right, and once the propagation has ceased, the stationary wind variables from Fig. 5.12 are recovered

We inject this spherical CME as a denser, faster, and hotter pulse than the wind, with number density of 4200, twice the density of the stationary wind at r_{min}, velocity $v_{CME} = 6.5 \times 10^5$ m/s, and temperature $T_{CME} = 10^6$ K, that will propagate across the domain and leave through the boundary at r_{max}, until the stationary wind is recovered. The state variables of the CME are injected during 3 hours, at r_{min} in the same way as wind variables, during the time window $t \in (800, 803)$, which is a domain after the stationary wind has formed (Fig. 5.13).

Diagnostics. Additional to the numerical solution of Euler equations, the useful information is the value of state variables at different space positions. For example, it is important to know the value of state variables at Earth, in terms of measurements made by probes flying around the Sun. In the spherically symmetric case, we cannot do much, but we can at least measure the state variables. Let us invent that the position of the Parker Solar Probe (PSP) is $r = 0.532801$ astronomical units and that Earth is at 0.985885 AU from the Sun at the moment the CME above is ejected (keep in mind that this is actually a spherical blast wave). Then one can interpolate the value of state variables at detecting positions, say at positions of the PSP and Earth with the solution of state variables at the cells near detecting positions. We show the state variables at these two detection positions in Fig. 5.14.

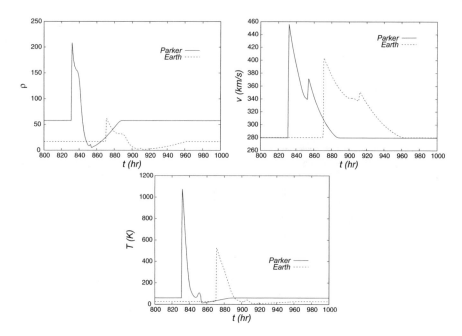

Fig. 5.14 Density, velocity, and temperature at the invented position of the PSP and Earth. The CME is launched at $t = 800$ and lasts 3 hours. It first reaches the position of the probe by $t \sim 830$, and about 40 hours later, it reaches Earth. After that, the stationary wind variables are again measured by the detectors

The Probe detects the high-density, high-velocity pulse, which leaves behind a hot region. About 40 hours later, the shock front is measured, with smaller density, velocity, and temperature at Earth. This simple exercise is an example of space weather; the properties of a CME are observed and measured near the Sun (with telescopes and probes); one should simulate a scenario, guess or find appropriate values of state variables ρ_{CME}, v_{CME}, and T_{CME} at r_{min} that reproduce those measurements (density, velocity, and temperature) and then solve Euler equations with such parameters in order to predict the state variables at Earth's position.

The code SolarWind.f90 can be used to reproduce all the results in this section. It includes the conversion of units and the tweaks needed to convert the shock-tube solver into the solar wind simulation, as well as the implementation of detectors. The code can be found as Electronic Supplementary Material described in Appendix B.

Exercise. (a) Using the code SolarWind.f90, extend the domain in such a way that it contains Pluto. (b) Find information of the solar wind density at

(continued)

various distances from the Sun at a given date, because the solar wind profile is nearly stationary but changes slightly with time. (c) Run the code for a number of combinations of temperature, density, and velocity at the inner boundary, and construct a catalog of stationary winds. (d) Find the one consistent with observations at various radii found in (b).

5.4 Application: 1+1 Relativistic Euler Equations

Relativistic Euler equations describe the evolution of a perfect fluid defined by the stress-energy tensor

$$T^{\mu\nu} = \rho_0 h u^\mu u^\nu + p\eta^{\mu\nu}, \tag{5.46}$$

where a volume element is characterized by the rest mass density ρ_0, specific enthalpy $h = 1 + e + p/\rho_0$, internal energy e, pressure p, and u^μ its four velocity, whereas $\eta^{\mu\nu} = \eta_{\mu\nu} = diag(-1, 1, 1, 1)$ is the Minkowski space-time metric.

Euler equations consist of the conservation of mass and the divergence-free constraint of the stress-energy tensor that, using Einstein summation notation, are, respectively,

$$\frac{\partial}{\partial x^\mu}(\rho_0 u^\mu) = 0, \tag{5.47}$$

$$\frac{\partial}{\partial x^\nu} T^{\mu\nu} = 0, \tag{5.48}$$

where $x^\mu = x^0, x^1, x^2, x^3$ or $x^\mu = x^0, x^i$ are the coordinates of the space-time, where $\mu = 0, 1, 2, 3$ label space-time coordinates and $i = 1, 2, 3$ label space coordinates only. Since x^0 is the coordinate time, we use indistinctly $x^0 = t$ in the calculations below. These equations can be cast in a flux balance set of laws that we construct based on the Valencia formulation [9, 10], for which we need to recall a few relations from special relativity. Considering the four-velocity of a fluid element has components $u^\mu = (u^0, u^i)$, the following relations hold:

$$u^0 = \frac{dt}{d\lambda} = W = \frac{1}{\sqrt{1 - v^i v_i}}, \quad v^i = \frac{u^i}{u^0} = \frac{dx^i}{dt} = \frac{u^i}{W},$$

where t is the coordinate time associated with a fluid element from a Eulerian reference frame, λ the proper time measured at its location, and v^i the components of its three-velocity and W its Lorentz factor.

Development of the conservation of mass (5.47) gives

$$\frac{\partial}{\partial x^\mu}(\rho_0 u^\mu) = \frac{\partial}{\partial x^0}(\rho_0 u^0) + \frac{\partial}{\partial x^i}(\rho_0 u^i) = \frac{\partial}{\partial x^0}(\rho_0 W) + \frac{\partial}{\partial x^i}(\rho_0 W v^i) = 0,$$

or equivalently

$$\frac{\partial \mathcal{D}}{\partial t} + \frac{\partial}{\partial x^i}(\mathcal{D} v^i) = 0, \tag{5.49}$$

where $\mathcal{D} = \rho_0 W$. Now, let us expand Eqs. (5.48), for which we need the components of the stress-energy tensor distinguished with time and spatial labels as follows:

$$T^{\mu\nu} = \rho_0 h u^\mu u^\nu + p \eta^{\mu\nu} \quad \Rightarrow$$
$$T^{00} = \rho_0 h (u^0)^2 + p \eta^{00} = \rho_0 h (u^0)^2 - p = \rho_0 h W^2 - p,$$
$$T^{0i} = T^{i0} = \rho_0 h u^0 u^i + p \eta^{i0} = \rho_0 h W^2 v^i,$$
$$T^{ij} = \rho_0 h u^i u^j + p \eta^{ij} = \rho_0 h v^i v^j W^2 + p \delta^{ij}. \tag{5.50}$$

With these expressions, we expand the summation in (5.48) for components of the stress-energy tensor with $\mu = i$, a spatial index:

$$\frac{\partial}{\partial x^\nu}(T^{i\nu}) = \frac{\partial}{\partial x^0}(T^{i0}) + \frac{\partial}{\partial x^j}(T^{ij})$$
$$= \frac{\partial}{\partial t}(\rho_0 h W^2 v^i) + \frac{\partial}{\partial x^j}(\rho_0 h v^i v^j W^2 + p \delta^{ij}) = 0,$$

or equivalently

$$\frac{\partial \mathcal{S}^i}{\partial t} + \frac{\partial}{\partial x^j}(\mathcal{S}^i v^j + p \delta^{ij}) = 0, \tag{5.51}$$

where $\mathcal{S}^i = \rho_0 h W^2 v^i$ is a new three-vector. For $\mu = 0$, the time coordinate, one has

$$\frac{\partial}{\partial x^\nu}(T^{0\nu}) = \frac{\partial}{\partial x^0}(T^{00}) + \frac{\partial}{\partial x^i}(T^{0i}) \quad \Rightarrow$$
$$\frac{\partial}{\partial t}(\rho_0 h W^2 - p) + \frac{\partial}{\partial x^i}(\rho_0 h W^2 v^i) = 0,$$

by adding (5.49)

5.4 Application: 1+1 Relativistic Euler Equations

$$\frac{\partial}{\partial t}(\rho_0 h W^2 - p - \mathcal{D}) + \frac{\partial}{\partial x^i}(\rho_0 h W^2 v^i - \mathcal{D}) = 0,$$

reduces to

$$\frac{\partial \tau}{\partial t} + \frac{\partial}{\partial x^i}(\mathcal{S}^i - \mathcal{D}v^i) = 0, \tag{5.52}$$

where $\tau = \rho_0 h W^2 - p - \mathcal{D}$ is another new variable. The set of Eqs. (5.49), (5.51), and (5.52) constitutes the relativistic Euler equations in flux balance law form:

$$\frac{\partial \mathcal{D}}{\partial t} + \frac{\partial}{\partial x^i}(\mathcal{D}v^i) = 0, \tag{5.53}$$

$$\frac{\partial \mathcal{S}^i}{\partial t} + \frac{\partial}{\partial x^j}(\mathcal{S}^i v^j + p\delta_{ij}) = 0, \tag{5.54}$$

$$\frac{\partial \tau}{\partial t} + \frac{\partial}{\partial x^i}(\mathcal{S}^i - \mathcal{D}v^i) = 0, \tag{5.55}$$

where $\mathcal{D}, \mathcal{S}^i, \text{and } \tau$ are the conservative variables. In these terms, an IVP for the 1+1 relativistic Euler equations is formulated as follows:

$$\begin{array}{ll}
\partial_t \mathcal{D} + \partial_{x^i}(\mathcal{D}v^i) = 0 & \\
\partial_t \mathcal{S}^i + \partial_{x^j}(\mathcal{S}^i v^j + p\delta_{ij}) = 0 & \\
\partial_t \tau + \partial_{x^i}(\mathcal{S}^i - \mathcal{D}v^i) = 0 & \mathcal{D} = \mathcal{D}(t, x^i),\ \mathcal{S}^i = \mathcal{S}^i(t, x^i), \\
 & \tau = \tau(t, x^i) \\
\mathcal{D},\ \mathcal{S}^i,\ \tau \text{ depend on } \rho_0,\ v^i,\ p,\ e, & \\
\quad \text{and } x^i = x, y, z,\ i = 1, 2, 3 & \\
D = [x_{min}, x_{max}] \times [y_{min}, y_{max}] & \\
\quad \times [z_{min}, z_{max}] \times [0, t_f] & \text{Domain} \\
\rho_0(0, x^i),\ v^i(0, x^i),\ p(0, x^i),\ e(0, x^i) & \text{Initial Conditions} \\
\mathcal{D}(t, \partial D) & \text{Boundary Conditions} \\
\mathcal{S}^i(t, \partial D) & \\
\tau(t, \partial D) & \\
\end{array}$$
(5.56)

The system of equations is a set of flux balance laws:

$$\frac{\partial \mathbf{u}}{\partial t} + \frac{\partial \mathbf{F}^i(\mathbf{u})}{\partial x^i} = 0, \tag{5.57}$$

where **u** is the state vector of conservative variables and **F** the vector of fluxes

$$\mathbf{u} = \begin{pmatrix} D \\ S^1 \\ S^2 \\ S^3 \\ \tau \end{pmatrix}, \quad \mathbf{F}^i = \begin{pmatrix} Dv^i \\ S^1 v^i + p\delta^{1i} \\ S^2 v^i + p\delta^{2i} \\ S^3 v^i + p\delta^{3i} \\ S^i - Dv^i \end{pmatrix}. \tag{5.58}$$

Notice that in the limit $v^i \ll 1$, the enthalpy $h \to 1$, and the conservative variables tend toward their nonrelativistic values:

$$\begin{aligned} D &\to \rho_0, \\ S^i &\to \rho_0 v^i, \\ \tau &\to \rho_0 E = \rho_0 e + \rho_0 v^2/2, \end{aligned} \tag{5.59}$$

and Eqs. (5.55) reduce to the classical Euler equations (5.17).

The solution of Eqs. (5.57) and (5.58) in three spatial dimensions is proposed as a project in Sect. 6.7.2. Meanwhile, in this chapter, we restrict to the 1+1 case. In the 1D case, the three-velocity has only one component that we choose to be v^x, so that the sum $v^i v_i = \eta_{ij} v^i v^j = (v^x)^2$. Consequently, there is only one component of S^i, specifically S^x. Also, the domain reduces to a two-dimensional set $D = [x_{min}, x_{max}] \times [0, t_f]$.

Now, like in the Newtonian case, the system is underdetermined and is closed using a caloric EoS $p = p(\rho, e)$, and like in the Newtonian case, one needs the dictionary between primitive and conservative variables that here we summarize. Conservative variables are defined by

$$\begin{aligned} D &= \rho_0 W = \frac{\rho_0}{\sqrt{1 - (v^x)^2}}, \\ S^x &= \rho_0 h W^2 v^x = \frac{\rho_0 h v^x}{1 - (v^x)^2}, \\ \tau &= \rho_0 h W^2 - p - \rho_0 W = \frac{\rho_0 h}{1 - (v^x)^2} - p - \frac{\rho_0}{\sqrt{1 - (v^x)^2}}. \end{aligned} \tag{5.60}$$

Now, the primitive variables in terms of the conservative variables read

$$\begin{aligned} v^x &= \frac{S^x}{\rho_0 h W^2} = \frac{S^x}{\tau + p + D}, \\ \rho_0 &= \frac{D}{W} = D\sqrt{1 - (v^x)^2}, \\ p &= p(\rho_0, e). \end{aligned} \tag{5.61}$$

5.4 Application: 1+1 Relativistic Euler Equations

We point out that pressure depends on density and internal energy. For the ideal gas EoS $p = \rho_0 e(\gamma - 1)$, developing the third of Eqs. (5.61), we have

$$p = \rho_0 e(\gamma - 1)$$
$$= (\gamma - 1)(\rho_0 h - \rho_0 - p) \quad \{h = 1 + e + p/\rho_0 \Rightarrow \rho_0 e = \rho_0 h - \rho_0 - p\}$$
$$= (\gamma - 1)\left(\frac{\tau + p + D}{W^2} - \rho_0 - p\right)$$
$$= \rho_0(\gamma - 1)\left[\frac{\tau + D(1 - W) + p(1 - W^2)}{DW}\right],$$
$$= (\gamma - 1)\left[\frac{\tau + D(1 - W) + p(1 - W^2)}{W^2}\right] \tag{5.62}$$

which means that pressure obeys a trascendental equation, *unlike* in the Newtonian case, where p could be written explicitly in terms of conservative variables.

In order to solve this equation, we consider p to be the root of $f(p) = p - \rho_0(\gamma - 1)e(D, \tau, W(v^x(p)))$, that is, W is a function $W(v^x)$, which in turn is a function of $v^x(p)$. According to the last of Eqs. (5.62),

$$W = \frac{1}{\sqrt{1 - (v^x)^2}} = \frac{1}{\sqrt{1 - \frac{S^{x2}}{(\tau + p + D)^2}}} = \frac{\tau + p + D}{\sqrt{(\tau + p + D)^2 - S^{x2}}}, \tag{5.63}$$

has to be substituted into (5.62) and then solve for p. The resulting transcendental equation $f(p) = 0$ is solved using the Newton-Raphson method, with $f(p)$ and its derivative $f'(p)$:

$$f(p) = p - (\gamma - 1)\left[\frac{\tau + D(1 - W) + p(1 - W^2)}{W^2}\right]$$
$$= -p + (\gamma - 1)\tau - (\gamma - 1)\frac{S^{x2}}{\tau + p + D}$$
$$- (\gamma - 1)D\left(-1 + \sqrt{1 - \frac{S^{x2}}{(\tau + p + D)^2}}\right),$$
$$f'(p) = -1 + (\gamma - 1)\frac{S^{x2}}{(\tau + p + D)^2} - (\gamma - 1)\frac{D}{\sqrt{1 - \frac{S^{x2}}{(\tau + p + D)^2}}}\frac{S^{x2}}{(\tau + p + D)^3}, \tag{5.64}$$

where the value of pressure at iteration k is $p_k = p_{k-1} - f(p)/f'(p)$, provided an initial guess for iteration zero, which can be the value of the pressure at previous time.

Like in the Newtonian case, one has to calculate the Jacobian matrix $d\mathbf{F}(\mathbf{u})/d\mathbf{u}$ of the system (5.57) and its eigenvalues. The eigenvalues of the Jacobian matrix are [10]

$$\lambda_1 = v^x, \quad \lambda_2 = \frac{v + c_s}{1 + vc_s}, \quad \lambda_3 = \frac{v - c_s}{1 - vc_s}, \quad (5.65)$$

where c_s is the speed of sound defined by

$$hc_s^2 = \chi + \left(\frac{p}{\rho_0^2}\right)\kappa, \quad \chi = \frac{\partial p}{\partial \rho_0}, \quad +\kappa = \frac{\partial p}{\partial e}.$$

In our case $p = \rho_0 e(\gamma - 1)$, thus

$$\chi = \frac{\partial p}{\partial \rho_0} = e(\gamma - 1) = \frac{p}{\rho_0},$$

$$\kappa = \frac{\partial p}{\partial e} = \rho_0(\gamma - 1),$$

and then, using that $h = 1 + e + p/\rho_0 = 1 + \frac{p}{\rho_0}\frac{\gamma}{\gamma-1}$ implies that the speed of sound is finally

$$c_s^2 = \frac{\chi + \kappa \frac{p}{\rho_0^2}}{h} = \frac{p(\gamma - 1)\gamma}{\rho_0(\gamma - 1) + p\gamma}. \quad (5.66)$$

Flux Formula. With all this information, it is possible to calculate the numerical fluxes at the intercell boundaries of the numerical domain, using like we did in the Newtonian case of Sect. 5.2, the HLLE flux formula (5.32):

$$\mathbf{F}^{HLLE} = \frac{\lambda_+ \mathbf{F}(\mathbf{u}^L) - \lambda_- \mathbf{F}(\mathbf{u}^R) + \lambda_+ \lambda_- (\mathbf{u}^R - \mathbf{u}_L)}{\lambda_+ - \lambda_-}, \quad (5.67)$$

where $\lambda_\pm$ are defined by

$$\lambda_+ = \max\left(0, \lambda_1^R, \lambda_2^R, \lambda_3^R, \lambda_1^L, \lambda_2^L, \lambda_3^L\right),$$

$$\lambda_- = \min\left(0, \lambda_1^R, \lambda_2^R, \lambda_3^R, \lambda_1^L, \lambda_2^L, \lambda_3^L\right). \quad (5.68)$$

and $\lambda_i^{L,R}$ are the eigenvalues (5.65) evaluated at the left and right cells from intercell boundaries. With these fluxes, it is straightforward to use the MoL and solve the IVP associated with the relativistic hydrodynamics equations.

Implementation. The implementation of the numerical solution is nearly the same as in the Newtonian case. Nevertheless, for the calculation of primitive variables in

5.4 Application: 1+1 Relativistic Euler Equations

Table 5.2 Initial conditions for two standard shock-tube test problems using $\gamma = 1.6$ which is appropriate for a relativistic fluid

Case	ρ_L	ρ_R	v_L^x	v_R^x	p_L	p_R
Mild shock	10.0	1.0	0.0	0.0	13.33	0.0
Strong shock	1.0	1.0	0.0	0.0	1000.0	0.01

the Newtonian case, one uses relations (5.38), whereas in the relativistic case it is necessary to solve the transcendental equation (5.62) for p and then use equations of (5.61) to calculate v^x and ρ_0 and with these two the internal energy e from the EoS.

Workhorse Case. Standard tests of relativistic hydrodynamics codes are again shock tubes, that is, the initial conditions define a Riemann problem.

Initial data are defined for the primitive variables ρ_0, p, v^x. With these variables, one calculates the Lorentz factor, internal energy, and specific enthalpy:

$$W = \frac{1}{\sqrt{1-(v^x)^2}},$$

$$e = \frac{p}{\rho_0(\gamma-1)},$$

$$h = 1 + e + \frac{p}{\rho_0},$$

and with these quantities, one constructs the conservative variables $(\mathcal{D}, \mathcal{S}^x, \tau)$ using (5.60). In Table 5.2, we indicate the initial conditions for two standard tests.

Mild Shock. In Fig. 5.15, we show a snapshot of the solution in the mild shock case using Godunov (5.35) and minmod (5.36) reconstructors. We also show the exact solution, constructed following the recipe in [7]. Notice that the exact solution captures exactly the discontinuities, whereas the numerical solutions introduce dissipation that smoothens out the discontinuities. It is also clear that using the minmod reconstructor, the solution is closer to the exact solution than using the Godunov reconstructor.

These numerical solutions were calculated using $N_x = 1000$ or equivalently $\Delta x = 0.001$, with $\Delta t = 0.25 \Delta x$.

Strong Shock. In this case, in order to approach the exact solution, one needs to use a higher resolution. For this, we set `resolution_label = 4`, that is, $N_x = 1000 \times 2^3$ or equivalently $\Delta x = 0.000125$, with $\Delta t = 0.25 \Delta x$.

The results at $t = 0.4$ are shown in Fig. 5.16. A particular property of a strong shock is that a thin shell density forms while the fluid of the high-pressure region brooms the fluid of the low-pressure zone. The performance of minmod versus Godunov reconstructors can be noticed precisely at this shell. The density using Godunov fails to reach the value of the density by more than 20%.

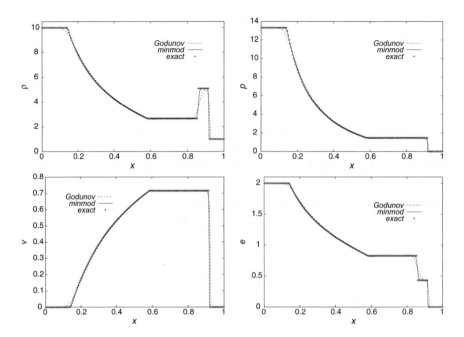

Fig. 5.15 Mild shock test at $t = 0.5$ using Godunov and minmod reconstructors. The exact solution indicates the correct location of the discontinuities

The implementation of this solution is written in the code `Relativistic Euler.f90`, which can be in the Electronic Supplementary Material.

The transcendental equation for p is solved using the Newton-Raphson method at each cell of the numerical domain. For this, we use expressions (5.64) and iterate the pressure:

$$p_k = p_{k-1} - \frac{f(p_{k-1})}{f'(p_{k-1})},$$

for $k = 1, 2, \ldots$, with the initial guess $p_{k=0} = p_i$, `press_guess`. The process is iterated until the value of the root converges within a tolerance. In the code, we use `pressK` for p_k, `pressK_p` for p_{k-1}, `fun_of_p` for $f(p_{k-1})$, and `funprime_of_p` for $f'(p_{k-1})$.

> **Exercise.** Using the recipe in [7], construct the exact solution to the relativistic shock-tube problem. Incorporate it to the program and calculate the error of the density.

5.5 Application: Spherical Accretion of a Fluid onto a Schwarzschild Black Hole

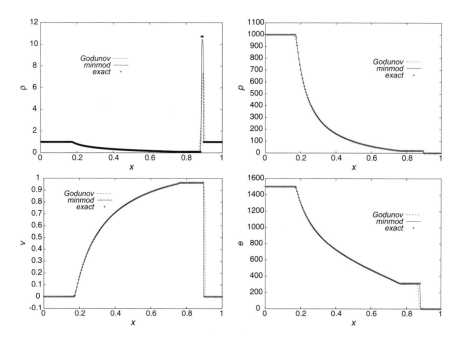

Fig. 5.16 Strong shock test at $t = 0.4$ using Godunov and minmod reconstructors. A property of a strong shock is that it forms a sharp thin shell of fluid

5.5 Application: Spherical Accretion of a Fluid onto a Schwarzschild Black Hole

Let us now bring relativistic hydrodynamics to the next level. The plan is to solve relativistic Euler equations on top of the background space-time of a Schwarzschild black hole for a radial flow. Like we did in Sect. 4.7, we use Eddington-Finkelstein coordinates to describe the Schwarzschild black hole.

Consider the matter to be a perfect fluid, whose stress-energy tensor for a general space-time with metric $g_{\mu\nu}$ is

$$T^{\mu\nu} = \rho_0 h u^\mu u^\nu + p g^{\mu\nu}, \tag{5.69}$$

where each volume element has rest mass density ρ_0, specific enthalpy $h = 1 + e + p/\rho_0$, internal energy e, pressure p, and four-velocity u^μ. Relativistic hydrodynamics equations consist of the conservation of mass and the divergence-free condition of the stress-energy tensor. For a general space-time, these laws read

$$\nabla_\mu(\rho_0 u^\mu) = 0, \tag{5.70}$$

$$\nabla_\mu T^{\mu\nu} = 0, \tag{5.71}$$

where ∇_μ is the covariant derivative consistent with the metric $g_{\mu\nu}$, where $x^\mu = x^0, x^1, x^2, x^3$ or $x^\mu = x^0, x^i$ are the coordinates of the space-time, $\mu = 0, 1, 2, 3$ labels space-time coordinates, and $i = 1, 2, 3$ labels space coordinates only. Since x^0 is the coordinate time, we use indistinctly $x^0 = t$ in the developments below.

These equations can be cast in a flux balance set of laws that we construct based on the Valencia formulation [10, 11] and the 3+1 decomposition of space-time (see, for example, [12, 13]). First we write the line element of the space-time metric in the 3 + 1 decomposition notation:

$$ds^2 = (-\alpha^2 + \beta^i \beta_i)dt^2 + 2\beta_i dt dx^i + \gamma_{ij} dx^i dx^j, \tag{5.72}$$

where α is the lapse function, β^i is the shift vector, and γ_{ij} is the three-metric of the spatial hypersurfaces of the space-time with i, j labeling spatial coordinates and $\beta_i = \gamma_{ij}\beta^j$. With this metric, one can construct expressions for the four-velocity of a fluid element $u^\mu = (u^0, u^i)$, in terms of the three-velocity

$$u^0 = \frac{dt}{d\lambda} = \frac{W}{\alpha} = \frac{1}{\alpha\sqrt{1 - v^i v_i}}, \qquad v^i = \frac{u^i}{u^0} = \frac{u^i}{W} + \frac{\beta^i}{\alpha},$$

where t is the coordinate time associated with a fluid element from a Eulerian reference frame, λ is the proper time measured at its location, W its Lorentz factor, and v^i are the components of its three-velocity according to the Eulerian observer itself. In the expressions above, $v^i v_i = \gamma_{ij} v^i v^j$.

With all these definitions and following [10, 11], one can write Eqs. (5.70)–(5.71) as a flux balance set of equations

$$\frac{\partial \mathbf{u}}{\partial t} + \frac{\partial \mathbf{F}^i(\mathbf{u})}{\partial x^i} = \mathbf{S}(\mathbf{u}), \tag{5.73}$$

with the following state vector $\mathbf{u}$, vector of fluxes $\mathbf{F}$, and vector of sources $\mathbf{S}$:

$$\mathbf{u} = \begin{pmatrix} D \\ S_j \\ \tau \end{pmatrix}, \quad \mathbf{F}^i(\mathbf{u}) = \begin{pmatrix} \alpha\left(v^i - \frac{\beta^i}{\alpha}\right) D \\ \alpha\left(v^i - \frac{\beta^i}{\alpha}\right) S_j + \alpha\sqrt{\gamma} p \delta^i_j \\ \alpha\left(v^i - \frac{\beta^i}{\alpha}\right) \tau + \alpha\sqrt{\gamma} p v^i \end{pmatrix},$$

$$\mathbf{S}(\mathbf{u}) = \begin{pmatrix} 0 \\ \alpha\sqrt{\gamma} T^{\mu\nu} g_{\nu\sigma} \Gamma^\sigma{}_{\mu j} \\ \alpha\sqrt{\gamma} (T^{\mu 0} \partial_\mu \alpha - \alpha T^{\mu\nu} \Gamma^0{}_{\mu\nu}) \end{pmatrix}, \tag{5.74}$$

for $i, j = 1, 2, 3$. Conservative variables are defined in terms of the primitive variables and geometric quantities by

5.5 Application: Spherical Accretion of a Fluid onto a Schwarzschild Black Hole

$$\mathcal{D} = \sqrt{\gamma}\rho_0 W,$$
$$\mathcal{S}_i = \sqrt{\gamma}\rho_0 h W^2 v_i,$$
$$\tau = \sqrt{\gamma}(\rho_0 h W^2 - p - \rho_0 W). \quad (5.75)$$

This is an underdetermined system of five equations for six primitive variables ρ, v^i, e, and p. In order to close it, we assume the fluid obeys an ideal gas equation of state

$$p = \rho_0 e(\Gamma - 1), \quad (5.76)$$

with $\Gamma = c_p/c_v$ the adiabatic index.

We now restrict to the case of a *spherically symmetric space-time* written in spherical coordinates (r, θ, ϕ, t). The most general 3+1 line element under these conditions is

$$ds^2 = (-\alpha^2 + \beta_r\beta^r)dt^2 + 2\beta_r dt dr + \gamma_{rr}dr^2 + \gamma_{\theta\theta}(d\theta^2 + \sin^2\theta d\phi^2), \quad (5.77)$$

where all the functions depend on (r, t).

From this metric, one has that if $\gamma = \det(\gamma_{ij})$, then $\sqrt{\gamma} = \sqrt{\gamma_{rr}\gamma_{\theta\theta}}\sin\theta$. Assuming there is only radial motion, then $v^i = (v^r, 0, 0)$, which implies $v_i v^i = \gamma_{rr}v^r v^r$. Consequently, the components of $\mathcal{S}_i$ are $\mathcal{S}_r = \sqrt{\gamma}\rho_0 h W^2 \gamma_{rr}v^r$, $\mathcal{S}_\theta = \sqrt{\gamma}\rho_0 h W^2 \gamma_{\theta\theta}v^\theta = 0$, and $\mathcal{S}_\phi = \sqrt{\gamma}\rho_0 h W^2 \gamma_{\phi\phi}v^\phi = 0$, the two later are zero due to the condition that $v^\theta = v^\phi = 0$.

Therefore, the system of Eqs. (5.73)–(5.74) becomes

$$\frac{\partial \mathbf{u}}{\partial t} + \frac{\partial F^r(\mathbf{u})}{\partial r} + \frac{\partial F^\theta(\mathbf{u})}{\partial \theta} + \frac{\partial F^\phi(\mathbf{u})}{\partial \phi} = \mathbf{S}(\mathbf{u}), \quad (5.78)$$

with the following state vector, flux vectors, and sources:

$$\mathbf{u} = \begin{pmatrix} \mathcal{D} \\ \mathcal{S}_r \\ \mathcal{S}_\theta \\ \mathcal{S}_\phi \\ \tau \end{pmatrix} = \begin{pmatrix} \mathcal{D} \\ \mathcal{S}_r \\ 0 \\ 0 \\ \tau \end{pmatrix} = \begin{pmatrix} \sqrt{\gamma}\rho_0 W \\ \sqrt{\gamma}\rho_0 h W^2 v_r \\ 0 \\ 0 \\ \sqrt{\gamma}(\rho_0 h W^2 - p - \rho_0 W) \end{pmatrix},$$

$$F^r(\mathbf{u}) = \begin{pmatrix} \alpha\left(v^r - \frac{\beta^r}{\alpha}\right)\mathcal{D} \\ \alpha\left(v^r - \frac{\beta^r}{\alpha}\right)\mathcal{S}_r + \alpha\sqrt{\gamma}p\delta^r_r \\ \alpha\left(v^r - \frac{\beta^r}{\alpha}\right)\mathcal{S}_\theta + \alpha\sqrt{\gamma}p\delta^r_\theta \\ \alpha\left(v^r - \frac{\beta^r}{\alpha}\right)\mathcal{S}_\phi + \alpha\sqrt{\gamma}p\delta^i_\phi \\ \alpha\left(v^r - \frac{\beta^r}{\alpha}\right)\tau + \alpha\sqrt{\gamma}pv^r \end{pmatrix} = \begin{pmatrix} \alpha\left(v^r - \frac{\beta^r}{\alpha}\right)\mathcal{D} \\ \alpha\left(v^r - \frac{\beta^r}{\alpha}\right)\mathcal{S}_r + \alpha\sqrt{\gamma}p \\ 0 \\ 0 \\ \alpha\left(v^r - \frac{\beta^r}{\alpha}\right)\tau + \alpha\sqrt{\gamma}pv^r \end{pmatrix},$$

$$F^\theta(\mathbf{u}) = \begin{pmatrix} \alpha\left(v^\theta - \frac{\beta^\theta}{\alpha}\right)\mathcal{D} \\ \alpha\left(v^\theta - \frac{\beta^\theta}{\alpha}\right)\mathcal{S}_r + \alpha\sqrt{\gamma}p\delta^\theta_r \\ \alpha\left(v^\theta - \frac{\beta^\theta}{\alpha}\right)\mathcal{S}_\theta + \alpha\sqrt{\gamma}p\delta^\theta_\theta \\ \alpha\left(v^\theta - \frac{\beta^\theta}{\alpha}\right)\mathcal{S}_\phi + \alpha\sqrt{\gamma}p\delta^\theta_\phi \\ \alpha\left(v^\theta - \frac{\beta^\theta}{\alpha}\right)\tau + \alpha\sqrt{\gamma}pv^\theta \end{pmatrix} = \begin{pmatrix} 0 \\ 0 \\ \alpha\sqrt{\gamma}p \\ 0 \\ 0 \end{pmatrix},$$

$$F^\phi(\mathbf{u}) = \begin{pmatrix} \alpha\left(v^\phi - \frac{\beta^\phi}{\alpha}\right)\mathcal{D} \\ \alpha\left(v^\phi - \frac{\beta^\phi}{\alpha}\right)\mathcal{S}_r + \alpha\sqrt{\gamma}p\delta^\phi_r \\ \alpha\left(v^\phi - \frac{\beta^\phi}{\alpha}\right)\mathcal{S}_\theta + \alpha\sqrt{\gamma}p\delta^\phi_\theta \\ \alpha\left(v^\phi - \frac{\beta^\phi}{\alpha}\right)\mathcal{S}_\phi + \alpha\sqrt{\gamma}p\delta^\phi_\phi \\ \alpha\left(v^\phi - \frac{\beta^\phi}{\alpha}\right)\tau + \alpha\sqrt{\gamma}pv^\phi \end{pmatrix} = \begin{pmatrix} 0 \\ 0 \\ 0 \\ \alpha\sqrt{\gamma}p \\ 0 \end{pmatrix},$$

$$\mathbf{S}(\mathbf{u}) = \begin{pmatrix} 0 \\ \alpha\sqrt{\gamma}T^{\mu\nu}g_{\nu\sigma}\Gamma^\sigma{}_{\mu r} \\ \alpha\sqrt{\gamma}T^{\mu\nu}g_{\nu\sigma}\Gamma^\sigma{}_{\mu\theta} \\ \alpha\sqrt{\gamma}T^{\mu\nu}g_{\nu\sigma}\Gamma^\sigma{}_{\mu\phi} \\ \alpha\sqrt{\gamma}(T^{\mu 0}\partial_\mu\alpha - \alpha T^{\mu\nu}\Gamma^0{}_{\mu\nu}) \end{pmatrix}. \tag{5.79}$$

Since F^θ and F^ϕ have a nonzero component each, it is worth discussing the implication in Eq. (5.78). Let us first see F^ϕ, notice that $\alpha\sqrt{\gamma}p = \alpha(t,r)\sqrt{\gamma_{rr}(t,r)}\gamma_{\theta\theta}(t,r)\sin\theta\ p(t,r)$, therefore $\frac{\partial F^\phi}{\partial \phi} = 0$, and thus the fourth term in (5.78) vanishes.

The third term $\frac{\partial F^\theta}{\partial \theta}$ in Eq. (5.78) is not as simple, and for its discussion, we have to first explicitly calculate the sources $\mathbf{S}$. Let us then calculate the nonzero stress-energy tensor components that will contribute to sources:

$$T^{00} = \frac{1}{\alpha^2}(\rho_0 h W^2 - p),$$

$$T^{r0} = \frac{\rho_0 h W^2}{\alpha}\left(v^r - \frac{\beta^r}{\alpha}\right) + \frac{\beta^r}{\alpha^2}p,$$

$$T^{rr} = \rho_0 h W^2 \left(v^r - \frac{\beta^r}{\alpha}\right)^2 + \left(\gamma^{rr} - \frac{\beta^r\beta^r}{\alpha^2}\right)p,$$

$$T^{\theta\theta} = \frac{p}{\gamma_{\theta\theta}},$$

$$T^{\phi\phi} = \frac{p}{\gamma_{\phi\phi}}. \tag{5.80}$$

5.5 Application: Spherical Accretion of a Fluid onto a Schwarzschild Black Hole

The other components are zero because $u^\theta = u^\phi = 0$ or $g^{0\theta} = g^{0\phi} = g^{r\theta} = g^{r\phi} = g^{\theta\phi} = 0$. Also in the sources, there are the Christoffel symbols of the space-time metric. The nonzero ones for the metric (5.77) are the following:

$$\Gamma^0_{\ 00} = \frac{(\dot{\gamma}_{rr} - 2\gamma_{rr}\beta^{r\prime} - \beta^r \gamma'_{rr})(\beta^r)^2 + 2\alpha(\dot{\alpha} + \alpha'\beta^r)}{2\alpha^2},$$

$$\Gamma^0_{\ 0r} = \frac{2\alpha\alpha' + \beta^r(\dot{\gamma}_{rr} - 2\gamma_{rr}\beta^{r\prime} - \beta^r \gamma'_{rr})}{2\alpha^2},$$

$$\Gamma^0_{\ rr} = \frac{\dot{\gamma}_{rr} - 2\gamma_{rr}\beta^{r\prime} - \beta^r \gamma'_{rr}}{2\alpha^2}, \quad \Gamma^0_{\ \theta\theta} = \frac{\dot{\gamma}_{\theta\theta} - \beta^r \gamma'_{\theta\theta}}{2\alpha^2}, \quad \Gamma^0_{\ \phi\phi} = \sin^2\theta \, \Gamma^0_{\ \theta\theta},$$

$$\Gamma^r_{\ 00} = \frac{1}{2}\left[\frac{\beta^r(\beta^r(\dot{\gamma}_{rr}\beta^r + 2\gamma_{rr}\dot{\beta}^r + 2\dot{\gamma}_{rr}\beta^r) - 2\alpha\dot{\alpha})}{\alpha^2}\right.$$

$$\left. + \left(\frac{1}{\gamma_{rr}} - \frac{(\beta^r)^2}{\alpha^2}\right)\left(-\gamma'_{rr}(\beta^r)^2 + 2(\dot{\gamma}_{rr} - \gamma_{rr}\beta^{r\prime})\beta^r + 2\alpha\alpha' + 2\gamma_{rr}\dot{\beta}^r\right)\right],$$

$$\Gamma^r_{\ 0r} = \frac{1}{2}\left(\frac{(-\dot{\gamma}_{rr} + 2\gamma_{rr}\beta^{r\prime} + \beta^r \gamma'_{rr})(\beta^r)^2}{\alpha^2} - \frac{2\alpha'\beta^r}{\alpha} + \frac{\dot{\gamma}_{rr}}{\gamma_{rr}}\right),$$

$$\Gamma^r_{\ rr} = \frac{1}{2}\left(\frac{\gamma'_{rr}}{\gamma_{rr}} + \frac{\beta^r(-\dot{\gamma}_{rr} + 2\gamma_{rr}\beta^{r\prime} + \beta^r \gamma'_{rr})}{\alpha^2}\right),$$

$$\Gamma^r_{\ \theta\theta} = \frac{1}{2}\left(\frac{\beta^r(\beta^r \gamma'_{\theta\theta} - \dot{\gamma}_{\theta\theta})}{\alpha^2} - \frac{\gamma'_{\theta\theta}}{\gamma_{rr}}\right),$$

$$\Gamma^r_{\ \phi\phi} = \sin^2\theta \, \Gamma^r_{\ \theta\theta}, \quad \Gamma^\theta_{\ 0\theta} = \frac{\dot{\gamma}_{\theta\theta}}{2\gamma_{\theta\theta}}, \quad \Gamma^\theta_{\ r\theta} = \frac{\gamma'_{\theta\theta}}{2\gamma_{\theta\theta}},$$

$$\Gamma^\theta_{\ \phi\phi} = -\cos\theta\sin\theta, \quad \Gamma^\phi_{\ 0\phi} = \frac{\dot{\gamma}_{\theta\theta}}{2\gamma_{\theta\theta}}, \quad \Gamma^\phi_{\ r\phi} = \frac{\gamma'_{\theta\theta}}{2\gamma_{\theta\theta}}, \quad \Gamma^\phi_{\ \theta\phi} = \cot\theta, \quad (5.81)$$

where, for the sake of compactness, we used prime and dot to indicate derivative with respect to r and t.

With the nonzero stress-energy tensor components (5.80) and the nonzero Christoffel symbols (5.81), the five entries of the source vector in (5.79) are

$$0 = 0$$

$$\alpha\sqrt{\gamma}T^{\mu\nu}g_{\nu\sigma}\Gamma^\sigma_{\ \mu r} = \alpha\sqrt{\gamma}(\, T^{00}[(-\alpha^2 + \gamma_{rr}(\beta^r)^2)\Gamma^0_{\ 0r} + \gamma_{rr}\beta^r \Gamma^r_{\ 0r}]$$

$$+ T^{r0}[(-\alpha^2 + \gamma_{rr}(\beta^r)^2)\Gamma^0_{\ rr} + \gamma_{rr}\Gamma^r_{\ 0r}$$

$$+ \gamma_{rr}\beta^r(\Gamma^r_{\ rr} + \Gamma^0_{\ 0r})]$$

$$+ T^{rr}[\gamma_{rr}\Gamma^r_{\ rr} + \gamma_{rr}\beta^r \Gamma^0_{\ rr}]$$

$$+T^{\theta\theta}\gamma_{\theta\theta}\Gamma^{\theta}{}_{\theta r}+T^{\phi\phi}\gamma_{\phi\phi}\Gamma^{\phi}{}_{\phi r}),$$

$$\alpha\sqrt{\gamma}T^{\mu\nu}g_{\nu\sigma}\Gamma^{\sigma}{}_{\mu\theta}=\alpha\sqrt{\gamma}\,p\cot\theta,$$

$$\alpha\sqrt{\gamma}T^{\mu\nu}g_{\nu\sigma}\Gamma^{\sigma}{}_{\mu\phi}=0,$$

$$\alpha\sqrt{\gamma}(T^{\mu 0}\partial_{\mu}\alpha-\alpha T^{\mu\nu}\Gamma^{0}{}_{\mu\nu})=\alpha\sqrt{\gamma}(\,T^{00}(\partial_{0}\alpha-\alpha\Gamma^{0}{}_{00})+T^{r0}(\partial_{r}\alpha-2\alpha\Gamma^{0}{}_{r0})$$
$$-\alpha(T^{rr}\Gamma^{0}{}_{rr}+T^{\theta\theta}\Gamma^{0}{}_{\theta\theta}+T^{\phi\phi}\Gamma^{0}{}_{\phi\phi})\,). \quad (5.82)$$

Now, we note the importance of F^{θ}, which accounts to the third term of the third equation in (5.78), because

$$\frac{\partial F^{\theta}}{\partial\theta}=\partial_{\theta}[\alpha(t,r)\sqrt{\gamma_{rr}(t,r)}\gamma_{\theta\theta}(t,r)\sin\theta\,p(t,r)]$$
$$=\alpha(t,r)\sqrt{\gamma_{rr}(t,r)}\gamma_{\theta\theta}(t,r)\cos\theta\,p(t,r).$$

On the other hand, the third entry of the sources (5.79), namely, the third of Eqs. (5.82), is precisely $\alpha\sqrt{\gamma_{rr}}\gamma_{\theta\theta}\sin\theta\cot\theta$, and therefore, these terms in the third of Eqs. (5.78) cancel out. Then **only three out of the five Eqs. (5.78)–(5.79), first, second, and fifth, are nontrivial.**

Therefore, one can define the following initial value problem for the evolution of the spherically symmetric flow of a perfect fluid on top of a spherically symmetric space-time written in the form of (5.78), which reduces to three evolution equations only:

$$\boxed{\begin{aligned}
&\partial_t \mathcal{D} + \partial_r \left(\alpha \left(v^r - \frac{\beta^r}{\alpha} \right) \mathcal{D} \right) = 0 \\
&\partial_t \mathcal{S}_r + \partial_r \left(\alpha \left(v^r - \frac{\beta^r}{\alpha} \right) \mathcal{S}_r + \alpha\sqrt{\gamma}\,p \right) = \alpha\sqrt{\gamma}T^{\mu\nu}g_{\nu\sigma}\Gamma^{\sigma}{}_{\mu r} \\
&\partial_t \tau + \partial_r \left(\alpha \left(v^r - \frac{\beta^r}{\alpha} \right) \mathcal{S}_r + \alpha\sqrt{\gamma}\,p \right) = \alpha\sqrt{\gamma}(T^{\mu 0}\partial_{\mu}\alpha - \alpha T^{\mu\nu}\Gamma^{0}{}_{\mu\nu}) \\
&\mathcal{D} = \mathcal{D}(r,r),\ \mathcal{S}_r = \mathcal{S}_r(r,t),\ \tau = \tau(r,t) \\[4pt]
&\text{where the variables } \mathcal{D},\ \mathcal{S}_r,\ \tau \text{ depend on } \rho_0,\ v^r,\ p,\ e,\ \text{through Eqs. (5.75)} \\[4pt]
&\mathcal{D} = [r_{min}, r_{max}] \times [0, t_f] \qquad\qquad \text{Domain} \\
&\rho_0(0,r),\ v^r(0,r),\ p(0,r),\ e(0,r) \qquad \text{Initial Conditions} \\
&\mathcal{D}(t, \partial\mathcal{D}),\ \mathcal{J}_r(t, \partial\mathcal{D}),\ \tau(t, \partial\mathcal{D}) \qquad \text{Boundary Conditions}
\end{aligned}}$$

$$(5.83)$$

5.5 Application: Spherical Accretion of a Fluid onto a Schwarzschild Black Hole

Characteristic Structure of Equations (5.83.) In order to use the finite volume approach with the HLLE flux formula as in previous sections, one needs the characteristic structure of the system of equations, at least the eigenvalues of the Jacobian matrix $\frac{d\mathbf{F}}{d\mathbf{u}}$ in Eqs. (5.78)–(5.79). We do not calculate them here; instead, we take the eigenvalues from equations (21) and (22) in [11], which are

$$\lambda_1 = \alpha v^r - \beta^r,$$

$$\lambda_2 = \frac{\alpha}{1 - v_r v^r c_s^2}$$
$$\times \left[v^r (1 - c_s^2) + c_s \sqrt{(1 - v_r v^r)[\gamma^{rr}(1 - v_r v^r c_s^2) - v^r v^r (1 - c_s^2)]} \right] - \beta^r,$$

$$\lambda_5 = \frac{\alpha}{1 - v_r v^r c_s^2}$$
$$\times \left[v^r (1 - c_s^2) - c_s \sqrt{(1 - v_r v^r)[\gamma^{rr}(1 - v_r v^r c_s^2) - v^r v^r (1 - c_s^2)]} \right] - \beta^r,$$

(5.84)

with c_s the speed of sound as in (5.66):

$$c_s^2 = \frac{\chi + \kappa \frac{p}{\rho_0^2}}{h} = \frac{p(\Gamma - 1)\Gamma}{\rho_0(\Gamma - 1) + p\Gamma}. \qquad (5.85)$$

The dictionary between primitive and conservative variables is in turn as follows: Conservative variables are defined by

$$D = \sqrt{\gamma}\rho_0 W = \frac{\rho_0}{\sqrt{1 - \gamma_{rr}(v^r)^2}},$$

$$S_r = \sqrt{\gamma}\rho_0 h W^2 v_r = \frac{\sqrt{\gamma}\rho_0 h \gamma_{rr} v^r}{1 - \gamma_{rr}(v^r)^2},$$

$$\tau = \sqrt{\gamma}(\rho_0 h W^2 - p - \rho_0 W) = \frac{\sqrt{\gamma}\rho_0 h}{1 - \gamma_{rr}(v^r)^2} - p - \frac{\rho_0}{\sqrt{1 - \gamma_{rr}(v^r)^2}}. \qquad (5.86)$$

The primitive variables in terms of the conservative variables read

$$v^r = \frac{S_r}{\sqrt{\gamma}\rho_0 h W^2 \gamma_{rr}} = \frac{S_r}{(\tau + \sqrt{\gamma} p + D)\gamma_{rr}},$$

$$\rho_0 = \frac{D}{W\sqrt{\gamma}} = \frac{D}{\sqrt{\gamma}}\sqrt{1 - \gamma_{rr}(v^r)^2},$$

$$p = p(\rho_0, e), \qquad (5.87)$$

where the latter is a caloric EoS, in which the pressure depends on density and internal energy. Developing the third of Eqs. (5.87), we have

$$p = \rho_0 e(\Gamma - 1)$$
$$= (\Gamma - 1)(\rho_0 h - \rho_0 - p) \quad \{\text{since } h = 1 + e + p/\rho_0 \Rightarrow \rho_0 e = \rho_0 h - \rho_0 - p\}$$
$$= (\Gamma - 1)\left(\frac{1}{W^2}\left(\frac{\tau}{\sqrt{\gamma}} + p + \frac{D}{\sqrt{\gamma}}\right) - \rho_0 - p\right) \quad \{\text{from (5.86)}\}$$
$$= \rho_0(\Gamma - 1)\left(\frac{\tau + p\sqrt{\gamma} + D - \rho_0\sqrt{\gamma}W^2 - p\sqrt{\gamma}W^2}{\rho_0\sqrt{\gamma}W^2}\right)$$
$$= \rho_0(\Gamma - 1)\left[\frac{\tau + D(1-W) + p\sqrt{\gamma}(1-W^2)}{DW}\right],$$
$$= (\Gamma - 1)\left[\frac{\tau + D(1-W) + p\sqrt{\gamma}(1-W^2)}{W^2}\right], \quad (5.88)$$

which is a trascendental equation for p and slightly more general than (5.62) due to geometric factors.

In order to make this equation more explicit, we consider p to be the root of the function $f(p) = p - \rho_0(\Gamma - 1)e$. According to the first of Eqs. (5.87),

$$W = \frac{1}{\sqrt{1 - \gamma_{rr}(v^r)^2}} = \frac{1}{\sqrt{1 - \frac{S_r^2}{\gamma_{rr}(\tau + \sqrt{\gamma}p + D)^2}}} = \frac{\sqrt{\gamma_{rr}}(\tau + \sqrt{\gamma}p + D)}{\sqrt{\gamma_{rr}(\tau + \sqrt{\gamma}p + D)^2 - S_r^2}},$$
$$(5.89)$$

which substituted into (5.88) defines a transcendental equation for p that we solve using the Newton-Raphson method, which locates the zeroes of the function $f(p)$:

$$f(p) = p - (\Gamma - 1)\left[\frac{\tau + D(1-W) + p\sqrt{\gamma}(1-W^2)}{W^2}\right],$$
$$= p - \frac{(\Gamma - 1)}{\sqrt{\gamma}}\left(1 - \frac{\gamma^{rr}S_r^2}{(D + p\sqrt{\gamma} + \tau)^2}\right).$$
$$\cdot \left(\tau + D\left(1 - \frac{1}{\sqrt{1 - \frac{\gamma^{rr}S_r^2}{(D+p\sqrt{\gamma}+\tau)^2}}}\right) + p\sqrt{\gamma}\left(1 + \frac{1}{(-1 + \frac{\gamma^{rr}S_r^2}{(D+p\sqrt{\gamma}+\tau)^2})}\right)\right)$$
$$(5.90)$$

5.5 Application: Spherical Accretion of a Fluid onto a Schwarzschild Black Hole

$$f'(p) = 1 + \frac{\left[\begin{array}{c}(\Gamma - 1)\gamma^{rr} S_r^2 (\gamma^{rr} S_r^2 + (\mathcal{D} + p\sqrt{\gamma} + \tau)) \\ \left(-\sqrt{\gamma} p - \tau + \mathcal{D}\left(-1 + \sqrt{1 - \frac{\gamma^{rr} S_r^2}{(\mathcal{D}+p\sqrt{\gamma}+\tau)^2}}\right)\right)\end{array}\right]}{(\mathcal{D} + p\sqrt{\gamma} + \tau)^2 (-\gamma^{rr} S_r^2 + (\mathcal{D} + p\sqrt{\gamma} + \tau)^2)} \tag{5.91}$$

The values of p are iterated with the relation $p_k = p_{k-1} - \frac{f(p)}{f'(p)}$, until a tolerance is achieved, for each cell in the domain. The initial guess we use to start the loop is $p_{k=0} = (p_i + p_{i+1})/2$, which is an average of the pressure at two neighboring cells at the previous time step.

Now, these results are valid for an **arbitrary** spherically symmetric space-time. We restrict now to the case of a *Schwarzschild black hole* described with Eddington-Finkelstein coordinates, whose line element is

$$ds^2 = -\left(1 - \frac{2M}{r}\right)dt^2 + \frac{4M}{r}dtdr + \left(1 + \frac{2M}{r}\right)dr^2 + r^2(d\theta^2 + \sin^2\theta d\phi^2), \tag{5.92}$$

where M is the mass of the black hole and (r, θ, ϕ, t) the coordinates used to describe the points of space-time, like we did for the wave equation in Sect. 4.7. Identification of (5.77) and (5.92) leads to the 3+1 gauge and metric functions:

$$\alpha = \frac{1}{\sqrt{1 + \frac{2M}{r}}},$$

$$\beta^i = (\beta^r, 0, 0) = \left(\frac{2M}{r}\frac{1}{1+\frac{2M}{r}}, 0, 0\right),$$

$$\gamma_{ij} = \mathrm{diag}(\gamma_{rr}, \gamma_{\theta\theta}, \sin^2\theta\gamma_{\theta\theta}) = \mathrm{diag}\left(1 + \frac{2M}{r}, r^2, r^2\sin^2\theta\right),$$

$$\beta_r = \gamma_{rr}\beta^r = \frac{2M}{r},$$

$$\sqrt{\gamma} = \sqrt{\gamma_{rr}\gamma_{\theta\theta}\gamma_{\phi\phi}} = r^2\sin\theta\sqrt{1 + \frac{2M}{r}}. \tag{5.93}$$

The nonzero Christoffel symbols for the EF space-time metric (5.92), needed to calculate the sources in (5.79), are the following:

$$\Gamma^0{}_{00} = \tfrac{2M^2}{r^3}, \qquad \Gamma^0{}_{0r} = \tfrac{M}{r^2}\left(1 + \tfrac{2M}{r}\right), \quad \Gamma^r{}_{0r} = -\tfrac{2M^2}{r^3}, \quad \Gamma^0{}_{rr} = \tfrac{2M}{r^2}\left(1 + \tfrac{M}{r}\right),$$

$$\Gamma^r{}_{rr} = -\tfrac{M}{r^2}\left(1 + \tfrac{2M}{r}\right), \quad \Gamma^\theta{}_{r\theta} = \tfrac{1}{r}, \qquad\qquad \Gamma^\phi{}_{r\phi} = \tfrac{1}{r}, \qquad \Gamma^0{}_{\theta\theta} = -2M,$$

$$\Gamma^0{}_{\phi\phi} = -2M\sin^2\theta, \quad \Gamma^r{}_{00} = \tfrac{M}{r^2}\left(1 - \tfrac{2M}{r}\right), \quad \Gamma^r{}_{\theta\theta} = 2M - r, \quad \Gamma^r{}_{\phi\phi} = (2M - r)\sin^2\theta,$$

$$\Gamma^\theta{}_{\phi\phi} = -\sin\theta\cos\theta, \qquad \Gamma^\phi{}_{\theta\phi} = \cot\theta.$$

This completes the information needed to simulate the dynamics of a perfect fluid being accreted spherically by a black hole.

Domain. We set the inner boundary at $r_{min} = M$ in order to allow the matter to enter the horizon, and the external boundary at $r_{max} = 51M$, sufficiently far from the horizon. We define the numerical domain $D_d = \{(r_i, t^n)\}$, where $r_i = r_{min} + i\Delta r$, $i = 0, 1, 2, \ldots, N_r$, $\Delta r = (r_{max} - r_{min})/N_r$ and $t^n = n\Delta t$; we choose as base resolution $N_r = 1000$ or equivalently $\Delta r = 0.05$. Time is discretized with resolution $\Delta t = 0.25\,\Delta r$. Moreover, in all the examples, we set the black hole mass to $M = 1$.

Initial Conditions. We consider the fluid to have initially a constant density $\rho(r, 0) = \rho_c$ and constant inward velocity $v^r(r, 0) = v^r_c < 0$.

Evolution. For the evolution, we use the RK3 integrator.

Boundary Conditions. Since we want to simulate a process of accretion, at the **outer boundary** (r_{max}, t), we maintain the inward flux at all times and impose time-independent Dirichlet conditions. Then we set the right-hand sides of $\mathcal{D}, \mathcal{S}_r,$ and τ to zero at this boundary.

At the **inner boundary**, the excision boundary (r_{min}, t), due to the light cone structure in Fig. 4.16, we extrapolate the values of the conservative variables as follows:

$$\mathcal{D}_0 = 3\mathcal{D}_1 - 3\mathcal{D}_2 + \mathcal{D}_3,$$
$$\mathcal{S}_0 = 3\mathcal{S}_1 - 3\mathcal{S}_2 + \mathcal{S}_3,$$
$$\tau_0 = 3\tau_1 - 3\tau_2 + \tau_3,$$

after each intermediate step of the evolution integration within the steps of the RK3 method (4.5). These formulas are obtained from the Lagrange interpolation formula with three points (see, for example, [14]).

With these initial and boundary conditions, the fluid dynamics will lead the fluid to a stationary inward flow. We illustrate this evolution process for two emblematic exact solutions, the stationary accretion of a pressureless fluid known as the Bondi solution [15] and the Michel stationary solution for the accretion of a perfect fluid with pressure [16].

5.5 Application: Spherical Accretion of a Fluid onto a Schwarzschild Black Hole

Accretion of a pressureless fluid assumes that the fluid accreted onto a black hole is dust, $p = 0$. There is an exact solution to this problem. In [17], the solution is developed for the Eddington-Finkelstein metric, and the density and velocity profiles are

$$\rho_{0,dust}(r) = \frac{-C}{r^2 \sqrt{\frac{2M}{r}}},$$

$$v^r_{dust}(r) = -\frac{1}{\sqrt{1 + \frac{r}{2M}} \left(1 + \sqrt{\frac{2M}{r} + \frac{2M}{r}}\right)}, \quad (5.94)$$

where C is a constant of integration. Since we set the initial density and velocity profiles to constant, we will set their values to $\rho_0 = \rho_{dust}(r_{max})$ and $v^r(r) = v^r_{dust}(r_{max})$, respectively. Following the example in [17], where the authors use $C = -0.195$, here, we show the results for $C = -0.5$, just to have a different case.

With all this information, instructions to initialize the variables are as follows:

- Set $\rho_0(r, 0) = \rho_{dust}(r_{max})$.
- Set $v^r(r, 0) = v^r_{dust}(r_{max})$.
- Set $p = 0$.
- Set $e = \frac{p}{\rho_0(\Gamma - 1)} = 0$.
- Calculate $h = 1 + e + p/\rho_c = 1$.
- Calculate W and then $\mathcal{D}$, $\mathcal{S}_r$, and τ using (5.75) and (5.93).
- Then start evolution.

For these initial conditions, the results of the evolution are in Fig. 5.17, where it is shown that at $t = 0$ the profiles of density and velocity are constant, and with time, the flow becomes consistent with the exact solution (5.94). This numerical solution shows that the exact solution behaves as an attractor in time provided some fixed

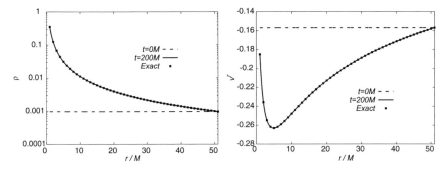

Fig. 5.17 Numerical solution at $t = 0$ and $t = 200M$ for ρ_0 and v^r, together with the exact pressureless solution for $C = -0.5$. The coordinate r is in units of M

values of density and velocity feed the inward flux from the external boundary at $r = r_{max}$.

Michel Accretion. This is the spherically symmetric stationary inflow of a perfect fluid with $p \neq 0$ onto a black hole [16]. Didactical descriptions of this solution can be found in references [17] and [18]. Michel solution is parametrized by two constants ρ_c and r_c, and unlike dust solution, Michel solution does not have a closed formula for the hydrodynamical variables.

Similar to what we did for the pressureless case, here, we set initially constant values for density, velocity, and pressure, which are those of the exact solution with $\Gamma = 4/3$, $\rho_c = 0.1$, and $r_c = 100$ evaluated at r_{max}. The system is initialized as follows:

- Set $\Gamma = 4/3$.
- Set $\rho_0(r, 0) = \rho_{Michel}(r_{max})$.
- Set $p(r, 0) = p_{Michel}(r_{max})$.
- Set $v^r(r, 0) = v^r_{Michel}(r_{max})$.
- Calculate $e = \frac{p}{\rho_c(\Gamma-1)}$.
- Calculate $h = 1 + e + p/\rho_c$.
- Calculate W and then $\mathcal{D}$, $\mathcal{S}_r$, and τ using (5.75) and (5.93).
- Then start evolution.

The results are shown in Fig. 5.18. We show the initial conditions for ρ_0 and v^r and their values at $t = 500M$. With evolution, the numerical solution resembles the exact Michel solution represented with points.

Another common setting can be written in terms of the asymptotic values of the speed of sound and velocity of the fluid. Assume that at r_{max}, we impose the density to $\rho_0(r_{max}) = \rho_0$ and the velocity magnitude to $|v_c|$. In order to properly define the velocity field v^i, we notice that

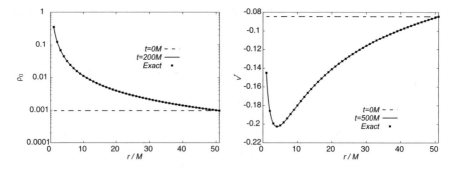

Fig. 5.18 Exact Michel solution and the numerical solution of ρ and v^r at initial time $t = 0$ and at $t = 500M$ when the flow has become stationary

5.5 Application: Spherical Accretion of a Fluid onto a Schwarzschild Black Hole

$$|v_c|^2 = v_c{}_i v_c^i = \gamma_{rr}(v_c^r)^2 \quad \Rightarrow \quad v_c^r = \pm \frac{|v_c|}{\sqrt{\gamma_{rr}}} \quad \Rightarrow \quad v^i = \left(-\frac{|v_c|}{\sqrt{\gamma_{rr}}}, 0, 0\right),$$
(5.95)

where we use the negative sign due to inward motion.

Now the pressure. For this, we use the speed of sound expression (5.85), from which pressure can be expressed in terms of a given value of the speed of sound c_s:

$$c_s^2 = \frac{p(\Gamma - 1)\Gamma}{\rho_0(\Gamma - 1) + p\Gamma} \quad \Rightarrow \quad p = \frac{\rho_0 c_s^2}{\Gamma - c_s^2 \frac{\Gamma}{\Gamma - 1}},$$
(5.96)

in fact one can choose to set either p or c_s for a given adiabatic index Γ. We choose c_s because one can estimate the Mach number of the fluid at r_{max} easily. Setting initial data is as follows:

- Set Γ.
- Set $\rho_0(r, 0)$ to a constant value.
- Set $c_s < \sqrt{\Gamma - 1}$ for it to be real.
- Use (5.96) to calculate p.
- Calculate $e = \frac{p}{\rho_c(\Gamma - 1)}$.
- Calculate $h = 1 + e + p/\rho_c$.
- Set $|v_c|$ and calculate v^i using (5.95)
- Calculate W and then $\mathcal{D}$, $\mathcal{S}_r$ and τ, using (5.75) and (5.93)
- Then start evolution.

We illustrate the solution using the values

$$\Gamma = 4/3,$$
$$\rho_0(r, 0) = 10^{-4},$$
$$c_s = 0.1 < \sqrt{\Gamma - 1},$$
$$|v_c| = 0.5.$$
(5.97)

Notice that velocities are in units of $c = 1$, which means that this case uses very high speeds.

We know the variables start being constant and evolve toward a stationary state. What remains time-independent is the value of the variables at r_{max}. We can formally estimate the Mach number of the flow at the outer boundary as follows:

$$\mathcal{M}_{r_{max}}^R = \frac{W |v_c(r_{max})|}{W_s \, c_s(r_{max})} = \frac{W}{W_s} \mathcal{M}_{r_{max}} = \frac{\sqrt{1 - |c_s(r_{max})|^2}}{\sqrt{1 - |v_c(r_{max})|^2}} \mathcal{M}_{r_{max}},$$

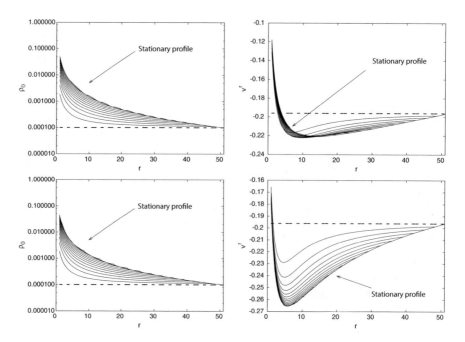

Fig. 5.19 Snapshots of ρ and v^r, showing the evolution toward a stationary flow. The dashed line indicates the values of ρ and v^r at initial time. The top plots correspond to $|v_c| = 0.5$, and the bottom ones correspond to $|v_c| = 0.2$

with $\mathcal{M}_\infty^R$ and $\mathcal{M}_\infty$ the relativistic and Newtonian Mach numbers, whereas W_s is the Lorentz factor for the speed of sound and W the Lorentz factor of the fluid velocity. Then for initial conditions (5.97), Mach numbers at r_{max} are

$$\mathcal{M}_{r_{max}}^R = 5.038, \qquad \mathcal{M}_{r_{max}} = 5.$$

In Fig. 5.19, we show the evolution of these initial conditions. The velocity profile is considerably different from that in the dust case and the Michel case in Fig. 5.18. For this reason we also show the results for $|v_c| = 0.2$ corresponding to $\mathcal{M}_{r_{max}}^R = 2.061$ and $\mathcal{M}_{r_{max}} = 2$, whose results are more similar to the Michel case in Fig. 5.18.

The code `BlackHoleAccretion.f90` is structured in the same order as that of relativistic Euler equations, uses the variables and shorthands described in this section, and is part of the Electronic Supplementary Material described in Appendix B.

A final comment is in turn as follows: All physical quantities can be translated from geometric to physical units. The mass of the black hole is an important quantity of space-time that scales all the fluid state variables. For translation of units, we recommend Appendix A of [19].

Exercise. The snapshots in Fig. 5.19 suggest that Michel solutions are attractors. Define the initial conditions as in the text, except that density can be $\rho_0(r, 0) = \rho_{Michel}(r_{max}) + Ae^{-(r-20)^2}$. Run simulations for different values of A, and show that in the end, the flow becomes again the same Michel solutions of Fig. 5.19. This would indicate that Michel solution, given consistent density and velocity values at the outer boundary, is an attractor. For this exercise, it may help revising Sect. 2.11.5.

5.6 How to Improve the Methods in This Chapter

It is an advantage to account with knowledge to implement the simple methods described in this chapter. On top of these various elements can improve.

One part is the reconstruction of variables at each cell. Here, we described and used Godunov and minmod reconstructors. Other options are the maxmode or superbee limiter, the monotonized-centered (MC) limiter, and the piecewise parabolic method (PPM) reconstructors. Higher-order reconstructors include the class of essentially non-oscillatory (ENO) method and its weighted version WENO reconstructors. These methods are well described in [19].

Another essential part is the calculation of numerical fluxes. In this chapter, we described the approximate Riemann solver HLLE which has been good enough for the examples developed here but constructed based on the propagation of only the two fastest modes, leaving all the information in between incomplete. More complete Riemann solvers can be programed with small adjustments to the codes we have constructed. For example, the HLLC can handle the contact discontinuities that may develop within the two fastest modes; Roe and Marquina solvers also prevent the violation of entropy conditions among others. These methods are described in [3, 6, 19].

5.7 Projects

With the methods described in this chapter, it is possible to carry out the following projects:

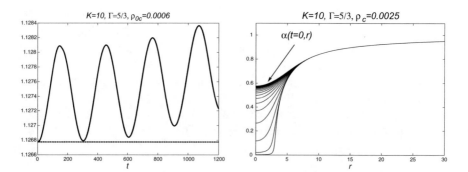

Fig. 5.20 On the left, we show the maximum of $\sqrt{\gamma_{rr}}$ as function of time (continuous line) and the value it should have in the continuum (dotted line), for a stable TOV configuration with $K = 10$, adiabatic index $\Gamma = 5/3$, and central density $\rho_{0c} = 0.0006$. On the right, we show snapshots of the lapse function α during the evolution of an unstable TOV star with polytropic constant $K = 10$, adiabatic index $\Gamma = 5/3$, and central density $\rho_{0c} = 0.0025$. The collapse of α illustrates the formation of a black hole

5.7.1 Evolution of a Relativistic TOV Star

In Sect. 2.11.7 from Chap. 2, we constructed stationary solutions of relativistic TOV stars. In Sect. 5.5, we described the methods needed to study the spherical accretion of a perfect fluid onto a Schwarzschild black hole.

A project consists in combining the two elements and evolve the equilibrium TOV stars within general relativistic hydrodynamics. In reference [20], the methods described here were used to evolve these configurations.

Relativistic Euler equations use exactly the same methods described in Sect. 5.5, but this time, Einstein equations are solved simultaneously, and thus geometry evolves in time. The paper uses a constrained evolution with an appropriate gauge, such that Einstein equations, for a spherically symmetric space-time described with spherical coordinates, reduce to ordinary differential equations that can be solved using the RK4 method for metric functions, at every time slice.

Two important results: One is that stable TOV stars oscillate and remain long-living, whereas a configuration of an unstable TOV star collapses and forms a black hole. These two results are at hand using the methods implemented so far in this book.

In Fig. 5.20, we reproduce some interesting results found in [20]. In the first plot, we show the maximum of the metric function $\sqrt{\gamma_{rr}}$ for a stable TOV star from Fig. 2.31 with $K = 10$, $\Gamma = 5/3$ and central density $\rho_{0c} = 0.0006$. In theory, these equilibrium configurations are stationary, and consequently, metric functions should remain time-independent, but in practice, there are numerical errors introduced in the construction of the configurations of Fig. 2.31 and also numerical errors during the evolution that add a permanent perturbation to the configuration. This type of perturbations serves to study the oscillation modes characterized by density and compactness of configurations. Then a first project would be to reproduce this result for various values of the central density ρ_{0c} in the stable branch of Fig. 2.31 and construct a diagram of M vs frequency of oscillation.

In the second plot of Fig. 5.20, we show the evolution of the lapse function α of an unstable TOV star in Fig. 2.31, with $K = 10$, $\Gamma = 5/3$, and central density $\rho_{0c} = 0.0025$. The black hole horizon is formed when this function collapses to zero and then an apparent horizon is formed as well as an event horizon as illustrated in [20]. A second project can be the reproduction of these results, along with the construction of a routine that calculates the growth of the black hole event horizon.

5.7.2 Nonlinear Michel Accretion

The Michel accretion problem from Sect. 5.5 can be generalized to the nonlinear case, where the space-time evolves in time. In this way, the flow does not only enter the black hole, but the black hole horizon grows.

The additional work consists in the implementation of the solution of Einstein equations for which a recipe can be found in [21]. In this reference, Einstein equations are written in the $3 + 1$ decomposition of space-time using the ADM formulation. Then evolution equations for the components of the three-metric of space-like hypersurfaces used to foliate the space-time and the components of their extrinsic curvature are coupled to the evolution of the perfect fluid from Sect. 5.5.

In [21], diagnostics tools can be found and can be easily programmed, including the location of the apparent horizon in time, its area and mass, and the accreted flow, not only through the integration of matter but also through the mass growth of the black hole itself. Methods are described also to construct the event horizon growth tracker.

References

1. B. Gustafsson, H.-O. Kreiss, J. Oliger, *Time Dependent Problems and Difference Methods.* (Wiley Interscience Series of Texts, Monographs, and Tracts, New York, 1995)
2. R.J. LeVeque, *Numerical Methods for Conservation Laws*, 2nd edn. (Birkhuser, Basel, 1992)
3. E.F. Toro, *Riemann Solvers and Numerical Methods for Fluid Dynamics*, 3rd edn. (Springer, Berlin, 2009)
4. A. Harten, P.D. Lax, B. van Leer, On upstream difference and Godunov-type schemes for hyperbolic conservation laws. SIAM Rev. **25**, 35 (1983). http://dx.doi.org/10.1137/1025002
5. B. Einfeldt, On Godunov-type methods for gas dynamics. SIAM J. Numer. Anal. **25**, 294 (1988). https://doi.org/10.1137/0725021
6. J.W. Thomas, *Numerical Partial Differential Equations: Conservation Laws and Elliptic Equations* (Springer, New York, 2013). Texts in Applied Mathematics
7. F.D. Lora-Clavijo, J.P. Cruz-Pérez, F.S. Guzmán, J.A. González, Exact solution of the 1D Riemann problem in Newtonian and relativistic hydrodynamics. Rev. Mex. Fis. E **59**, 28–50 (2013). arXiv:1303.3999 [astro-ph.HE]. https://doi.org/10.48550/arXiv.1303.3999

8. J.A. González-Esparza, A. Lara, E. Pérez-Tijerina, A. Santillán, N. Gopalswamy, A numerical study on the acceleration and transit time of coronal mass ejections in the interplanetary medium. J. Geophys. Res. **108**, 1039 (2003)
9. J.M. Martí, E. Müller, Numerical hydrodynamics in special relativity. Liv. Rev. Relativity **2**, 3 (1999). arXiv:astro-ph/9906333. https://doi.org/10.12942/lrr-1999-3
10. J.A. Font, J.M. Ibañez, A. Marquina, J.M. Martí, Multidimensional relativistic hydrodynamics: Characteristic fields and modern high-resolution shock-capturing schemes. Astron. Astrophys. **282**, 304–314 (1994). https://ui.adsabs.harvard.edu/abs/1994A&A...282..304F
11. F. Banyouls, J.A. Font, J. Ma. Ibañez, L. Ma. Marí, J.A. Miralles, Numerical 3+1 general relativistic hydrodynamics: a local characteristic approach. Astrophys. J. **476**, 221–231 (1997). https://iopscience.iop.org/article/10.1086/303604
12. M. Alcubierre, *Introduction to 3+1 Numerical Relativity* (Oxford Science Publications, Oxford, 2008)
13. T.W. Baumgarte, S.L. Shapiro, *Numerical Relativity: Solving Einstein's Equations on the Computer* (Cambridge University, Cambridge, 2010)
14. W.H. Press, S.A Teukolsky, W.T. Vettering, B.P. Flannery, *Numerical Recipes in Fortran 77: The Art of Scientific Computing*, 2nd edn. (Cambridge University, Cambridge, 2003)
15. H. Bondi, On spherically symmetrical accretion. Mon. Not. R. Astron. Soc. **112**, 195204 (1952). https://doi.org/10.1093/mnras/112.2.195
16. F.C. Michel, Accretion of matter by condensed objects. Astrophys. Space Sci. **15**, 153160 (1972). https://doi.org/10.1007/BF00649949
17. P. Papadopoulos, J.A. Font, Relativistic hydrodynamics around black holes and horizon adapted coordinate systems. Phys. Rev. **D 58**, 024005 (1998). arXiv:gr-qc/9803087. https://doi.org/10.48550/arXiv.gr-qc/9803087
18. M. Gracia-Linares, F.S. Guzmán, Astrophys. J. **812**, 23 (2015). arXiv:1510.05947 [astro-ph.HE]. https://doi.org/10.48550/arXiv.1510.05947
19. L. Rezzolla, O. Zanotti, *Relativistic Hydrodynamics* (Oxford University, Oxford, 2013)
20. F.S. Guzmán, F.D. Lora-Clavijo, M.D. Morales, Revisiting spherically symmetric relativistic hydrodynamics. Rev. Mex. Fis. **E 58**, 84–98 (2012). arXiv:1212.1421 [gr-qc]. https://doi.org/10.48550/arXiv.1212.1421
21. F.S. Guzmán, A. Romero-Amezcua, I. Alvarez-Ríos, Spherical accretion of a perfect fluid onto a black hole. Rev. Mex. Fis. E **18**, 020206, 1–24 (2021). https://doi.org/10.31349/RevMexFisE.18.020206

Chapter 6
Initial Value Problems in 3+1 and 2+1 Dimensions

Abstract In this chapter, we present the solution of physics problems defined on a space-time domain with three and two spatial dimensions. The objective is to leave the reader in a position from which it will be possible to tackle state-of-the-art problems. The three representative problems in three spatial dimensions involve the wave equation, Schrödinger equation, and Euler equations. At the end of the chapter, we present examples with the 2+1 diffusion equation.

Keywords 3+1 Partial differential equations · Wave equation · Schrödinger equation · Diffusion equation

In this chapter, we present the solution of physics problems defined on a space-time domain with three and two spatial dimensions. The objective is to leave the reader in a position from which it will be possible to tackle state-of-the-art problems. The three representative problems in three spatial dimensions involve the wave equation, Schrödinger equation, and Euler equations. At the end of the chapter, we present examples with the 2+1 diffusion equation.

6.1 General Problem

In previous chapters, when using finite volume or finite difference discretizations, we defined the same numerical domain. We proceed in the same manner here. In this section, we define a general domain that will be used during the whole chapter and also develop various general formulas that will be helpful.

We will define a general IVP on the domain $D = [x_{min}, x_{max}] \times [y_{min}, y_{max}] \times [z_{min}, z_{max}] \times [0, t_f]$ and the discrete domain D_d in a more general but similar way as before. The discrete version of the domain will be written in Cartesian coordinates and defined as $D_d = \{(x_i, y_j, z_k, t^n)\}$, where

$$x_i = x_{min} + i \Delta x,$$

© The Author(s), under exclusive license to Springer Nature Switzerland AG 2023
F. S. Guzmán, *Numerical Methods for Initial Value Problems in Physics*,
https://doi.org/10.1007/978-3-031-33556-3_6

$$y_j = y_{min} + j\Delta y,$$
$$z_k = z_{min} + k\Delta z,$$
$$t^n = n\Delta t, \tag{6.1}$$

and labels take values $i = 0, \ldots, N_x$, $j = 0, \ldots, N_y$, $k = 0, \ldots, N_z$, and spatial resolutions are defined by $\Delta x = (x_{max} - x_{min})/N_x$, $\Delta y = (y_{max} - y_{min})/N_y$, $\Delta z = (z_{max} - z_{min})/N_z$. Time resolution is set to $\Delta t = C \min(\Delta x, \Delta y, \Delta z)$ in order to take the smallest value in terms of spatial resolutions. This is formal, and in the examples here, we will always use $\Delta x = \Delta y = \Delta z$.

6.1.1 Expressions for Partial Derivatives

In problems involving three spatial dimensions and time, 3+1 dimensions from now on, one has to define discrete versions of partial derivatives. For this, we will denote a generic function $f = f(x, y, z, t)$ at the arbitrary point $(x_i, y_j, z_k, t^n) \in D_d$ by $f_{i,j,k}^n$.

In 3+1 dimensions, one can have four partial derivatives, whose discrete versions are extensions of the expressions we derived in Sect. 3.1 using Taylor series expansions, this time along each of the spatial direction and time. Here, we summarize the expressions for the various first-order partial derivatives of the generic function f.

First-order derivatives and second-order accurate, centered at (x_i, y_j, z_k, t^n), follow directly from expressions (3.5) and (3.15):

$$\left.\frac{\partial f}{\partial x}\right|_{(x_i, y_j, z_k, t^n)} = \frac{f_{i+1,j,k}^n - f_{i-1,j,k}^n}{2\Delta x} + \mathcal{O}(\Delta x^2), \tag{6.2}$$

$$\left.\frac{\partial f}{\partial y}\right|_{(x_i, y_j, z_k, t^n)} = \frac{f_{i,j+1,k}^n - f_{i,j-1,k}^n}{2\Delta y} + \mathcal{O}(\Delta y^2), \tag{6.3}$$

$$\left.\frac{\partial f}{\partial z}\right|_{(x_i, y_j, z_k, t^n)} = \frac{f_{i,j,k+1}^n - f_{i,j,k-1}^n}{2\Delta z} + \mathcal{O}(\Delta z^2), \tag{6.4}$$

$$\left.\frac{\partial f}{\partial t}\right|_{(x_i, y_j, z_k, t^n)} = \frac{f_{i,j,k}^{n+1} - f_{i,j,k}^{n-1}}{2\Delta t} + \mathcal{O}(\Delta t^2), \tag{6.5}$$

whereas forward and backward one-sided formulas (3.7), (3.8), (3.16), and (3.17) are generalized. For example, the expressions for first-order partial derivatives with respect to x

6.1 General Problem

$$\left.\frac{\partial f}{\partial x}\right|_{(x_i,y_j,z_k,t^n)} = -\frac{f^n_{i+2,j,k} - 4f^n_{i+1,j,k} + 3f^n_{i,j,k}}{2\Delta x} + \mathcal{O}(\Delta x^2), \quad (6.6)$$

$$\left.\frac{\partial f}{\partial x}\right|_{(x_i,y_j,z_k,t^n)} = \frac{f^n_{i-2,j,k} - 4f^n_{i-1,j,k} + 3f^n_{i,j,k}}{2\Delta x} + \mathcal{O}(\Delta x^2), \quad (6.7)$$

and the derivatives with respect to y and z are analogous, where, instead of the index i, indices j and k need to be shifted in each term of the expressions.

Among the one-sided expressions for the derivative with respect to t, we need the backward version, which is a generalization of (3.17):

$$\left.\frac{\partial f}{\partial t}\right|_{(x_i,y_j,z_k,t^n)} = \frac{f^{n-2}_{i,j,k} - 4f^{n-1}_{i,j,k} + 3f^n_{i,j,k}}{2\Delta t} + \mathcal{O}(\Delta t^2), \quad (6.8)$$

for all i, j, and k.

Centered second-order derivatives, second-order accurate follow from (3.9) and (3.18):

$$\left.\frac{\partial^2 f}{\partial x^2}\right|_{(x_i,y_j,z_k,t^n)} = \frac{f^n_{i+1,j,k} - 2f^n_{i,j,k} + f^n_{i-1,j,k}}{\Delta x^2} + \mathcal{O}(\Delta x^2), \quad (6.9)$$

$$\left.\frac{\partial^2 f}{\partial y^2}\right|_{(x_i,y_j,z_k,t^n)} = \frac{f^n_{i,j+1,k} - 2f^n_{i,j,k} + f^n_{i,j-1,k}}{\Delta y^2} + \mathcal{O}(\Delta y^2), \quad (6.10)$$

$$\left.\frac{\partial^2 f}{\partial z^2}\right|_{(x_i,y_j,z_k,t^n)} = \frac{f^n_{i,j,k+1} - 2f^n_{i,j,k} + f^n_{i,j,k-1}}{\Delta z^2} + \mathcal{O}(\Delta z^2), \quad (6.11)$$

$$\left.\frac{\partial^2 f}{\partial t^2}\right|_{(x_i,y_j,z_k,t^n)} = \frac{f^{n+1}_{i,j,k} - 2f^n_{i,j,k} + f^{n-1}_{i,j,k}}{\Delta t^2} + \mathcal{O}(\Delta t^2), \quad (6.12)$$

which will be useful too.

Each of the examples of IVP in this chapter will have its own initial and boundary conditions that we will define. These problems will be the generalization of problems solved in 1+1 dimensions from previous chapters.

6.1.2 Programming

In 3+1 evolution problems, the solution will be constructed via the time evolution of functions that depend on three independent spatial variables. The collection of all values of the generic *grid function* $f^n_{i,j,k}$ defined at $(x_i, y_j, z_k, t^n) \in D_d$ will be stored in a three-dimensional array f, whereas at time t^{n-1} in the array f_p:

```
real(kind=8), allocatable, dimension(:,:,:) :: f,f_p
```

and the allocation of memory would read

```
allocate(f(0:Nx,0:Ny,0:Nz),f_p(0:Nx,0:Ny,0:Nz)).
```

For ease of programming, we will also declare spatial coordinates x, y, z as three-dimensional arrays, which will also facilitate the output.

Output. Unlike in 1+1 problems, where a grid function is one-dimensional, it was easy to visualize the numerical solution because the output was the whole collection of entries, whereas in 3+1 problems one cannot easily visualize all the components of the arrays. We write simple routines that produce the projections of a grid function along each of the three axes x, y, z or the projection of a function on the xy, xz, yz planes.

In the problems of this chapter, we will output projections of the function $f^n_{i,j,k}$ along the axes that pass through the center of the domain, which are $f^n_{i,N_y/2,N_z/2}$ for $i = 0, \ldots, N_x$ along the x-axis, $f^n_{N_x/2,j,N_z/2}$ for $j = 0, \ldots, N_y$ along the y-axis, and $f^n_{N_x/2,N_y/2,k}$ for $k = 0, \ldots, N_z$ along the z-axis.

The projection of $f^n_{i,j,k}$ on the plane parallel to the xy-plane that passes through the center of the domain is also useful and can be done by saving the 2D array $f^n_{i,j,N_z/2}$ for $i = 0, \ldots, N_x$ and $j = 0, \ldots, N_y$. Similarly, the projection of $f^n_{i,j,k}$ on the plane passing through the center of the domain parallel to the yz-plane is $f^n_{N_x/2,j,k}$ for $j = 0, \ldots, N_y$ and $k = 0, \ldots, N_z$ and the projection of $f^n_{i,j,k}$ on a plane that passes through the center of the domain and is parallel to the xz-plane is $f^n_{i,N_y/2,k}$ for $i = 0, \ldots, N_x$ and $k = 0, \ldots, N_z$.

These projections will allow us to have sufficient information for the analysis of the numerical solutions.

Exercise. At the faces of the domain $x = x_{min}$ and $x = x_{max}$, centered finite differencing formulas (6.3) and (6.4), for first-order partial derivatives with respect to y and z directions, can be used. However, partial derivatives along x need one-sided formulas. Write down such formulas based on the construction of (3.7) and (3.8). Write also the formulas for first-order partial derivatives with respect to y and z, to be used at the faces $y = y_{min}$, $y = y_{max}$ and $z = z_{min}$, $z = z_{max}$, respectively.

6.2 The 3+1 Wave Equation Using a Simple Discretization

We now solve the 3+1 wave equation in Cartesian coordinates. This problem can be solved using either the simple finite differences method of Sect. 3.2, the implicit method of Sect. 3.3, or the method of lines of Sect. 4.4. As done before, we first define the IVP in the continuum, construct its exact solution, and then proceed to construct the numerical solution with different methods.

The IVP in the continuum is defined as follows:

$$\begin{array}{ll}\frac{\partial^2 \phi}{\partial t^2} - \frac{\partial^2 \phi}{\partial x^2} - \frac{\partial^2 \phi}{\partial y^2} - \frac{\partial^2 \phi}{\partial z^2} = 0, & \phi = \phi(x, y, z, t) \\ D = [x_{min}, x_{max}] \times [y_{min}, y_{max}] \\ \quad \times [z_{min}, z_{max}] \times [0, t_f] & Domain \\ \phi(x, y, z, 0) = \phi_0(x, y, z) & Initial\ Conditions \\ \dot\phi(x, y, z, 0) = \dot\phi_0(x, y, z) \\ \phi(\partial D, t) & Boundary\ Conditions \end{array} \quad (6.13)$$

Like in the 1+1 case, we impose Cauchy boundary conditions at the side $(x, y, z, 0)$ of the domain, corresponding to a time-symmetric Gaussian pulse centered at the origin:

$$\phi_0(x, y, z) = A e^{-(x^2+y^2+z^2)/\sigma^2}, \quad (6.14)$$

$$\dot\phi_0(x, y, z) = 0. \quad (6.15)$$

Notice that the initial pulse is spherically symmetric, because $\phi_0(x, y, z)$ depends on $r = \sqrt{x^2 + y^2 + z^2}$. We choose this profile because it is the generalization of the 1+1 problem, for which we can calculate the exact solution to compare with and also illustrate the properties of a spherical wave.

6.2.1 The Exact Solution

For the exact solution of this particular IVP, we rewrite the equation in spherical coordinates, knowing that the wave function with the spherical Gaussian initial conditions will propagate spherically is such that $\phi(x, y, z, t) = \phi(r, t)$:

$$\frac{\partial^2 \phi}{\partial t^2} - \frac{\partial^2 \phi}{\partial r^2} - \frac{2}{r}\frac{\partial \phi}{\partial r} = 0, \quad (6.16)$$

by defining the auxiliary variable $\varphi = r\phi$, this equation transforms into the following one:

$$\frac{1}{r}\left[\frac{\partial^2 \varphi}{\partial t^2} - \frac{\partial^2 \varphi}{\partial r^2}\right] = 0. \quad (6.17)$$

Let us find a solution *for* $r > 0$. This equation has a general solution of the type $\varphi = g(r+t) + h(r-t)$ with g and h generic functions of class C^2. Now, let us introduce time-symmetry $\dot{\phi} = 0$ at initial time, which implies that $\dot{\varphi} = 0$, which in turn implies that $\dot{\varphi} = g' - h' = 0$, whose simplest solution is $g = h$. If we propose an initial form of $g(\xi) = \frac{A}{2}\xi e^{-\xi^2/\sigma^2}|_{t=0}$, where $\xi = r \pm t$, we have the following solution for all times:

$$\varphi(r,t) = \frac{A}{2}(r+t)e^{-(r+t)^2/\sigma^2} + \frac{A}{2}(r-t)e^{-(r-t)^2/\sigma^2}. \tag{6.18}$$

Then the solution for the original ϕ and its initial profile read

$$\phi(r,t) = \frac{A}{2}\frac{(r+t)}{r}e^{-(r+t)^2/\sigma^2} + \frac{A}{2}\frac{(r-t)}{r}e^{-(r-t)^2/\sigma^2}, \tag{6.19}$$

$$\phi(r,0) = Ae^{-r^2/\sigma^2}. \tag{6.20}$$

where $r = \sqrt{x^2 + y^2 + z^2}$.

Now the solution at $r = 0$. Notice that the factor in brackets of (6.17) is zero if one substitutes (6.18), which would possibly compensate the singularity when $r \to 0$. One can check that solution (6.19) is regular at $r = 0$. For this, we expand the solution (6.19) around $r = 0$:

$$\phi(r \gtrsim 0, t) = -\frac{2A}{\sigma^2}t^2 e^{-t^2/\sigma^2} + Ae^{-t^2/\sigma^2}$$

$$+ \frac{A}{\sigma^4}\left[-\sigma^2 + 4t^2 - \frac{4}{3}\frac{t^4}{\sigma^2}\right]e^{-t^2/\sigma^2}r^2 + \mathcal{O}(r^4), \Rightarrow$$

$$\phi(0,t) = -\frac{2A}{\sigma^2}t^2 e^{-t^2/\sigma^2} + Ae^{-t^2/\sigma^2}. \tag{6.21}$$

which shows that the result is regular at $r = 0$. The exact solution is therefore the expression (6.19) for $r > 0$ and (6.21) for $r = 0$. This is the exact solution in the whole domain $\mathbb{R}^3 \times \mathbb{R}$ and is used here to compare with the numerical solutions to be constructed in a finite domain.

6.2.2 Numerical Solution Using a Simple Discretization

We denote the value of ϕ at the generic point (x_i, y_j, z_k, t^n) of the numerical domain D_d by $\phi_{i,j,k}^n$. The discrete version of the *initial conditions* is written exactly in the same manner as in the 1+1 case. The wave function at t^0 is the Gaussian pulse itself (6.14). However, the time-symmetric condition on the time derivative (6.15) has to be implemented using a virtual time $t^{-1} = t^0 - \Delta t$, where we define

6.2 The 3+1 Wave Equation Using a Simple Discretization

appropriate values of $\phi_{i,j,k}^{-1}$ as in Eq. (3.29). For this, we construct $\phi_{i,j,k}^{-1}$ as a Taylor series expansion along the backward time direction and obtain

$$\phi_{i,j,k}^{-1} = \phi_{i,j,k}^{0} - \Delta t \frac{\partial \phi}{\partial t}\bigg|_{(x_i,y_j,z_k,t^0)} + \frac{\Delta t^2}{2} \frac{\partial^2 \phi}{\partial t^2}\bigg|_{(x_i,y_j,z_k,t^0)} + \mathcal{O}(\Delta t^3),$$

time symmetry $\frac{\partial \phi}{\partial t}\bigg|_{(x_i,y_j,z_k,t^0)} = 0$ implies

$$\phi_{i,j,k}^{-1} = \phi_{i,j,k}^{0} + \frac{\Delta t^2}{2} \frac{\partial^2 \phi}{\partial t^2}\bigg|_{(x_i,y_j,z_k,t^0)} + \mathcal{O}(\Delta t^3)$$

$$= \phi_{i,j,k}^{0} + \frac{\Delta t^2}{2} \left[\frac{\partial^2 \phi}{\partial x^2} + \frac{\partial^2 \phi}{\partial y^2} + \frac{\partial^2 \phi}{\partial z^2} \right]\bigg|_{(x_i,y_j,z_k,t^0)} + \mathcal{O}(\Delta t^3).$$

Therefore, using formulas (6.9), (6.10), and (6.11) and that the resolution is the same along the three spatial directions $\Delta x = \Delta y = \Delta z$, the initial conditions are set to

$$\phi_{i,j,k}^{0} = A e^{-(x_i^2 + y_j^2 + z_k^2)/\sigma^2}, \tag{6.22}$$

$$\phi_{i,j,k}^{-1} = \phi_{i,j,k}^{0}$$

$$+ \frac{\Delta t^2}{2 \Delta x^2} \left[\phi_{i+1,j,k}^{0} + \phi_{i-1,j,k}^{0} + \phi_{i,j+1,k}^{0} + \phi_{i,j-1,k}^{0} \right.$$

$$\left. + \phi_{i,j,k+1}^{0} + \phi_{i,j,k-1}^{0} - 8\phi_{i,j,k}^{0} \right]$$

$$+ \mathcal{O}(\Delta x^2, \Delta y^2, \Delta z^2, \Delta t^3),$$

for all interior points of D_d, except boundaries, where one-sided second-order derivative operators of the type (6.6)–(6.7) have to be used at each face.

The *evolution* is carried out using the discrete version of the wave equation, which is constructed using formulas (6.9), (6.10), (6.11), and (6.12) into the wave equation (6.13) and then solving for $\phi_{i,j,k}^{n+1}$:

$$\phi_{i,j,k}^{n+1} = \frac{\Delta t^2}{\Delta x^2} (\phi_{i+1,j,k}^{n} - 2\phi_{i,j,k}^{n} + \phi_{i-1,j,k}^{n})$$

$$+ \frac{\Delta t^2}{\Delta y^2} (\phi_{i,j+1,k}^{n} - 2\phi_{i,j,k}^{n} + \phi_{i,j-1,k}^{n})$$

$$+ \frac{\Delta t^2}{\Delta z^2} (\phi_{i,j,k+1}^{n} - 2\phi_{i,j,k}^{n} + \phi_{i,j,k-1}^{n}) + 2\phi_{i,j,k}^{n} - \phi_{i,j,k}^{n-1}$$

$$+ \mathcal{O}(\Delta x^2, \Delta y^2, \Delta z^2, \Delta t^2). \tag{6.23}$$

Notice that this is a generalization of formula (3.27) for the 1+1 wave equation and serves to calculate the solution ϕ at time t^{n+1} in terms of information at times t^n and t^{n-1}. Also notice that this formula does not serve to calculate the solution at the faces of the spatial domain, which will be determined by the boundary conditions to be used.

Outgoing Wave Boundary Conditions. The initial conditions used here correspond to a spherical wave, whereas the boundary is the faces of a cube; thus, the wave front is not plane and is not parallel to any of the faces. This interesting situation is very common in many problems, and here, we explain the implementation of practical wave boundary conditions for a spherical front on a cubic boundary. Knowing the pulse is spherically symmetric, notice that the wave operator in (6.16) can be factorized, like in the 1+1 case, for outgoing and ingoing spherical pulses assuming the origin of coordinates is at $r = 0$:

$$\left(\frac{\partial}{\partial t} - \frac{\partial}{\partial r} - \frac{1}{r}\right)\left(\frac{\partial}{\partial t} + \frac{\partial}{\partial r} + \frac{1}{r}\right)\phi = 0. \tag{6.24}$$

At the boundary, we want to set the inward mode to zero with the intention to avoid signals to be reflected back from the boundary toward the interior of the domain, for which we have to solve the following equation at the boundary surface:

$$\frac{\partial \phi}{\partial t} + \frac{\partial \phi}{\partial r} + \frac{\phi}{r} = 0, \tag{6.25}$$

a condition that needs to be translated into Cartesian coordinates. Using the chain rule, the derivative of ϕ with respect to each coordinate is $\frac{\partial \phi}{\partial x^k} = \frac{\partial \phi}{\partial r}\frac{\partial r}{\partial x^k} = \frac{x^k}{r}\frac{\partial \phi}{\partial x^k}$, with $r = \sqrt{x^2 + y^2 + z^2}$ and $x^k = x, y, z$, which implies that $\frac{\partial \phi}{\partial r} = \frac{r}{x^k}\frac{\partial \phi}{\partial x^k}$ for $x^k \neq 0$, which is true at any face of the cube. Here, x^k is the coordinate whose axis is normal to the face one desires to impose the boundary condition onto, which reads

$$\frac{\partial \phi}{\partial t} + \frac{r}{x^k}\frac{\partial \phi}{\partial x^k} + \frac{\phi}{r} = 0, \tag{6.26}$$

and r is evaluated from the origin of coordinates to each point of the cube's face perpendicular to the x^k-axis one deals with.

Explicitly, for the face of the boundary on the plane $x = x_{max}$, the discrete version of Eq. (6.26), using the second-order accurate backward one-sided first-order time derivative (6.8) and second-order accurate backward first-order space derivative (6.7), for all points (x_{N_x}, y_j, z_k, t^n), and the radius evaluated at each point of the face $r_{N_x, j, k} = \sqrt{x_{N_x}^2 + y_j^2 + z_k^2}$, one obtains

6.2 The 3+1 Wave Equation Using a Simple Discretization

$$\frac{\phi_{N_x,j,k}^{n-1} - 4\phi_{N_x,j,k}^{n} + 3\phi_{N_x,j,k}^{n+1}}{2\Delta t} + \frac{r_{N_x,j,k}}{x_{N_x,j,k}} \frac{\phi_{N_x-2,j,k}^{n+1} - 4\phi_{N_x-1,j,k}^{n+1} + 3\phi_{N_x,j,k}^{n+1}}{2\Delta x}$$

$$+ \frac{\phi_{N_x,j,k}^{n+1}}{r_{N_x,j,k}} = 0,$$

for all j, k. One solves for $\phi_{N_x,j,k}^{n+1}$, which is the value of the wave function at each point of the boundary plane $x = x_{max} = x_{N_x}$:

$$\phi_{N_x,j,k}^{n+1} = \frac{1}{\left(\frac{3}{2\Delta t} + \frac{3}{2\Delta x}\frac{r_{N_x,j,k}}{x_{N_x,j,k}} + \frac{1}{r_{N_x,j,k}}\right)}$$

$$\times \left[\frac{4\phi_{N_x,j,k}^{n} - \phi_{N_x,j,k}^{n-1}}{2\Delta t} - \frac{r_{N_x,j,k}}{x_{N_x,j,k}}\frac{\phi_{N_x-2,j,k}^{n+1} - 4\phi_{N_x-1,j,k}^{n+1}}{2\Delta x}\right]. \quad (6.27)$$

A similar calculation leads to the value of the wave function at the face where $x = x_{min}$, that is, for all points (x_0, y_j, z_k, t^n) of the numerical domain:

$$\phi_{0,j,k}^{n+1} = \frac{1}{\left(\frac{3}{2\Delta t} - \frac{3}{2\Delta x}\frac{r_{0,j,k}}{x_{0,j,k}} + \frac{1}{r_{0,j,k}}\right)} \left[\frac{4\phi_{0,j,k}^{n} - \phi_{0,j,k}^{n-1}}{2\Delta t} + \frac{r_{0,j,k}}{x_{0,j,k}}\frac{\phi_{2,j,k}^{n+1} - 4\phi_{1,j,k}^{n+1}}{2\Delta x}\right],$$

$$(6.28)$$

where $r_{0,j,k} = \sqrt{x_0^2 + y_j^2 + z_k^2}$.

Similar expressions have to be constructed for the faces on the planes $y = y_{max}$, $y = y_{min}$, $z = z_{max}$ and $z = z_{min}$ of the spatial boundary of D_d.

Let Us Workout an Example. Consider the domain $D = [-1, 1] \times [-1, 1] \times [-1, 1] \times [0, t_f]$, with $x_{max} = y_{max} = z_{max} = -x_{min} = -y_{min} = -z_{min} = 1$ and $t_f = 2$. The discrete domain D_d is defined according to (6.1) with $N_x = N_y = N_z = 100$ or equivalently $\Delta x = \Delta y = \Delta z = 0.02$ and $\Delta t = 0.5\Delta x$. The parameters of the initial Gaussian (6.14) are set to $A = 1$ and $\sigma = 0.1$

Snapshots of the numerical solution are shown in Fig. 6.1. Notice that unlike the 1+1 case, the amplitude of the pulse decays as a spherical wave $\sim 1/r$. The comparison of the numerical solution with the exact solution (6.19) and (6.21) appears in Fig. 6.2, where also the second-order convergence according to formula (3.21) is shown, in agreement with the accuracy of the operators and time integration.
The code that produces these results `WaveSimple3p1.f90` is found in the Electronic Supplementary Material.

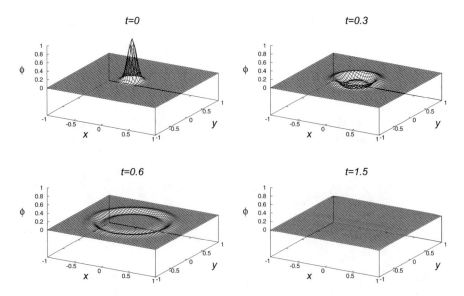

Fig. 6.1 Numerical solution ϕ of the 3+1 wave equation projected on the xy-plane at times $t = 0, 0.3, 0.6$, and 1.5. Notice that the wave gets off the domain showing the outgoing wave boundary conditions work fine. Since the pulse is spherically symmetric, the projection on the yz and zx planes has the same results

6.2.3 Example with Sources

Let us consider a nonspherically symmetric scenario. For this, we will slightly complicate the wave equation and add a source. The IVP to solve is based on the problem (6.13) with the following wave equation:

$$\frac{\partial^2 \phi}{\partial t^2} - \frac{\partial^2 \phi}{\partial x^2} - \frac{\partial^2 \phi}{\partial y^2} - \frac{\partial^2 \phi}{\partial z^2} = S(x, y, z, t), \tag{6.29}$$

where S is a source. We will construct a source consisting of two scalar charges orbiting around each other defined by

$$S = A_s e^{-[(x-x_0)^2 + (y-y_0)^2 + z^2]/\sigma_s^2} + A_s e^{-[(x+x_0)^2 + (y+y_0)^2 + z^2]/\sigma_s^2} \tag{6.30}$$

which is the sum of two Gaussians of amplitude A_s and width σ_s, centered at the points $(x_0, y_0, 0)$ and $(-x_0, -y_0, 0)$. We make the charges orbit on the circle of radius r_s with angular frequency ω by assuming $x_0 = r_s \cos(\omega t)$ and $y_0 = r_s \sin(\omega t)$.

6.2 The 3+1 Wave Equation Using a Simple Discretization

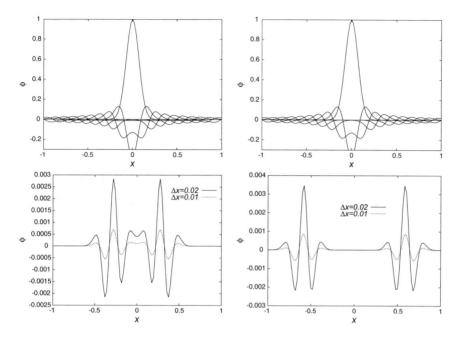

Fig. 6.2 At the top, we show the solutions, numerical on the left and exact on the right at various times, projected along the x−axis. At the bottom, we show the error along the x−axis at $t = 0.3$ and $t = 0.6$. Additionally, we show the error of the numerical solution using double resolution, and the result indicates second-order convergence, in agreement with the error of the methods

The *discrete version* of this equation leads to the following expression for the wave function at inner points of D_d:

$$\begin{aligned}\phi_{i,j,k}^{n+1} = {}& \frac{\Delta t^2}{\Delta x^2}(\phi_{i+1,j,k}^n - 2\phi_{i,j,k}^n + \phi_{i-1,j,k}^n) \\ &+ \frac{\Delta t^2}{\Delta y^2}(\phi_{i,j+1,k}^n - 2\phi_{i,j,k}^n + \phi_{i,j-1,k}^n) \\ &+ \frac{\Delta t^2}{\Delta z^2}(\phi_{i,j,k+1}^n - 2\phi_{i,j,k}^n + \phi_{i,j,k-1}^n) + 2\phi_{i,j,k}^n - \phi_{i,j,k}^{n-1} \\ &+ S_{i,j,k}\Delta t^2 + \mathcal{O}(\Delta x^2, \Delta y^2, \Delta z^2, \Delta t^2),\end{aligned} \quad (6.31)$$

which is Eq. (6.23) plus the contribution of the source.

The *initial conditions* are $\phi(x, y, z, 0) = S(x, y, z, 0)$, and time symmetry is assumed. For the construction of the wave function at the auxiliary time t^{-1}, we follow the second expressions of (6.22).

We impose *outgoing wave boundary conditions* following the recipe above, from (6.26) to (6.28).

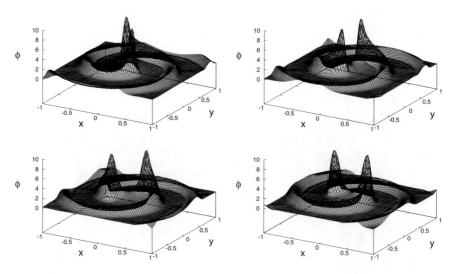

Fig. 6.3 Wave function at time $t = 3, 6, 9, 12$. The result is that the two orbiting sources produce a tail left behind

For this case, consider the same domain of the previous example, that is, $D = [-1, 1] \times [-1, 1] \times [-1, 1] \times [0, t_f]$, with $x_{max} = y_{max} = z_{max} = -x_{min} = -y_{min} = -z_{min} = 1$ and $t_f = 2$. The discrete domain D_d is defined according to (6.1) with $N_x = N_y = N_z = 100$, or equivalently $\Delta x = \Delta y = \Delta z = 0.02$ and $\Delta t = 0.5 \Delta x$. The parameters of the source in (6.30) are set to $A_s = 5 \times 10^3$, $\sigma = 0.05$, $r_s = 0.3$, and $\omega = 5$.

Snapshots of the solution are shown in Fig. 6.3. The sources orbit around the origin and leave a trace behind. In pioneering works, this type of numerical experiment was once considered a toy model for the generation of gravitational waves by binary objects [1], prior to the numerical solution of the binary black hole problem in 2005 [2].

In the code, the source is incorporated easily within the loop that updates the array of ϕ^{n+1} as follows:

```
! Update of the wave function
do i=1,Nx-1
do j=1,Ny-1
do k=1,Nz-1
   source(i,j,k) = amp_source * exp( -((x(i,j,k)-rad
*cos(omega*t))*))*2 &
         + (y(i,j,k)-rad*sin(omega*t))**2 &
         + z(i,j,k)**2) / sigma**2 ) &
         + amp_source * exp( -((x(i,j,k)+rad*cos(omega*t))**2 &
         + (y(i,j,k)+rad*sin(omega*t))**2 &
         + z(i,j,k)**2) / sigma**2 )
   phi(i,j,k) = dt**2 * ( phi_p(i+1,j,k) - 2.0d0*phi_p(i,j,k) +
phi_p(i-1,j,k)) / dx**2 &
```

```
               + dt**2 *  ( phi_p(i,j+1,k)  -  2.0d0*phi_p(i,j,k) +
phi_p(i,j-1,k)) / dy**2 &
               + dt**2 *  ( phi_p(i,j,k+1)  -  2.0d0*phi_p(i,j,k) +
phi_p(i,j,k-1)) / dz**2 &
               + 2.0d0 * phi_p(i,j,k) - phi_pp(i,j,k) + dt**2*source (i,j,k)
       end do
     end do
   end do
```

6.3 The 3+1 Wave Equation Using the Method of Lines

The IVP problem is the same as that in (6.13). In order to apply the method of lines, one has to convert the wave equation into a set of first-order PDEs. For that, we follow the same strategy as in Sect. 4.4, which consists in defining first-order variables:

$$\pi = \frac{\partial \phi}{\partial t}, \quad \psi_x = \frac{\partial \phi}{\partial x}, \quad \psi_y = \frac{\partial \phi}{\partial y}, \quad \psi_z = \frac{\partial \phi}{\partial z}. \tag{6.32}$$

In terms of these variables, the problem (6.13) is redefined as follows:

$$\boxed{\begin{aligned}
&\partial_t \pi = \partial_x \psi_x + \partial_y \psi_y + \partial_z \psi_z \\
&\partial_t \psi_x = \partial_x \pi \\
&\partial_t \psi_y = \partial_y \pi \\
&\partial_t \psi_z = \partial_z \pi \\
&\partial_t \phi = \pi, \qquad \phi = \phi(x,y,z,t), \ \psi_x = \psi_x(x,y,z,t) \\
&\qquad\qquad \psi_y = \psi_y(x,y,z,t), \ \psi_z = \psi_z(x,y,z,t), \ \pi = \pi(x,y,z,t) \\
&D = [x_{min}, x_{max}] \times [y_{min}, y_{max}] \times [z_{min}, z_{max}] \times [0, t_f], \ Domain \\
&\phi(x,y,z,0) = \phi_0(x,y,z) \\
&\psi_x(x,y,z,0) = \psi_{x0}(x,y,z) \\
&\psi_y(x,y,z,0) = \psi_{y0}(x,y,z) \\
&\psi_z(x,y,z,0) = \psi_{z0}(x,y,z) \\
&\pi(x,y,z,0) = \pi_0(x,y,z) \qquad\qquad\qquad Initial\ Conditions \\
&\phi(\partial D, t), \ \psi_x(\partial D, t), \ \psi_y(\partial D, t), \ \psi_z(\partial D, t), \ \pi(\partial D, t) \quad Boundary\ Conditions
\end{aligned}}$$

(6.33)

It is a problem of four unknowns $\psi_x, \psi_y, \psi_z,$ and π with the constraints $\psi_x = \partial_x \phi$, $\psi_y = \partial_y \phi$, and $\psi_z = \partial_z \phi$. The reconstruction of the original unknown ϕ is achieved via the integration of the equation $\partial_t \phi = \pi$.

All five variables are defined in the numerical domain D_d, and the *semi-discrete* version of the evolution equations in problem (6.38), at the inner points of the numerical domain D_d, using second-order accurate expressions for first-order spatial derivatives in (6.2)–(6.4), read:

$$\partial_t \pi = \frac{\psi_{x\,i+1,j,k} - \psi_{x\,i-1,j,k}}{2\Delta x} + \frac{\psi_{y\,i,j+1,k} - \psi_{y\,i,j-1,k}}{2\Delta y}$$
$$+ \frac{\psi_{z\,i,j,k+1} - \psi_{z\,i,j,k-1}}{2\Delta z} + \mathcal{O}(\Delta x^2, \Delta y^2, \Delta z^2),$$
$$\partial_t \psi_x = \frac{\pi_{i+1,j,k} - \pi_{i-1,j,k}}{2\Delta x} + \mathcal{O}(\Delta x^2),$$
$$\partial_t \psi_y = \frac{\pi_{i,j+1,k} - \pi_{i,j-1,k}}{2\Delta y} + \mathcal{O}(\Delta y^2),$$
$$\partial_t \psi_z = \frac{\pi_{i,j,k+1} - \pi_{i,j,k-1}}{2\Delta z} + \mathcal{O}(\Delta z^2),$$
$$\partial_t \phi = \pi_{i,j,k}. \tag{6.34}$$

The integration of the system at the boundary has to be implemented based on boundary conditions.

Time Integration. For time integration, the second- (4.4) or third-order (4.5) accurate Runge-Kutta integrators suffice for the solution of this problem.

6.3.1 The Time-Symmetric Gaussian Pulse

This example uses the exact solution (6.19) and (6.21) to compare with. In terms of the first-order variables, the Cauchy initial conditions (6.14)–(6.15) become Dirichlet for all variables at the cube $(x, y, z, 0)$:

$$\phi_0(x, y, z) = A e^{-(x^2+y^2+z^2)/\sigma^2},$$
$$\psi_{x0}(x, y, z) = -\frac{2x}{\sigma^2} \phi_0(x, y, z),$$
$$\psi_{y0}(x, y, z) = -\frac{2y}{\sigma^2} \phi_0(x, y, z),$$
$$\psi_{z0}(x, y, z) = -\frac{2z}{\sigma^2} \phi_0(x, y, z),$$
$$\pi_0(x, y, z) = 0, \tag{6.35}$$

where time symmetry is implemented simply through the value of π_0.

6.3 The 3+1 Wave Equation Using the Method of Lines

Outgoing wave boundary conditions need the Eq. (6.26) to be solved at the faces of the spatial boundary of D_d. This is achieved as follows: (i) impose the condition (6.26) on ϕ by defining the appropriate right-hand side for the evolution of ϕ at the faces of D_d; (ii) with the values of ϕ, calculate the space derivatives $\partial \phi / \partial x^i = \psi_i$, $i = x, y, z$ using one-sided formulas (6.6)–(6.7) and the equivalent formulas for the faces perpendicular to y and z axes, at each face; and (iii) finally calculate $\pi = \partial \phi / \partial t$ from (6.26) at all faces.

Specifically, the boundary condition at the faces where $x = x_{min}$ and $x = x_{max}$ is implemented as follows:

(i) The semi-discrete version of (6.26), which uses formulas (6.6)–(6.7) at these two faces, reads :

$$\partial_t \phi|_{0,j,k} = \frac{1}{2} \frac{\phi_{2,j,k} - 4\phi_{1,j,k} + 3\phi_{0,j,k}}{\Delta x} \frac{r_{0,j,k}}{x_{0,j,k}} - \frac{\phi_{0,j,k}}{r_{0,j,k}},$$

$$\partial_t \phi|_{N_x,j,k} = -\frac{1}{2} \frac{\phi_{N_x-2,j,k} - 4\phi_{N_x-1,j,k} + 3\phi_{N_x,j,k}}{\Delta x} \frac{r_{N_x,j,k}}{x_{N_x,j,k}} - \frac{\phi_{N_x,j,k}}{r_{N_x,j,k}},$$

for all j, k, that one programs in the calculation of the right-hand side of the evolution equation for ϕ.

(ii) Now that $\phi_{0,j,k}$ and $\phi_{N_x,j,k}$ are known, the value of ψ_x at the faces can be calculated using the one-sided formulas (6.6) and (6.7):

$$\psi_{x\,0,j,k} = -\frac{1}{2} \frac{\phi_{2,j,k} - 4\phi_{1,j,k} + 3\phi_{0,j,k}}{\Delta x},$$

$$\psi_{x\,N_x,j,k} = \frac{1}{2} \frac{\phi_{N_x-2,j,k} - 4\phi_{N_x-1,j,k} + 3\phi_{N_x,j,k}}{\Delta x}.$$

(iii) Finally, using directly condition (6.26), one obtains π as follows:

$$\pi_{0,j,k} = -\frac{r_{0,j,k}}{x_{0,j,k}} \psi_{x\,0,j,k} - \frac{\phi_{0,j,k}}{r_{0,j,k}},$$

$$\pi_{N_x,j,k} = -\frac{r_{N_x,j,k}}{x_{N_x,j,k}} \psi_{x\,N_x,j,k} - \frac{\phi_{N_x,j,k}}{r_{N_x,j,k}}.$$

The implementation of this boundary condition on the other four faces of the boundary is analogous.

The solution for the initial Gaussian pulse with the same parameters as in the previous section is shown in Fig. 6.4. The results are similar to those obtained with the simple discretization in Fig. 6.1, except for a bump at the center that remains in the domain even if the pulse has gone out of that zone.

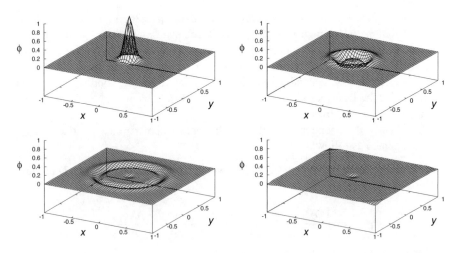

Fig. 6.4 Numerical solution ϕ of the 3+1 wave equation projected on the xy-plane at times $t = 0, 0.3, 0.6$, and 1.5. This time, we use the method of lines, and the plots can be compared with those in Fig. 6.1. Notice the bump left at the center of the domain, which is a remnant that converges to zero with resolution

In Fig. 6.5, we show the projection of the solution and the error along the x-axis, which should be compared with results in Fig. 6.2. The convergence test using two resolutions $\Delta x = 0.02, 0.01$ shows that the error bump at the center converges to zero with resolution.

The effort invested in solving the wave equation with the method of lines seems to be not retributed, since the solution shows glitches, the error is bigger in magnitude than with the simple discretization, and moreover, one has to define four extra variables to solve the system of equations, which in turn represents a cost in memory and number of operations. Nevertheless, the use of the method of lines pays once the systems of equations become more complex as we will show later in this chapter.

6.3.2 Example with a Plane Wave

A special case is the evolution of a plane wave of the type $\phi = \exp[i(\mathbf{k} \cdot \mathbf{x} - \omega t)]$, where $\mathbf{k} = (k_x, k_y, k_z)$ is the wave vector, ω the angular frequency of the wave, and $\mathbf{x} = (x, y, z)$ the position vector in space. From the wave equation $\partial^2 \phi / \partial t^2 - \nabla^2 \phi = 0$, the substitution of ϕ implies the dispersion relation $\omega^2 = \mathbf{k} \cdot \mathbf{k} = k_x^2 + k_y^2 + k_z^2$.

Domain. We solve the equation in the domain $D = [-1, 1] \times [-1, 1] \times [-1, 1] \times [0, t_f]$, with $x_{max} = y_{max} = z_{max} = -x_{min} = -y_{min} = -z_{min} = 1$ and $t_f = 2$. The discrete domain D_d is defined according to (6.1) with $N_x = N_y = N_z = 100$ or equivalently $\Delta x = \Delta y = \Delta z = 0.02$, and $\Delta t = 0.5 \Delta x$.

6.3 The 3+1 Wave Equation Using the Method of Lines

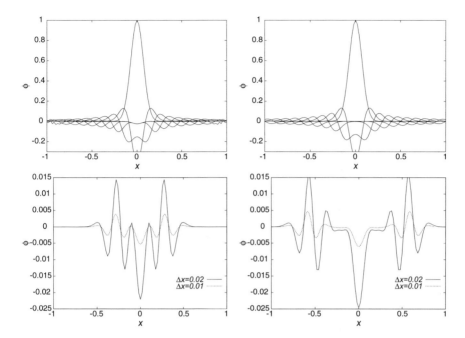

Fig. 6.5 Solution ϕ using the Method of Lines. At the top, we show the solutions, numerical on the left and exact on the right, projected along the x−axis. At the bottom, we show the error along the x−axis at $t = 0.3$ and $t = 0.6$. Additionally, we show the error of the numerical solution using double resolution, and the result indicates second-order convergence, in agreement with the order of accuracy of the discrete version of space derivative operators and time integration

Boundary Conditions. In this case, we use periodic boundary conditions, so that the wave can travel through the domain indefinitely.

Unlike in the 1+1 problems, for example, in Sect. 4.4 for the wave equation, where periodic boundary conditions involve only two points and the calculation of the right-hand sides is simple, as seen in Eqs. (4.16) and (4.17), in the 3+1 case, the spatial boundary consists of six faces, and these boundary conditions are not as easy to implement. A common solution consists in the definition of **ghost points**, which are extra points outside of D_d where the evolution is carried out like in the interior points, but the values of variables at these points are identified with those at the opposite face. Explicitly, for any grid function u, memory is allocated for points (x_i, y_j, z_k, t^n) for $i = -1, 0, \ldots, N_x, N_x + 1$, $j = -1, 0, \ldots, N_y, N_y + 1$ and $k = -1, 0, \ldots, N_z, N_z + 1$, with an extra cubic surface of points outside of D_d. These ghost points allow the calculation of space derivatives up to the faces of D_d, $x = x_0, x = x_{N_x}, y = y_0, y = y_{N_y}, z = z_0$, and $z = z_{N_z}$. Then, after each time step within the time integrator, RK2 or RK3, one identifies the values of u at the ghost faces with those at the opposite face. Explicitly,

$$u_{0,j,k} = u_{N_x,j,k} \qquad u_{N_x+1,j,k} = u_{1,j,k}$$
$$u_{i,0,k} = u_{i,N_y,k} \qquad u_{i,N_y+1,k} = u_{i,1,k} \qquad (6.36)$$
$$u_{i,j,0} = u_{i,j,N_z} \qquad u_{i,j,N_z+1} = u_{i,j,1}$$

which automatically changes the topology of the domain from a cube into a three-torus.

Initial Conditions. We choose the phase of ϕ such that at initial time it is real and expressions for all the variables at initial time read

$$\phi_0(x,y,z) = \cos(k_x x + k_y y + k_z z - \omega t)|_{t=0},$$
$$\psi_{x0}(x,y,z) = -k_x \sin(k_x x + k_y y + k_z z - \omega t)|_{t=0},$$
$$\psi_{y0}(x,y,z) = -k_y \sin(k_x x + k_y y + k_z z - \omega t)|_{t=0},$$
$$\psi_{z0}(x,y,z) = -k_z \sin(k_x x + k_y y + k_z z - \omega t)|_{t=0},$$
$$\pi_0(x,y,z) = \omega \sin(k_x x + k_y y + k_z z - \omega t)|_{t=0}. \qquad (6.37)$$

As an example, we set initial conditions for a wave that travels along a diagonal of the xy-plane, for which we use a wave number with $k_x = k_y = 2\pi$, $k_z = 0$. The results of the evolution are shown in Fig. 6.6.

The code MoLWave3p1.f90, listed below, solves the 3+1 wave equation using the method of lines, for the Gaussian pulse and the plane wave. The code illustrates

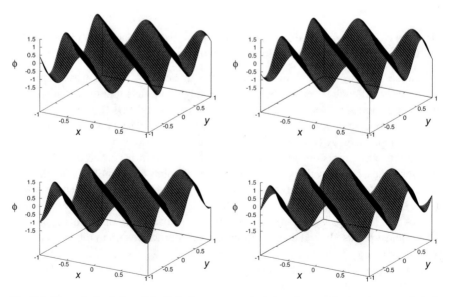

Fig. 6.6 Numerical solution of the 3+1 plane wave projected on the xy-plane at times $t = 2, 4, 6,$ and 8

6.4 The 3+1 Schrödinger Equation Using the ADI Scheme

the programming of the methods in 3+1 dimensions, along with the implementation of outgoing wave and periodic boundary conditions, and is available as Electronic Supplementary Material.

> **Exercise.** Construct the exact solution for a stationary wave in a cubic box. Then modify the code, with special care on the boundary conditions to be used, and verify that the solution is truly stationary during the evolution.

6.4 The 3+1 Schrödinger Equation Using the ADI Scheme

In Cartesian coordinates, the general three-dimensional Schrödinger equation reads

$$\begin{aligned}
&i\frac{\partial \Psi(\mathbf{x},t)}{\partial t} = -\frac{1}{2}\nabla^2 \Psi(\mathbf{x},t) + V(\mathbf{x})\Psi(\mathbf{x},t) \qquad & \Psi = \Psi(x,y,z,t) \\
&D = [x_{min}, x_{max}] \times [y_{min}, y_{max}] \times [z_{min}, z_{max}] \\
&\qquad \times [0, t_f] & Domain \\
&\Psi(x,y,z,0) = \Psi_0(x,y,z) & Initial\ Conditions \\
&\Psi(\partial D, t) & Boundary\ Conditions
\end{aligned} \tag{6.38}$$

The solution can be constructed with extensions of the strategies used for the one-dimensional case. In this section, we describe the extension of the implicit Crank-Nicolson method, for which we consider $\Psi^n_{i,j,k}$ is the wave function at point $(x_i, y_j, z_k, t^n) \in D_d$. Generalizing the development in Eqs. (3.72)–(3.74) to three spatial dimensions, the method for the evolution of the wave function from time t^n to time t^{n+1} is constructed as follows:

$$\Psi^{n+1}_{i,j,k} = \frac{e^{-i\frac{1}{2}\hat{H}\Delta t}}{e^{i\frac{1}{2}\hat{H}\Delta t}} \Psi^n_{i,j,k} \Rightarrow$$

$$\Psi^{n+1}_{i,j,k} \simeq \frac{(1 - \frac{1}{2}i\hat{H}\Delta t)}{(1 + \frac{1}{2}i\hat{H}\Delta t)} \Psi^n_{i,j,k} \Rightarrow$$

$$\left(1 + \frac{1}{2}i\hat{H}\Delta t\right)\Psi^{n+1}_{i,j,k} = \left(1 - \frac{1}{2}i\hat{H}\Delta t\right)\Psi^n_{i,j,k} \Rightarrow$$

$$\Psi^{n+1}_{i,j,k} + \frac{1}{2}i\hat{H}\Psi^{n+1}_{i,j,k}\Delta t = \Psi^n_{i,j,k} - \frac{1}{2}i\hat{H}\Psi^n_{i,j,k}\Delta t \Rightarrow$$

$$i\frac{\Psi^{n+1}_{i,j,k} - \Psi^n_{i,j,k}}{\Delta t} = \frac{1}{2}\left[\hat{H}\Psi^{n+1}_{i,j,k} + \hat{H}\Psi^n_{i,j,k}\right]. \tag{6.39}$$

278   6 Initial Value Problems in 3+1 and 2+1 Dimensions

This formula is Schrödinger equation with the spatial part averaged at times t^n and t^{n+1}, the essence of the Crank-Nicolson method. Notice also that the time derivative in (6.39) is second-order accurate centered at the point $(x_i, y_j, z_k, t^{n+1/2})$.

ADI Method. The Alternating Direction Implicit (ADI) method assumes that the evolution from t^n to t^{n+1} can be implemented by applying the Hamiltonian operator

$$\hat{H} = -\frac{1}{2}\left(\frac{\partial^2}{\partial x^2} + \frac{\partial^2}{\partial y^2} + \frac{\partial^2}{\partial z^2}\right) + V$$

in four successive steps, one for each of the partial second-order derivatives along each direction and one for the potential as follows:

$$\left(1 - \frac{i}{4}\Delta t \frac{\partial^2}{\partial x^2}\right) R_{i,j,k} = \left(1 + \frac{i}{4}\Delta t \frac{\partial^2}{\partial x^2}\right) \Psi_{i,j,k}^n$$

$$\left(1 - \frac{i}{4}\Delta t \frac{\partial^2}{\partial y^2}\right) S_{i,j,k} = \left(1 + \frac{i}{4}\Delta t \frac{\partial^2}{\partial y^2}\right) R_{i,j,k}$$

$$\left(1 - \frac{i}{4}\Delta t \frac{\partial^2}{\partial z^2}\right) T_{i,j,k} = \left(1 + \frac{i}{4}\Delta t \frac{\partial^2}{\partial z^2}\right) S_{i,j,k}$$

$$\left(1 + \frac{i}{2}\Delta t\, V\right) \Psi_{i,j,k}^{n+1} = \left(1 - \frac{i}{2}\Delta t\, V\right) T_{i,j,k} \qquad (6.40)$$

where $R_{i,j,k}$, $S_{i,j,k}$, and $T_{i,j,k}$ are auxiliary arrays that store the values of the wave function after applying the derivative operator along each of the spatial directions.

The implementation of the ADI method is based on the fact that each line in (6.40) defines a tridiagonal system as found for the one-dimensional case in Sect. 3.5. In particular for the interior points of the numerical domain, using centered discrete expressions for the second-order partial derivatives, these tridiagonal systems are explicitly

$$(-\alpha) R_{i-1,j,k} + (1 + 2\alpha) R_{i,j,k} + (-\alpha) R_{i+1,j,k} = (\alpha)\Psi_{i-1,j,k}^n$$
$$+ (1 - 2\alpha)\Psi_{i,j,k}^n + (\alpha)\Psi_{i+1,j,k}^n,$$

$$(-\alpha) S_{i,j-1,k} + (1 + 2\alpha) S_{i,j,k} + (-\alpha) S_{i,j+1,k} = (\alpha) R_{i,j-1,k}$$
$$+ (1 - 2\alpha) R_{i,j,k} + (\alpha) R_{i,j+1,k},$$

$$(-\alpha) T_{i,j,k-1} + (1 + 2\alpha) T_{i,j,k} + (-\alpha) T_{i,j,k+1} = (\alpha) S_{i,j,k-1}$$
$$+ (1 - 2\alpha) S_{i,j,k} + (\alpha) S_{i,j,k+1},$$

$$\Psi_{i,j,k}^{n+1} = \frac{1 - \beta V_{i,j,k}}{1 + \beta V_{i,j,k}} T_{i,j,k}, \qquad (6.41)$$

where $\alpha = \frac{1}{4}i\frac{\Delta t}{\Delta x^2}$ and $\beta = \frac{1}{2}i\Delta t$.

6.4 The 3+1 Schrödinger Equation Using the ADI Scheme

These formulas are valid in general for inner points, whereas the formulas at the faces of the domain depend on the boundary conditions imposed. In what follows, we work out the case of the particle in a box and the particle in the harmonic oscillator potential that illustrate two types of boundary conditions.

6.4.1 Two Exact Solutions

Before describing the construction of numerical solutions, we present the exact solution for the two basic problems, the particle in a box and the particle in a harmonic oscillator potential.

These two solutions are stationary, which allows us to start their construction together, noticing that in the two cases the potential is time-independent and Schrödinger equation reads

$$i\frac{\partial \Psi(\mathbf{x}, t)}{\partial t} = -\frac{1}{2}\nabla^2 \Psi(\mathbf{x}, t) + V(\mathbf{x})\Psi(\mathbf{x}, t), \tag{6.42}$$

that can be separated assuming $\Psi(\mathbf{x}, t) = \psi(\mathbf{x})T(t)$, which implies that

$$i\frac{1}{T}\frac{dT}{dt} = -\frac{1}{2\psi}\nabla^2\psi + V(\mathbf{x}) = E,$$

where the separation constant is the energy E like in the 1+1 case, which allows one to determine the time dependence of the wave function $T = T_0 e^{-iE(t-t_0)}$ and the spatial wave function ψ obeys the stationary Schrödinger equation:

$$-\frac{1}{2}\nabla^2\psi + V(\mathbf{x})\psi = E\psi. \tag{6.43}$$

Without loss of generality, one can set $T_0 = 1$ because the solution will be later normalized and $t_0 = 0$ as the initial time for the evolution of the wave function. From now on, the particular solutions will depend on the potential $V(\mathbf{x})$.

6.4.1.1 Particle in a 3D Box

We will use a unitary square box, where the potential is defined by

$$V(\mathbf{x}) = \begin{cases} 0, & \mathbf{x} \in [0, 1] \times [0, 1] \times [0, 1] \\ \infty, & \text{otherwise,} \end{cases} \tag{6.44}$$

which reduces Eq. (6.43) to

$$\nabla^2 \psi + 2E\psi = \left(\frac{\partial^2 \psi}{\partial x^2} + \frac{\partial^2 \psi}{\partial y^2} + \frac{\partial^2 \psi}{\partial z^2}\right) + 2E\psi = 0, \quad (6.45)$$

with $\psi = 0$ at the boundaries of the box $[0, 1] \times [0, 1] \times [0, 1]$. The solution is constructed by a further separation of variables assuming $\psi(x, y, x) = X(x)Y(y)Z(z)$ and finding $X_{n_x}(x) = \sqrt{2}\sin(n_x\pi x)$, $Y_{n_y}(y) = \sqrt{2}\sin(n_y\pi y)$, and $Z_{n_z}(z) = \sqrt{2}\sin(n_z\pi z)$, with n_x, n_y, and n_z integers. Applying the boundary conditions and the normalization condition

$$N = \int_0^1 \int_0^1 \int_0^1 \psi^* \psi \, dx dy dz = 1, \quad (6.46)$$

the solution is

$$\psi(x, y, z) = \sqrt{8} \sin(n_x \pi x) \sin(n_y \pi y) \sin(n_z \pi z), \quad (6.47)$$

where n_x, n_y, and n_z are integers and the energy is quantized

$$E = \frac{\pi^2}{2}(n_x^2 + n_y^2 + n_z^2). \quad (6.48)$$

Therefore, the time-dependent solution Ψ reads

$$\Psi(\mathbf{x}, t) = \psi(\mathbf{x})T(t) = \sqrt{8}e^{-it\pi^2(n_x^2+n_y^2+n_z^2)/2} \sin(n_x\pi x) \sin(n_y\pi y), \sin(n_z\pi z) \quad (6.49)$$

whose real part can be used to measure the error of the numerical solution.

Numerical Solution. As an example, we solve the problem with $n_x = n_y = n_z = 3$, in the domain $[0, 1]^3$ using resolution with $N_x = N_y = N_z = 100$ and $\Delta t = C\Delta x^2$ with $C = 1$. Initial conditions for the wave function are the expression (6.49) with $t = 0$. In Fig. 6.7, we show the evolution of the wave function and $\rho = |\Psi|^2$, which remains nearly time-independent.

Diagnostics of the solution includes the conservation of N in (6.46) and the value of the total energy E. In order to evaluate $E = \langle \hat{H} \rangle$, one can use the expectation value of the kinetic energy operator:

$$\langle \hat{H} \rangle = \langle \hat{K} \rangle = \langle \Psi | -\frac{1}{2}\nabla^2 |\Psi\rangle = -\frac{1}{2}\int_0^1 \int_0^1 \int_0^1 \Psi^*(\nabla^2 \Psi) \, dx dy dz. \quad (6.50)$$

In Fig. 6.8, we show the number of particles N calculated using (6.46), the total energy $E = \langle \hat{H} \rangle$ calculated using (6.50), and the central value of ρ (technically the infinity norm of ρ, $||\rho||_\infty$) as functions of time. For comparison, the exact values

6.4 The 3+1 Schrödinger Equation Using the ADI Scheme

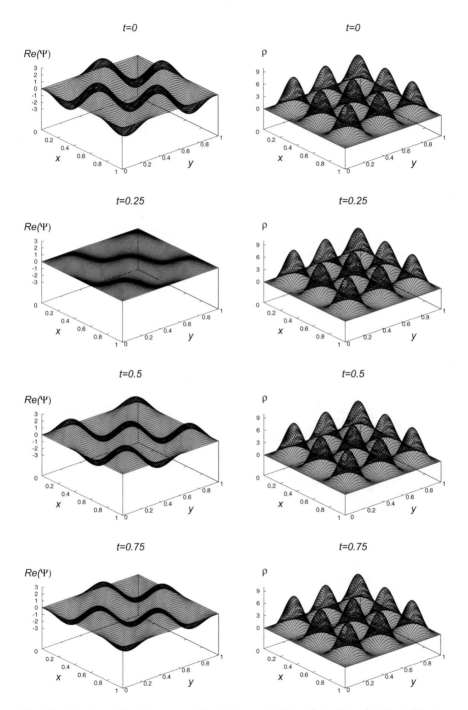

Fig. 6.7 Projection of Re(Ψ) and ρ on the plane $z = 1/2$ at various times, for the particle in a three-dimensional box with $n_x = n_y = n_z = 3$

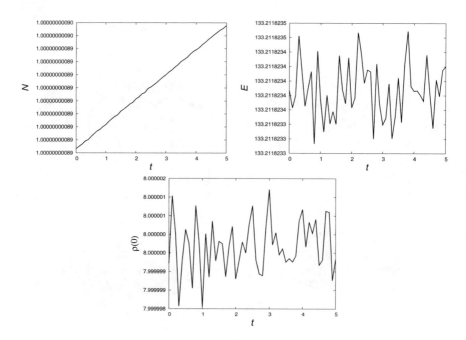

Fig. 6.8 Number of particles N, energy $E = \langle \hat{H} \rangle$, and the central value of ρ as function of time, for the particle in a three-dimensional box with $n_x = n_y = n_z = 3$

are, respectively, $N = 1$ from (6.46), $E = \pi^2(3 \times 3^2)/2 \simeq 133.23966$ from (6.48), and $\rho(\mathbf{0}, t) = 8$ from (6.49).

6.4.2 Particle in a Harmonic Oscillator Potential

In this case, the potential is $V(\mathbf{x}) = \frac{1}{2}\kappa(x^2+y^2+z^2)$, and the stationary Schrödinger equation (6.43) reads

$$\left(\frac{\partial^2 \psi}{\partial x^2} + \frac{\partial^2 \psi}{\partial y^2} + \frac{\partial^2 \psi}{\partial z^2}\right) + \frac{1}{2}\kappa(x^2 + y^2 + z^2) + 2E\psi = 0. \tag{6.51}$$

As we did in the 1D case, we define auxiliary constants $a = \sqrt{\kappa}$ and $b = 2E$. Definition of new variables $\xi = \sqrt{a}x, \eta = \sqrt{a}y, \text{and}\, \zeta = \sqrt{a}z$ leads to the equation

$$\left(\frac{\partial^2 \psi}{\partial \xi^2} + \frac{\partial^2 \psi}{\partial \eta^2} + \frac{\partial^2 \psi}{\partial \zeta^2}\right) + \left(\frac{b}{a} - \xi^2 - \eta^2 - \zeta^2\right)\psi = 0, \tag{6.52}$$

6.4 The 3+1 Schrödinger Equation Using the ADI Scheme

which can be separated via $\psi(\xi, \eta, \zeta) = X(\xi)Y(\eta)Z(\zeta)$, into three Hermite equations whose solutions are

$$X_{n_x}(\xi) = \sqrt{\frac{1}{\sqrt{\pi} 2^{n_x} n_x!}} e^{-\xi^2/2} H_{n_x}(\xi),$$

$$Y_{n_y}(\eta) = \sqrt{\frac{1}{\sqrt{\pi} 2^{n_y} n_y!}} e^{-\eta^2/2} H_{n_y}(\eta),$$

$$Z_{n_z}(\zeta) = \sqrt{\frac{1}{\sqrt{\pi} 2^{n_z} n_z!}} e^{-\zeta^2/2} H_{n_z}(\zeta),$$

where the energy is quantized as follows:

$$E_{n_x, n_y, n_z} = n_x + n_y + n_z + \frac{3}{2}. \tag{6.53}$$

Therefore, the fully time-dependent solution reads

$$\psi(x, y, z, t) = \sqrt{\frac{1}{\pi^{3/2} 2^{n_x + n_y + n_z} n_x! n_y! n_z!}} e^{-iE_{n_x,n_y,n_z}t}$$

$$\times e^{-(x^2+y^2+z^2)/2} H_{n_x}(x) H_{n_y}(y) H_{n_z}(z), \tag{6.54}$$

which can be used to initialize the evolution and to calculate the error of numerical solutions.

Numerical Solution. As an example, we solve the problem with $n_x = n_y = n_z = 2$, in the domain $[-5, 5]^3$ using resolution with $N_x = N_y = N_z = 100$ and $\Delta t = C \Delta x^2$ with $C = 1$. Initial conditions use the exact solution (6.54) with $t = 0$. In Fig. 6.9, we show the evolution of the wave function and that ρ remains nearly time-independent as expected.

Unlike (6.50), the expectation value of the Hamiltonian involves now the potential energy

$$E = \langle \Psi | \hat{H} | \Psi \rangle$$

$$= \langle \Psi | -\frac{1}{2}\nabla^2 + V | \Psi \rangle \int_0^1 \int_0^1 \int_0^1 \left[\Psi^* \left(-\frac{1}{2}\nabla^2 \Psi \right) + V \Psi^* \Psi \right] dx dy dz. \tag{6.55}$$

In Fig. 6.10, we show the number of particles N calculated using (6.46) and the total energy $E = \langle \Psi | \hat{H} | \Psi \rangle$ calculated using (6.55). For reference, the exact values are $N = 1$ and the energy, according to (6.53), is $E = 2 + 2 + 2 + 3/2 = 7.5$.

The code that implements these methods and solutions Schroedinger3p1.f90 can be found in Appendix B.

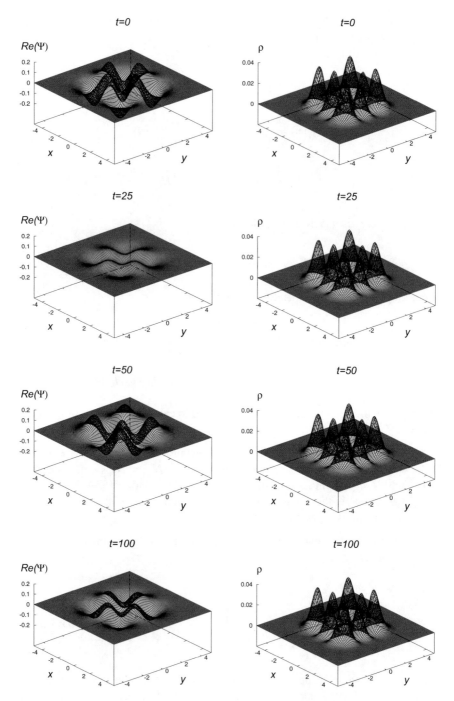

Fig. 6.9 Projection of Re(Ψ) and ρ on the plane $z = 1/2$ at various times for the three-dimensional harmonic oscillator with $n_x = n_y = n_z = 2$

6.5 3D Hydrodynamics

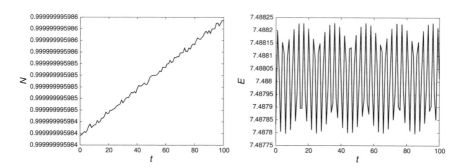

Fig. 6.10 Number of particles N and energy E as function of time, for the three-dimensional harmonic oscillator with $n_x = n_y = n_z = 2$

Exercise. Another way to monitor the energy is to track the oscillation of the wave function, for example, saving the central value of $\text{Re}(\Psi)$ as function of time and then calculating its Fourier transform (FT). The FT should have a dominant peak of frequency $E = n_x + n_y + n_z + 3/2$ for the harmonic oscillator potential. Show this is the case for the example worked out in the text with $n_x = n_y = n_z = 2$.

6.5 3D Hydrodynamics

The dynamics of a compressible inviscid fluid is ruled by the three-dimensional Euler equations. The IVP in this case reads

$$\begin{aligned}
& \partial_t \rho + \nabla \cdot (\rho \mathbf{v}) = 0 \\
& \partial_t (\rho \mathbf{v}) + \nabla \cdot (\mathbf{v} \otimes \rho \mathbf{v}) + \nabla p = -\rho \nabla \phi \\
& \partial_t E + \nabla \cdot [(E+p)\mathbf{v}] = -\rho \mathbf{v} \cdot \nabla \phi, \qquad \rho = \rho(\mathbf{x},t),\, p = p(\mathbf{x},t), \\
& \qquad\qquad\qquad\qquad\qquad\qquad\qquad\qquad\qquad \mathbf{v} = \mathbf{v}(\mathbf{x},t),\, E = E(\mathbf{x},t) \\
& D = [x_{min}, x_{max}] \times [y_{min}, y_{max}] \\
& \qquad \times [z_{min}, z_{max}] \times [0, t_f] \qquad\qquad Domain \\
& \rho(\mathbf{x},0) = \rho_0(\mathbf{x}),\, p(\mathbf{x},0) = p_0(\mathbf{x}), \qquad Initial\ Conditions \\
& \mathbf{v}(\mathbf{x},0) = \mathbf{v}_0(\mathbf{x}),\, E(\mathbf{x},0) = E_0(\mathbf{x}) \\
& \rho(\partial D, t),\, p(\partial D, t),\, \mathbf{v}(\partial D, t),\, E(\partial D, t) \quad Boundary\ Conditions
\end{aligned} \qquad (6.56)$$

where each volume element of the fluid has density ρ, velocity $\mathbf{v}$, pressure p, internal energy e, and total energy $E = \rho(\frac{1}{2}v^2 + e)$. This is a five-equations system for six unknowns ρ, $\mathbf{v}$, p, and e or E and is closed using an equation of state (EoS). We have written additionally the contribution of an external force represented by the potential ϕ, acting on each fluid element. In general, all variables involved depend on time and space.

The strategy to solve this IVP is essentially the generalization of that used for hydrodynamics in 1+1 dimensions, for a bigger system of equations:

- The system of equations in (6.56) for the primitive variables $\rho, p, \mathbf{v}, and\, E$ is transformed into a set of flux balance equations for conservative variables $u_1, u_2, u_3, u_4, and\, u_5$.
- Initial conditions are set for this IVP on primitive variables and then define initial conditions for the conservative variables.
- Evolve the equations written in flux conservative form using the finite volume method using suitable boundary conditions.
- Recover the primitive variables for the next evolution step and output.

We tackle the problem with no symmetries in Cartesian coordinates. To start with, notice that the second equation in (6.56) represents the conservation of momentum density along the three spatial directions, and it is worth expanding its flux. Considering that the velocity field has components $\mathbf{v} = (v_x, v_y, v_z)$

$$\mathbf{v} \otimes \rho\mathbf{v} = \begin{bmatrix} v_x \\ v_y \\ v_z \end{bmatrix} [\rho v_x, \rho v_y, \rho v_z] = \rho \begin{bmatrix} v_x v_x & v_x v_y & v_x v_z \\ v_y v_x & v_y v_y & v_y v_z \\ v_z v_x & v_z v_y & v_z v_z \end{bmatrix}, \quad (6.57)$$

and therefore the second of Eq. (6.56) is expanded into the following three equations:

$$\partial_t(\rho v_x) + \nabla \cdot (\rho(v_x v_x \hat{x} + v_x v_y \hat{y} + v_x v_z \hat{z})) + \partial_x p = -\rho \partial_x \Phi,$$
$$\partial_t(\rho v_y) + \nabla \cdot (\rho(v_y v_x \hat{x} + v_y v_y \hat{y} + v_y v_z \hat{z})) + \partial_y p = -\rho \partial_y \Phi,$$
$$\partial_t(\rho v_z) + \nabla \cdot (\rho(v_z v_x \hat{x} + v_z v_y \hat{y} + v_z v_z \hat{z})) + \partial_z p = -\rho \partial_z \Phi,$$

where $\hat{x}, \hat{y}, and\, \hat{z}$ are the canonical basis vectors of $\mathbb{R}^3$. Expanding the divergence operators in these equations and those of mass and energy conservation in (6.56) is explicitly the following system:

$$\partial_t \rho + \partial_x(\rho v_x) + \partial_y(\rho v_y) + \partial_z(\rho v_z) = 0,$$
$$\partial_t(\rho v_x) + \partial_x(\rho v_x v_x) + \partial_y(\rho v_x v_y) + \partial_z(\rho v_x v_z) + \partial_x p = -\rho \partial_x \Phi,$$
$$\partial_t(\rho v_y) + \partial_x(\rho v_y v_x) + \partial_y(\rho v_y v_y) + \partial_z(\rho v_y v_z) + \partial_y p = -\rho \partial_y \Phi,$$
$$\partial_t(\rho v_z) + \partial_x(\rho v_z v_x) + \partial_y(\rho v_z v_y) + \partial_z(\rho v_z v_z) + \partial_z p = -\rho \partial_z \Phi,$$
$$\partial_t E + \partial_x((E+p)v_x) + \partial_y((E+p)v_y) + \partial_z((E+p)v_z) = -\rho\,(v_x \partial_x \Phi$$
$$+ v_y \partial_y \Phi + v_z \partial_z \Phi\bigr).$$
$$(6.58)$$

6.5 3D Hydrodynamics

These equations suggest the definition of the conservative variables that we accommodate as

$$\mathbf{u} = \begin{bmatrix} \rho \\ \rho v_x \\ \rho v_y \\ \rho v_z \\ E \end{bmatrix} := \begin{bmatrix} u_1 \\ u_2 \\ u_3 \\ u_4 \\ u_5 \end{bmatrix} \tag{6.59}$$

and that the system is a set of flux balance law equations, which is a generalization of Eq. (5.18):

$$\frac{\partial \mathbf{u}}{\partial t} + \frac{\partial (\mathbf{F}_x)}{\partial x} + \frac{\partial (\mathbf{F}_y)}{\partial y} + \frac{\partial (\mathbf{F}_z)}{\partial z} = \mathbf{S}, \tag{6.60}$$

where $\mathbf{F}_x, \mathbf{F}_y,$ and $\mathbf{F}_z$ are five-component flux vectors along each direction, a generalization of Eq. (5.19), along with the source vector:

$$\mathbf{F}_x = \begin{bmatrix} \rho v_x \\ \rho v_x v_x + p \\ \rho v_y v_x \\ \rho v_z v_x \\ (u_5 + p) v_x \end{bmatrix}, \quad \mathbf{F}_y = \begin{bmatrix} \rho v_y \\ \rho v_x v_y \\ \rho v_y v_y + p \\ \rho v_z v_y \\ (u_5 + p) v_y \end{bmatrix},$$

$$\mathbf{F}_z = \begin{bmatrix} \rho v_z \\ \rho v_x v_z \\ \rho v_y v_z \\ \rho v_z v_z + p \\ (u_5 + p) v_z \end{bmatrix}, \quad \mathbf{S} = \begin{bmatrix} 0 \\ -\rho \partial_x \Phi \\ -\rho \partial_y \Phi \\ -\rho \partial_z \Phi \\ -\rho (v_x \partial_x \Phi + v_y \partial_y \Phi + v_z \partial_z \Phi) \end{bmatrix}. \tag{6.61}$$

For the application of the MoL, one writes the semi-discrete version of the system (6.60):

$$\frac{\partial \bar{\mathbf{u}}}{\partial t} = -\frac{(\bar{\mathbf{F}}_x)^{n+1/2}_{i+1/2,j,k} - (\bar{\mathbf{F}}_x)^{n+1/2}_{i-1/2,j,k}}{\Delta x} - \frac{(\bar{\mathbf{F}}_y)^{n+1/2}_{i,j+1/2,k} - (\bar{\mathbf{F}}_y)^{n+1/2}_{i,j-1/2,k}}{\Delta y}$$

$$-\frac{(\bar{\mathbf{F}}_z)^{n+1/2}_{i,j,k+1/2} - (\bar{\mathbf{F}}_z)^{n+1/2}_{i,j,k-1/2}}{\Delta z} + \bar{\mathbf{S}}^{n+1/2}_{i,j,k}, \tag{6.62}$$

which is a generalization of the one-dimensional case in Eq. (5.16). **Notice** that the flux $\mathbf{F}_x$ needs to be known at the left and right from an intercell face on the plane $x = x_i$, $\mathbf{F}_y$ needs to be known at the left and right from an intercell face on the plane $y = y_j$, and $\mathbf{F}_z$ needs to be known at the left and right from an intercell face on the plane $z = z_k$. This **observation** allows one to construct the fluxes to the left

288   6 Initial Value Problems in 3+1 and 2+1 Dimensions

L and right R from an intercell boundary as we did for the one-dimensional case: by considering flux transfer along one of the three directions at a time as we do below.

Following the one-dimensional case, the construction of fluxes aside an intercell boundary requires writing the system of equations using the chain rule and define Jacobian matrices. The generalization of (5.20) for the 3D case reads

$$\frac{\partial \mathbf{u}}{\partial t} + \frac{d\mathbf{F}_x}{d\mathbf{u}}\frac{\partial \mathbf{u}}{\partial x} + \frac{d\mathbf{F}_y}{d\mathbf{u}}\frac{\partial \mathbf{u}}{\partial y} + \frac{d\mathbf{F}_z}{d\mathbf{u}}\frac{\partial \mathbf{u}}{\partial z} = \mathbf{S}, \qquad (6.63)$$

where $d\mathbf{F}_x/d\mathbf{u}$, $d\mathbf{F}_y/d\mathbf{u}$, and $d\mathbf{F}_z/d\mathbf{u}$ are three 5×5 matrices, whereas in the one-dimensional case in (5.21) there was only one 3×3 Jacobian matrix. Each of these three matrices has its own characteristic structure needed for the construction of the numerical fluxes in (6.62).

As we did for one-dimensional problems, we construct numerical fluxes using the HLLE formula, which needs the eigenvalues of these Jacobian matrices. This requires writing the flux vectors in terms of conservative variables, which in turn needs closing the system with an EoS. We again use the ideal gas EoS $p = (\gamma - 1)\rho e$, which fixes the total energy to $u_5 = E = \rho(\frac{1}{2}v^2 + e) = \frac{1}{2}\rho v^2 + \frac{p}{\gamma-1}$. The flux vectors in (6.61) are then

$$\mathbf{F}_x = \begin{bmatrix} \rho v_x \\ \rho v_x v_x + p \\ \rho v_y v_x \\ \rho v_z v_x \\ (u_5 + p)v_x \end{bmatrix} = \begin{bmatrix} u_2 \\ \frac{1}{2}(3-\gamma)\frac{u_2^2}{u_1} + (\gamma-1)(u_5 - \frac{1}{2}\frac{u_3^2+u_4^2}{u_1}) \\ \frac{u_2 u_3}{u_1} \\ \frac{u_2 u_4}{u_1} \\ \gamma\frac{u_2 u_5}{u_1} - \frac{1}{2}(\gamma-1)(u_2^2 + u_3^2 + u_4^2)\frac{u_2}{u_1^2} \end{bmatrix}, \qquad (6.64)$$

$$\mathbf{F}_y = \begin{bmatrix} \rho v_y \\ \rho v_x v_y \\ \rho v_y v_y + p \\ \rho v_z v_y \\ (u_5 + p)v_y \end{bmatrix} = \begin{bmatrix} u_3 \\ \frac{u_3 u_2}{u_1} \\ \frac{1}{2}(3-\gamma)\frac{u_3^2}{u_1} + (\gamma-1)(u_5 - \frac{1}{2}\frac{u_2^2+u_4^2}{u_1}) \\ \frac{u_3 u_4}{u_1} \\ \gamma\frac{u_3 u_5}{u_1} - \frac{1}{2}(\gamma-1)(u_2^2 + u_3^2 + u_4^2)\frac{u_3}{u_1^2} \end{bmatrix}, \qquad (6.65)$$

$$\mathbf{F}_z = \begin{bmatrix} \rho v_z \\ \rho v_x v_z \\ \rho v_y v_z \\ \rho v_z v_z + p \\ (u_5 + p)v_z \end{bmatrix} = \begin{bmatrix} u_4 \\ \frac{u_4 u_2}{u_1} \\ \frac{u_4 u_3}{u_1} \\ \frac{1}{2}(3-\gamma)\frac{u_4^2}{u_1} + (\gamma-1)(u_5 - \frac{1}{2}\frac{u_2^2+u_3^2}{u_1}) \\ \gamma\frac{u_4 u_5}{u_1} - \frac{1}{2}(\gamma-1)(u_2^2 + u_3^2 + u_4^2)\frac{u_4}{u_1^2} \end{bmatrix}. \qquad (6.66)$$

The eigenvalues of the Jacobian matrices $d\mathbf{F}_x/d\mathbf{u}$, $d\mathbf{F}_y/d\mathbf{u}$, and $d\mathbf{F}_z/d\mathbf{u}$ are needed for the implementation of the HLLE flux formula and are

6.5 3D Hydrodynamics

eigenvalues of $\dfrac{d\mathbf{F}_x}{d\mathbf{u}}$, $\lambda_1 = v_x - c_s$, $\lambda_2 = v_x$, $\lambda_3 = v_x$, $\lambda_4 = v_x$, λ_5
$$= v_x + c_s \tag{6.67}$$

eigenvalues of $\dfrac{d\mathbf{F}_y}{d\mathbf{u}}$, $\lambda_1 = v_y - c_s$, $\lambda_2 = v_y$, $\lambda_3 = v_y$, $\lambda_4 = v_y$, λ_5
$$= v_y + c_s \tag{6.68}$$

eigenvalues of $\dfrac{d\mathbf{F}_z}{d\mathbf{u}}$, $\lambda_1 = v_z - c_s$, $\lambda_2 = v_z$, $\lambda_3 = v_z$, $\lambda_4 = v_z$, λ_5
$$= v_z + c_s \tag{6.69}$$

where $c_s = \sqrt{\gamma p/\rho}$ is the speed of sound for the ideal gas.

The following step may seem **excessively** developed; however, it is essential for the careful implementation of the method in 3D, and we specify the needed steps here.

Fluxes x. Calculation of $(\bar{\mathbf{F}}_x)_{i+1/2,j,k}$ in Eq. (6.62) uses the following HLLE formula for all $i = 0, \ldots, N_x - 1$

$$(\bar{\mathbf{F}}_x)^{HLLE}_{i+1/2,j,k} = \frac{\lambda_+ \mathbf{F}_x(\mathbf{u}^L_{i+1/2,j,k}) - \lambda_- \mathbf{F}_x(\mathbf{u}^R_{i+1/2,j,k}) + \lambda_+ \lambda_- (\mathbf{u}^R_{i+1/2,j,k} - \mathbf{u}^L_{i+1/2,j,k})}{\lambda_+ - \lambda_-}, \tag{6.70}$$

where λ_+ and λ_- are generalized from (5.33) to

$$\lambda_+ = \max(0, \lambda^R_1, \lambda^R_2, \lambda^R_3, \lambda^R_4, \lambda^R_5, \lambda^L_1, \lambda^L_2, \lambda^L_3, \lambda^L_4, \lambda^L_5),$$
$$\lambda_- = \min(0, \lambda^R_1, \lambda^R_2, \lambda^R_3, \lambda^R_4, \lambda^R_5, \lambda^L_1, \lambda^L_2, \lambda^L_3, \lambda^L_4, \lambda^L_5), \tag{6.71}$$

with the eigenvalues of $d\mathbf{F}_x/d\mathbf{u}$ in (6.67) at left and right from the intercell boundary at the plane $x = x_i$:

$$\lambda^L_1 = v_{xL} - c_{sL}, \quad \lambda^L_2 = v_{xL}, \quad \lambda^L_3 = v_{xL}, \quad \lambda^L_4 = v_{xL}, \quad \lambda^L_5 = v_{xL} + c_{sL},$$

$$\lambda^R_1 = v_{xR} - c_{sR}, \quad \lambda^R_2 = v_{xR}, \quad \lambda^R_3 = v_{xR}, \quad \lambda^R_4 = v_{xR}, \quad \lambda^R_5 = v_{xR} + c_{sR},$$

where

$$c_{sL} = \sqrt{p_L \gamma/\rho_L},$$

$$c_{sR} = \sqrt{p_R \gamma/\rho_R}. \tag{6.72}$$

On the other hand, in Eq. (6.70), the flux is evaluated at $\mathbf{u}^L_{i+1/2,j,k}$, $\mathbf{u}^R_{i+1/2,j,k}$, which are the values of the conservative variables reconstructed at the left L and right R cells from the interface at $x = x_i$. For the examples in this book, we use the Godunov reconstructor (5.35) that along the x–direction for a variable w reads

$$w^L_{i+1/2,j,k} = w_{i,j,k},$$
$$w^R_{i+1/2,j,k} = w_{i+1,j,k}, \tag{6.73}$$

or the minmod reconstructor (5.36) that now reads

$$w^L_{i+1/2,j,k} = w_{i,j,k} + \sigma_i(x_{i+1/2} - x_i),$$
$$w^R_{i+1/2,j,k} = w_{i+1,j,k} + \sigma_{i+1}(x_{i+1/2} - x_{i+1}), \tag{6.74}$$
$$\sigma_i = \text{minmod}(m_{i-1/2}, m_{i+1/2})$$

$$\text{minmod}(a, b) = \begin{cases} a & \text{if } |a| < |b| \text{ and } ab > 0 \\ b & \text{if } |a| > |b| \text{ and } ab > 0 \\ 0 & \text{if } ab < 0. \end{cases}$$

$$m_{i+1/2} = \frac{w_{i+1,j,k} - w_{i,j,k}}{x_{i+1} - x_i},$$
$$m_{i-1/2} = \frac{w_{i,j,k} - w_{i-1,j,k}}{x_i - x_{i-1}}.$$

Notice that w is any variable, either primitive or conservative. Our strategy here is that we reconstruct the primitive variables ρ^L, v^L_x, v^L_y, v^L_z, p^L, e^L, E^L, ρ^R, v^R_x, v^R_y, v^R_z, p^R, e^R, and E^R, and from these, we calculate the conservative variables $\mathbf{u}^L_{i+1/2,j,k} = (u^L_1, u^L_2, u^L_3, u^L_4, u^L_5)$ and $\mathbf{u}^R_{i+1/2,j,k} = (u^R_1, u^R_2, u^R_3, u^R_4, u^R_5)$ needed in the flux formula (6.70). Finally, the flux vector $\mathbf{F}_x$, according to (6.64) evaluated at $\mathbf{u}^L_{i+1/2,j,k}$, $\mathbf{u}^R_{i+1/2,j,k}$, reads

$$\mathbf{F}_x(\mathbf{u}^L_{i+1/2,j,k}) = \begin{bmatrix} u^L_2 \\ \frac{1}{2}(3-\gamma)\frac{(u^L_2)^2}{u^L_1} + (\gamma-1)(u^L_5 - \frac{1}{2}\frac{(u^L_3)^2+(u^L_4)^2}{u^L_1}) \\ \frac{u^L_2 u^L_3}{u^L_1} \\ \frac{u^L_2 u^L_4}{u^L_1} \\ \gamma \frac{u^L_2 u^L_5}{u^L_1} - \frac{1}{2}(\gamma-1)((u^L_2)^2 + (u^L_3)^2 + (u^L_4)^2)\frac{u^L_2}{(u^L_1)^2} \end{bmatrix},$$
$$\tag{6.75}$$

$$\mathbf{F}_x(\mathbf{u}^R_{i+1/2,j,k}) = \begin{bmatrix} u^R_2 \\ \frac{1}{2}(3-\gamma)\frac{(u^R_2)^2}{u^R_1} + (\gamma-1)(u^R_5 - \frac{1}{2}\frac{(u^R_3)^2+(u^R_4)^2}{u^R_1}) \\ \frac{u^R_2 u^R_3}{u^R_1} \\ \frac{u^R_2 u^R_4}{u^R_1} \\ \gamma \frac{u^R_2 u^R_5}{u^R_1} - \frac{1}{2}(\gamma-1)((u^R_2)^2 + (u^R_3)^2 + (u^R_4)^2)\frac{u^R_2}{(u^R_1)^2} \end{bmatrix},$$
$$\tag{6.76}$$

which are the expressions used in the HLLE flux formula (6.70).

6.5 3D Hydrodynamics

Fluxes y. Calculation of $(\bar{\mathbf{F}}_y)_{i,j+1/2,k}$ in Eq. (6.62) uses the following HLLE formula for all $j = 0, \ldots, N_y - 1$

$$(\bar{\mathbf{F}}_y)^{HLLE}_{i,j+1/2,k} = \frac{\lambda_+ \mathbf{F}_y(\mathbf{u}^L_{i,j+1/2,k}) - \lambda_- \mathbf{F}_y(\mathbf{u}^R_{i,j+1/2,k}) + \lambda_+ \lambda_- (\mathbf{u}^R_{i,j+1/2,k} - \mathbf{u}^L_{i,j+1/2,k})}{\lambda_+ - \lambda_-}, \tag{6.77}$$

where λ_+ and λ_- are

$$\lambda_+ = \max(0, \lambda_1^R, \lambda_2^R, \lambda_3^R, \lambda_4^R, \lambda_5^R, \lambda_1^L, \lambda_2^L, \lambda_3^L, \lambda_4^L, \lambda_5^L),$$
$$\lambda_- = \min(0, \lambda_1^R, \lambda_2^R, \lambda_3^R, \lambda_4^R, \lambda_5^R, \lambda_1^L, \lambda_2^L, \lambda_3^L, \lambda_4^L, \lambda_5^L), \tag{6.78}$$

with the eigenvalues of $d\mathbf{F}_y/d\mathbf{u}$ in (6.68) at left and right from the intercell boundary at the plane $y = y_j$:

$$\lambda_1^L = v_{yL} - c_{sL}, \quad \lambda_2^L = v_{yL}, \quad \lambda_3^L = v_{yL}, \quad \lambda_4^L = v_{yL}, \quad \lambda_5^L = v_{yL} + c_{sL},$$

$$\lambda_1^R = v_{yR} - c_{sR}, \quad \lambda_2^R = v_{yR}, \quad \lambda_3^R = v_{yR}, \quad \lambda_4^R = v_{yR}, \quad \lambda_5^R = v_{yR} + c_{sR},$$

where

$$c_{sL} = \sqrt{p_L \gamma / \rho_L},$$

$$c_{sR} = \sqrt{p_R \gamma / \rho_R}. \tag{6.79}$$

In this case, $\mathbf{u}^L_{i,j+1/2,k}$, $\mathbf{u}^R_{i,j+1/2,k}$ are the values of the variables reconstructed at the left L and right R cells from the interface at $y = y_j$. The Godunov reconstructor (5.35) along the y-direction for a variable w, which as described above can represent either a primitive or a conservative variable, reads

$$w^L_{i,j+1/2,k} = w_{i,j,k},$$
$$w^R_{i,j+1/2,k} = w_{i,j+1,k}, \tag{6.80}$$

or the minmod reconstructor (5.36) along the y-direction

$$w^L_{i,j+1/2,k} = w_{i,j,k} + \sigma_j (y_{j+1/2} - y_j),$$
$$w^R_{i,j+1/2,k} = w_{i,j+1,k} + \sigma_{j+1}(y_{j+1/2} - y_{j+1}), \tag{6.81}$$
$$\sigma_j = \text{minmod}(m_{j-1/2}, m_{j+1/2})$$

$$\text{minmod}(a, b) = \begin{cases} a & \text{if } |a| < |b| \text{ and } ab > 0 \\ b & \text{if } |a| > |b| \text{ and } ab > 0 \\ 0 & \text{if } ab < 0. \end{cases}$$

$$m_{j+1/2} = \frac{w_{i,j+1,k} - w_{i,j,k}}{y_{j+1} - y_j},$$

$$m_{j-1/2} = \frac{w_{i,j,k} - w_{i,j-1,k}}{y_j - y_{j-1}}.$$

Notice that w is any variable, either primitive or conservative. Again, our strategy here is that we reconstruct the primitive variables ρ^L, v_x^L, v_y^L, v_z^L, p^L, e^L, E^L, ρ^R, v_x^R, v_y^R, v_z^R, p^R, e^R, and E^R, and from these, we calculate the conservative ones $\mathbf{u}_{i,j+1/2,k}^L = (u_1^L, u_2^L, u_3^L, u_4^L, u_5^L)$ and $\mathbf{u}_{i,j+1/2,k}^R = (u_1^R, u_2^R, u_3^R, u_4^R, u_5^R)$ needed in the flux formula (6.77). The flux vector $\mathbf{F}_y$, according to (6.65) evaluated at $\mathbf{u}_{i,j+1/2,k}^L$, $\mathbf{u}_{i,j+1/2,k}^R$, reads

$$\mathbf{F}_y(\mathbf{u}_{i,j+1/2,k}^L) = \begin{bmatrix} u_3^L \\ \frac{u_3^L u_2^L}{u_1^L} \\ \frac{1}{2}(3-\gamma)\frac{(u_3^L)^2}{u_1^L} + (\gamma - 1)(u_5^L - \frac{1}{2}\frac{(u_2^L)^2 + (u_4^L)^2}{u_1^L}) \\ \frac{u_3^L u_4^L}{u_1^L} \\ \gamma \frac{u_3^L u_5^L}{u_1^L} - \frac{1}{2}(\gamma - 1)((u_2^L)^2 + (u_3^L)^2 + (u_4^L)^2)\frac{u_3^L}{(u_1^L)^2} \end{bmatrix},$$

(6.82)

$$\mathbf{F}_y(\mathbf{u}_{i,j+1/2,k}^R) = \begin{bmatrix} u_3^R \\ \frac{u_3^R u_2^R}{u_1^R} \\ \frac{1}{2}(3-\gamma)\frac{(u_3^R)^2}{u_1^R} + (\gamma - 1)(u_5^R - \frac{1}{2}\frac{(u_2^R)^2 + (u_4^R)^2}{u_1^R}) \\ \frac{u_3^R u_4^R}{u_1^R} \\ \gamma \frac{u_3^R u_5^R}{u_1^R} - \frac{1}{2}(\gamma - 1)((u_2^R)^2 + (u_3^R)^2 + (u_4^R)^2)\frac{u_3^R}{(u_1^R)^2} \end{bmatrix},$$

(6.83)

which are the expressions needed in the HLLE flux formula (6.77).

Fluxes z. Finally, the calculation of $(\bar{\mathbf{F}}_z)_{i,j,k+1/2}$ in Eq. (6.62) uses the following HLLE formula for all $k = 0, \ldots, N_z - 1$:

$$(\bar{\mathbf{F}}_z)_{i,j,k+1/2}^{HLLE} = \frac{\lambda_+ \mathbf{F}_z(\mathbf{u}_{i,j,k+1/2}^L) - \lambda_- \mathbf{F}_z(\mathbf{u}_{i,j,k+1/2}^R) + \lambda_+ \lambda_- (\mathbf{u}_{i,j,k+1/2}^R - \mathbf{u}_{i,j,k+1/2}^L)}{\lambda_+ - \lambda_-},$$

(6.84)

6.5 3D Hydrodynamics

where λ_+ and λ_- are

$$\lambda_+ = \max(0, \lambda_1^R, \lambda_2^R, \lambda_3^R, \lambda_4^R, \lambda_5^R, \lambda_1^L, \lambda_2^L, \lambda_3^L, \lambda_4^L, \lambda_5^L),$$
$$\lambda_- = \min(0, \lambda_1^R, \lambda_2^R, \lambda_3^R, \lambda_4^R, \lambda_5^R, \lambda_1^L, \lambda_2^L, \lambda_3^L, \lambda_4^L, \lambda_5^L), \quad (6.85)$$

with the eigenvalues of $d\mathbf{F}_z/d\mathbf{u}$ in (6.69) at left and right from the intercell boundary at the plane $z = z_k$:

$$\lambda_1^L = v_{zL} - c_{sL}, \quad \lambda_2^L = v_{zL}, \quad \lambda_3^L = v_{zL}, \quad \lambda_4^L = v_{zL}, \quad \lambda_5^L = v_{zL} + c_{sL},$$

$$\lambda_1^R = v_{zR} - c_{sR}, \quad \lambda_2^R = v_{zR}, \quad \lambda_3^R = v_{zR}, \quad \lambda_4^R = v_{zR}, \quad \lambda_5^R = v_{zR} + c_{sR},$$

where (6.86)

$$c_{sL} = \sqrt{p_L \gamma / \rho_L},$$

$$c_{sR} = \sqrt{p_R \gamma / \rho_R}.$$

Now, $\mathbf{u}^L_{i,j,k+1/2}$, $\mathbf{u}^R_{i,j,k+1/2}$ are the values of the variables reconstructed at the left L and right R cells from the interface at $z = z_k$. Godunov reconstructor (5.35) along the z-direction for a variable w, which can represent either a primitive or a conservative variable, reads

$$w^L_{i,j,k+1/2} = w_{i,j,k},$$
$$w^R_{i,j,k+1/2} = w_{i,j,k+1}, \quad (6.87)$$

or the minmod reconstructor (5.36), which along the z-direction reads

$$w^L_{i,j,k+1/2} = w_{i,j,k} + \sigma_j (z_{k+1/2} - z_k),$$
$$w^R_{i,j,k+1/2} = w_{i,j,k+1} + \sigma_{k+1} (z_{k+1/2} - z_{k+1}), \quad (6.88)$$
$$\sigma_k = \mathrm{minmod}(m_{k-1/2}, m_{k+1/2})$$

$$\mathrm{minmod}(a,b) = \begin{cases} a & \text{if } |a| < |b| \text{ and } ab > 0 \\ b & \text{if } |a| > |b| \text{ and } ab > 0 \\ 0 & \text{if } ab < 0. \end{cases}$$

$$m_{k+1/2} = \frac{w_{i,j,k+1} - w_{i,j,k}}{z_{k+1} - z_k},$$

$$m_{k-1/2} = \frac{w_{i,j,k} - w_{i,j,k-1}}{z_k - z_{k-1}}.$$

Notice that w is any variable, either primitive or conservative. Our strategy here is that we reconstruct the primitive variables ρ^L, v_x^L, v_y^L, v_z^L, p^L, e^L, E^L, ρ^R, v_x^R, v_y^R, v_z^R, p^R, e^R, and E^R, and from these, we calculate the conservative variables $\mathbf{u}_{i,j,k+1/2}^L = (u_1^L, u_2^L, u_3^L, u_4^L, u_5^L)$ and $\mathbf{u}_{i,j,k+1/2}^R = (u_1^R, u_2^R, u_3^R, u_4^R, u_5^R)$ needed in the flux formula (6.84). The vector of fluxes $\mathbf{F}_z$, according to (6.66) at $\mathbf{u}_{i,j,k+1/2}^L$, $\mathbf{u}_{i,j,k+1/2}^R$, reads

$$\mathbf{F}_z(\mathbf{u}_{i,j,k+1/2}^L) = \begin{bmatrix} u_4^L \\ \frac{u_4^L u_2^L}{u_1^L} \\ \frac{u_4^L u_3^L}{u_1^L} \\ \frac{1}{2}(3-\gamma)\frac{(u_4^L)^2}{u_1^L} + (\gamma-1)(u_5^L - \frac{1}{2}\frac{(u_2^L)^2+(u_3^L)^2}{u_1^L}) \\ \gamma\frac{u_4^L u_5^L}{u_1^L} - \frac{1}{2}(\gamma-1)((u_2^L)^2 + (u_3^L)^2 + (u_4^L)^2)\frac{u_4^L}{(u_1^L)^2} \end{bmatrix}, \tag{6.89}$$

$$\mathbf{F}_z(\mathbf{u}_{i,j,k+1/2}^R) = \begin{bmatrix} u_4^R \\ \frac{u_4^R u_2^R}{u_1^R} \\ \frac{u_4^R u_3^R}{u_1^R} \\ \frac{1}{2}(3-\gamma)\frac{(u_4^R)^2}{u_1^R} + (\gamma-1)(u_5^R - \frac{1}{2}\frac{(u_2^R)^2+(u_3^R)^2}{u_1^R}) \\ \gamma\frac{u_4^R u_5^R}{u_1^R} - \frac{1}{2}(\gamma-1)((u_2^R)^2 + (u_3^R)^2 + (u_4^R)^2)\frac{u_4^R}{(u_1^R)^2} \end{bmatrix}, \tag{6.90}$$

which are the expressions needed in the HLLE flux formula (6.84).

Boundary Conditions. Two types of boundary conditions are common in 3D hydrodynamics, namely, outflow and periodic boundary conditions.

Outflow Boundary Conditions. These boundary conditions are provided by Eq. (5.39) for 1+1 problems. In 3+1 problems, one generalizes these formulas to be implemented on each face of the spatial domain, that is, for every conservative variable u_m, $m = 1, 2, 3, 4, 5$, one defines trivial Riemann problems on cells at the boundary faces:

$$\mathbf{u}_{0,j,k} = \mathbf{u}_{1,j,k}, \quad \mathbf{u}_{N_x,j,k} = \mathbf{u}_{N_x-1,j,k}$$
$$\mathbf{u}_{i,0,k} = \mathbf{u}_{i,1,k}, \quad \mathbf{u}_{i,N_y,k} = \mathbf{u}_{i,N_y-1,k}$$
$$\mathbf{u}_{i,j,0} = \mathbf{u}_{i,j,1}, \quad \mathbf{u}_{i,j,N_z} = \mathbf{u}_{i,j,N_z-1} \tag{6.91}$$

with the aim of preventing boundary cells to interfere with the dynamics of the interior of the numerical domain.

6.5 3D Hydrodynamics

Periodic boundary conditions are implemented following the indications in Eq. (6.36).

Wrapping up the above elements, the method needed to construct the solution of the IVP has the following steps:

1. Set initial conditions for primitive variables of the IVP (6.56). Then use the definition (6.59) to initialize conservative variables.
2. Evolve a time step from time t^n to t^{n+1} using the MoL using the semi-discrete version of the evolution equations (6.62). This will need the evaluation of sources and fluxes as follows:

    (a) Calculate numerical fluxes $\mathbf{F}_x^{HLLE}$ using the recipe in the sequence from (6.70) to (6.76).
    (b) Calculate numerical fluxes $\mathbf{F}_y^{HLLE}$ using the recipe in the sequence from (6.77) to (6.83).
    (c) Calculate numerical fluxes $\mathbf{F}_z^{HLLE}$ using the recipe in the sequence from (6.84) to (6.90).
    (d) Calculate the sources $\mathbf{S}$ if any.
    (e) Implement the method of lines on the semi-discrete formula (6.62) using the Heun-RK2 or RK3 methods.
    (f) Implement boundary conditions on conservative variables.
    (g) Calculate primitive variables for the next step and output

$$\rho = u_1,$$
$$v_x = \frac{u_2}{u_1},$$
$$v_y = \frac{u_3}{u_1},$$
$$v_z = \frac{u_4}{u_1},$$
$$p = (\gamma - 1)\left[u_5 - \frac{1}{2}\frac{u_2^2 + u_3^2 + u_4^2}{u_1}\right]. \tag{6.92}$$

3. Repeat step 2 until $t = t_f$.

The method is applied to the following traditional example problems.

6.5.1 2D Spherical Blast Wave

Two-dimensional problems can be solved using the 3D code by suppressing one of the dimensions by hand into the code as we illustrate here. We define the domain to be $D = [x_{min}, x_{max}] \times [y_{min}, y_{max}] \times [z_{min}, z_{max}] \times [0, t_f] = [-0.5, 0.5] \times$

$[-0.5, 0.5] \times [-0.01, -0.01] \times [0, 0.5]$ that we cover with $N_x = N_y = 100$ and $N_z = 2$ or equivalently spatial resolution $\Delta x = 0.01$ and time resolution given by $\Delta t = 0.125 \Delta x$. We use outflow boundary conditions on the cubic surface.

We use the adiabatic index $\gamma = 5/3$ and outflow boundary conditions along the three directions. Notice that these conditions along the z-direction, being covered with two cells only, avoid leaking fluxes through the faces z_{min} and z_{max} of the domain. The initial conditions for the wave are the following:

$$(\rho_0(\mathbf{x}), p_0(\mathbf{x}), \mathbf{v}_0(\mathbf{x})) = \begin{cases} (1, 1, \mathbf{0}), & \text{if } \sqrt{x^2 + y^2} \leq 0.1 \\ (0.125, 0.1, \mathbf{0}), & \text{if } \sqrt{x^2 + y^2} > 0.1 \end{cases}. \quad (6.93)$$

The evolution of these initial conditions triggers an initial explosion with a front shock expanding outward. A few snapshots are shown in Fig. 6.11. Notice that the

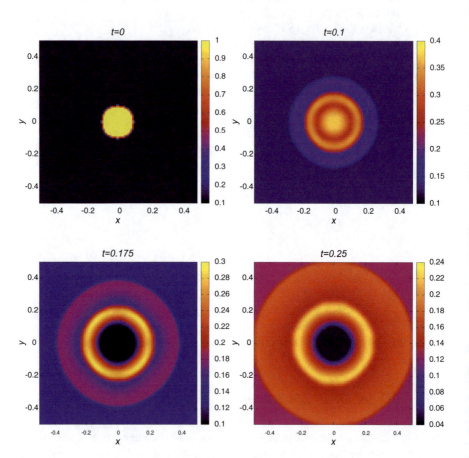

Fig. 6.11 Density ρ of the numerical solution of the 2D spherical blast wave problem on the xy-plane at times $t = 0, 0.1, 0.175$, and 0.25

6.5 3D Hydrodynamics

use of Cartesian coordinates limits the accuracy in setting initial data in this case and the high-density circle is only a *lego circle*. Afterward, the wave front smoothens and becomes more circular.

6.5.2 3D Spherical Blast Wave

We now evolve the blast wave in 3D in order to notice the differences with the 2D problem above. In this case, we define the domain to be $D = [x_{min}, x_{max}] \times [y_{min}, y_{max}] \times [z_{min}, z_{max}] \times [0, t_f] = [-0.5, 0.5] \times [-0.5, 0.5] \times [-0.5, 0.5] \times [0, 0.5]$ that we cover with $N_x = N_y = N_z = 100$ or equivalently resolution $\Delta x = 0.01$ with time resolution $\Delta t = 0.125 \Delta x$. We use outflow boundary conditions on the cubic surface.

We use the same adiabatic index $\gamma = 5/3$ and the following initial conditions:

$$(\rho_0(\mathbf{x}), p_0(\mathbf{x}), \mathbf{v}_0(\mathbf{x})) = \begin{cases} (1, 1, \mathbf{0}), & \text{if } \sqrt{x^2 + y^2 + z^2} \leq 0.1 \\ (0.125, 0.1, \mathbf{0}), & \text{if } \sqrt{x^2 + y^2 + z^2} > 0.1 \end{cases}. \quad (6.94)$$

The evolution of these initial conditions triggers the initial explosion, and a few snapshots are shown in Fig. 6.12. Differences between the propagation of a spherical shock wave in 3D and a circular one in 2D are the amplitude of density seen in the color map and the velocity of the wave front. In the 2D case, the ejected matter fills an area, whereas in the 3D it fills a volume.

6.5.3 2D Kelvin-Helmholtz Instability

This is an example of how a fluid develops an instability due to a perturbed counter flux scenario. We solve this problem on the domain $D = [x_{min}, x_{max}] \times [y_{min}, y_{max}] \times [z_{min}, z_{max}] \times [0, t_f] = [-.5, 0.5] \times [-0.5, 0.5] \times [-0.01, -0.01] \times [0, 2]$ that we cover with $N_x = N_y = 100$ and $N_z = 2$ or equivalently spatial resolution $\Delta x = 0.01$ and time resolution $\Delta t = 0.125 \Delta x$.

We use the adiabatic index $\gamma = 1.4$ and periodic boundary conditions. These boundary conditions are important because the instability develops after the fluid crosses the domain various times. The initial conditions correspond to a lane of fluid moving to the right, above and below which there is a denser counterflow moving to the left as follows:

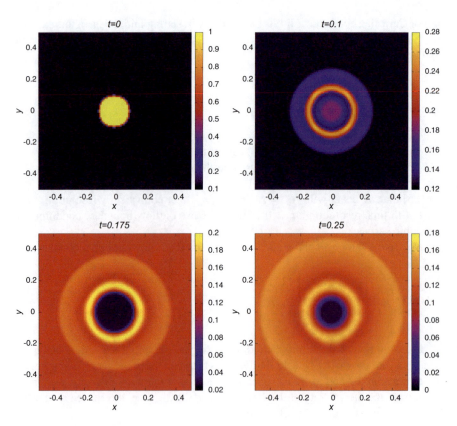

Fig. 6.12 Density ρ of the numerical solution of the 3D spherical blast wave problem on the xy-plane at times $t = 0, 0.1, 0.175$, and 0.25. These results are to be compare with those in Fig. 6.11

$$(\rho_0(\mathbf{x}), p_0(\mathbf{x}), \mathbf{v}_0(\mathbf{x})) = \begin{cases} (1, 2.5, (0.5 + \delta, \delta, 0)), & \text{if } \sqrt{x^2 + y^2} \leq 0.1 \\ (2, 2.5, (-0.5 + \delta, \delta, 0)), & \text{if } \sqrt{x^2 + y^2} > 0.1 \end{cases}$$
(6.95)

where the perturbation of the velocity field has the form $\delta = 0.1 \cos(4\pi x) \sin(4\pi y)$.

The evolution of these initial conditions triggers the instability, and a few snapshots are shown in Fig. 6.13, where the typical curly shapes of this problem can be seen.

The code `Euler3p1.f90` serves to solve these exercises and illustrates the implementation of the solution of Euler equations according to the theory in this section. The code is available as Electronic Supplementary Material.

6.5 3D Hydrodynamics

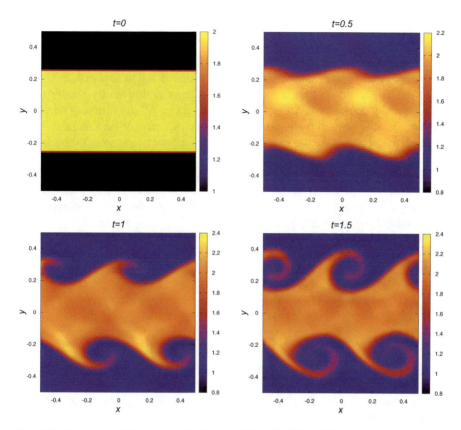

Fig. 6.13 Density ρ of the numerical solution of the 2D Kelvin-Helmholtz problem on the xy-plane at times $t = 0, 0.5, 1$, and 1.5

Exercise. Construct a potential ϕ that represents a constant, vertical gravitational field. Code it as a source according to Eq. (6.56) in the appropriate arrays source1, source2, source3, source 4, and source 5 of the code above. Then study the effects on the Kelvin-Helmholtz instability.

Exercise. Add gravity to the equations and implement the Rayleigh-Taylor instability problem within the code, following the recipe used by the ATHENA code [3].

6.6 Diffusion Equation in 2+1 Dimensions

In the previous section, the example of the 2D blast wave shows how to solve problems in 2+1 dimensions using a code designed for hydrodynamics in 3+1 domains. The reduction to actual 2+1 domains is straightforwardly obtained from 3D restricting operators to act only on the xy-plane. All the formulas developed in Sect. 6.1 can be used by ignoring dependency of functions on z and the label k of discrete operators. We use examples with the diffusion equation to show how to implement these ideas.

There are different degrees of complexity of the diffusion equation, which may include spatial dependence of the diffusion coefficient and proliferation terms useful in the study of populations. We solve three different scenarios with an increasing complexity. **First**, consider the following IVP:

$$\begin{array}{ll} \frac{\partial u}{\partial t} = \kappa \left(\frac{\partial^2 u}{\partial x^2} + \frac{\partial^2 u}{\partial y^2} \right), & u = u(x, y, t) \\ D = [x_{min}, x_{max}] \times [y_{min}, y_{max}] \times [0, t_f] & Domain \\ u(x, y, 0) =_0 (x, y) & Initial\ Conditions \\ u(\partial D, t) & Boundary\ Conditions \end{array} \quad (6.96)$$

where κ is a constant. This is the simplest and typical diffusion equation and is the generalization of the case we solved in Sect. 3.4 from one to two spatial dimensions. We use the domain $D = [0, 1] \times [0, 1] \times [0, 0.1]$ and zero Dirichlet boundary conditions $u(0, y) = u(1, y) = u(x, 0) = u(x, 1) = 0$.

This IVP has exact solution. Using separation of variables $u(x, y, t) = T(t)X(x)Y(y)$ leads to the set of three ODEs:

$$\frac{1}{T}\frac{dT}{dt} = -K, \quad \frac{1}{X}\frac{d^2 X}{dx^2} = -a, \quad \frac{1}{Y}\frac{d^2 Y}{dy^2} = -b \quad (6.97)$$

with K, a, and b positive constants and the constraint $a + b = K/\kappa$. Imposing the boundary conditions, the solution to these equations is $T = T_0 e^{-K(t-t_0)}$, $X = A\sin(n_x \pi x)$, $Y = B\sin(n_y \pi y)$, with n_x, n_y integers constrained by the condition $(n_x^2 + n_y^2)\pi^2 = K/\kappa$, and A, B arbitrary amplitudes. Setting $t_0 = 0$ and grouping all coefficients into a single one, the exact solution is

$$u^e(x, y, t) = C e^{-\kappa(n_x^2 + n_y^2)\pi^2 t} \sin(n_x \pi x) \sin(n_y \pi y), \quad (6.98)$$

that we use to set initial conditions and monitor the numerical solution. We solve the problem numerically using the MoL, on the discrete domain D_d defined with $N_x = N_y = 100$ or equivalently $\Delta x = 0.01$ and $\Delta t = 0.125 \Delta x^2$. The semi-discrete version of the evolution equation reads

6.6 Diffusion Equation in 2+1 Dimensions

$$\frac{\partial u}{\partial t} = \kappa \left(\frac{u_{i+1,j}^n - 2u_{i,j}^n + u_{i-1,j}^n}{\Delta x^2} + \frac{u_{i,j+1}^n - 2u_{i,j}^n + u_{i,j+1}^n}{\Delta y^2} \right). \tag{6.99}$$

We solve the problem with $C = \kappa = 1$, $n_x = 2$, and $n_y = 3$. In Fig. 6.14, we show two snapshots of the numerical solution, where diffusion can be noticed in the amplitude of u. For diagnostics, we also calculate the maximum of u as function of time that we compare with the time modulation function $T(t) = e^{-(n_x^2 + n_y^2)\pi^2 t}$.

A **second** problem involves a space-dependent diffusion coefficient. This case has various applications where the evolution process can be faster or slower in terms of the properties of the domain, like in the evolution of brain tumors for example in [4] or spread of populations, for example [5]. The IVP is formulated as follows:

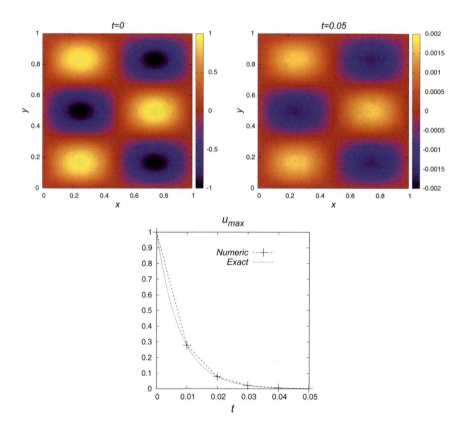

Fig. 6.14 Snapshots of u at times $t = 0$, and $t = 0.05$ for the initial conditions with the exact solution (6.98), considering the parameters $C = \kappa = 1$, $n_x = 2$, and $n_y = 3$. Also shown is the maximum u_{max} of the solution in the domain as function of time and compared with the exact time solution $T = e^{-(2^2 + 3^2)\pi^2 t}$

$$\begin{array}{ll} \frac{\partial u}{\partial t} = \nabla \cdot (\kappa \nabla u), & u = u(x, y, t), \ \kappa = \kappa(x, y) \\ D = [x_{min}, x_{max}] \times [y_{min}, y_{max}] \times [0, t_f] & Domain \\ u(x, y, 0) = u_0(x, y) & Initial\ Conditions \\ u(\partial D, t) & Boundary\ Conditions \end{array}$$
(6.100)

We again illustrate its solution considering the domain $D = [0, 1] \times [0, 1] \times [0, 0.1]$ and Dirichlet boundary conditions $u(0, y) = u(1, y) = u(x, 0) = u(x, 1) = 0$. Since $\kappa = \kappa(x, y)$ is space-dependent, the equation for u is

$$\frac{\partial u}{\partial t} = \nabla \cdot (\kappa \nabla u) \tag{6.101}$$

$$= \nabla \kappa \cdot \nabla u + \kappa \nabla^2 u$$

$$= \frac{\partial \kappa}{\partial x} \frac{\partial u}{\partial x} + \frac{\partial \kappa}{\partial y} \frac{\partial u}{\partial y} + \kappa \left(\frac{\partial^2 u}{\partial x^2} + \frac{\partial^2 u}{\partial y^2} \right). \tag{6.102}$$

In order to implement the MoL, we write the semi-discrete version of this equation using second-order accurate space derivatives of u and κ, for the interior points of the discrete domain D_d:

$$\frac{\partial u}{\partial t} = \frac{\kappa_{i+1,j} - \kappa_{i-1,j}}{2\Delta x} \frac{u^n_{i+1,j} - u^n_{i-1,j}}{2\Delta x} + \frac{\kappa_{i,j+1} - \kappa_{i,j-1}}{2\Delta y} \frac{u^n_{i,j+1} - u^n_{i,j-1}}{2\Delta y}$$

$$+ \kappa_{i,j} \left(\frac{u^n_{i+1,j} - 2u^n_{i,j} + u^n_{i-1,j}}{\Delta x^2} + \frac{u^n_{i,j+1} - 2u^n_{i,j} + u^n_{i,j+1}}{\Delta y^2} \right). \tag{6.103}$$

As example, we use a diffusion coefficient of the form

$$\kappa(x, y) = 2 \tanh\left(\frac{x - 0.5}{0.05}\right) + 3, \tag{6.104}$$

a smooth version of a step function that brings the values $\kappa \sim 1$ for $x < 0.5$ to $\kappa \sim 5$ for $x > 0.5$ and would show how the diffusion coefficient affects the evolution at each half domain.

For the numerical solution, we use the discrete domain D_d defined with $N_x = N_y = 100$ or equivalently $\Delta x = 0.01$ and $\Delta t = 0.125 \Delta x^2$. As initial conditions, we use the Gaussian function $u_0(x, y) = e^{-[(x-0.5)^2 + (y-0.5)^2]/0.05}$. The results at various times are shown in Fig. 6.15, which illustrate how the right side of the domain diffuses the signal more rapidly than the left half.

The **third** problem is a generalization of the predator-prey problem of Sect. 2.11.1, which models the interaction of two species but this time with the intervention of the space through diffusion. This is a two-species ecological system whose implementation follows the recipe in [6]. This model assumes there are two

6.6 Diffusion Equation in 2+1 Dimensions

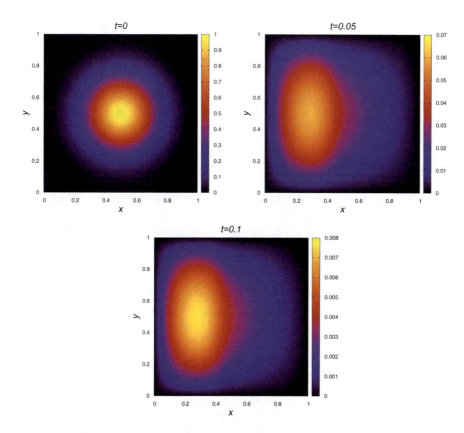

Fig. 6.15 Snapshots of u at times $t = 0, 0.05,$ and 0.1 for the initial condition $u_0(x, y) = e^{-[(x-0.5)^2+(y-0.5)^2]/0.05}$. The right half of the domain has a five times bigger diffusion coefficient than at the left, and consequently the initial data are diffused more rapidly

species described by populations u_1 and u_2, whose evolution is governed by the following IVP:

$$
\begin{array}{ll}
\frac{\partial u_1}{\partial t} = u_1(1-u_1) - \frac{u_1 u_2}{u_1+\alpha u_2} + d_1 \nabla^2 u_1 & \\
\frac{\partial u_2}{\partial t} = \delta u_2 \left(1 - \beta \frac{u_2}{u_1}\right) + d_2 \nabla^2 u_2 & u_1 = u_1(x,y,t),\ u_2 = u_2(x,y,t) \\
D = [x_{min}, x_{max}] \times [y_{min}, y_{max}] \times [0, t_f] & \textit{Domain} \\
u_1(x, y, 0) = u_{1,0}(x, y), & \\
\quad u_2(x, y, 0) = u_{2,0}(x, y) & \textit{Initial Conditions} \\
u_1(\partial D, t),\ u_2(\partial D, t) & \textit{Boundary Conditions}
\end{array}
$$

(6.105)

Notice that population u_1 grows according to the logistic model and interacts with u_2 through the crossed nonlinear term in the first equation. This species diffuses in space with coefficient d_1. Population u_2 interacts with species u_1 through coefficients δ and β and diffuses spatially with coefficient d_2.

As shown in [6], this system of equations has an equilibrium point at

$$u_1^* = 1 - \frac{1}{\alpha + \beta}, \quad u_2^* = \frac{u_1^*}{\beta}, \qquad (6.106)$$

that helps to design a couple of evolution scenarios. We solve the system using the MoL, with second-order accurate semi-discrete version of the equations in the domain $D = [0, 200] \times [0, 200] \times [0, 2000]$. The discrete domain D_d uses $N_x = N_y = 200$ or equivalently $\Delta x = \Delta y = 1$ and $\Delta t = 0.01 \Delta x^2$.

Initial conditions are chosen near the equilibrium values $u_{1,0} = u_1^* + noise$ and $u_{2,0} = u_2^*$, where $noise = 0.001 * random$, where $random$ is a random number with values in the range [0, 1]. We present two examples with parameters $\alpha = 0.4$, $\beta = 5/4$, $d_1 = 1$, and $d_2 = 7$ and two values $\delta = 0.6$ and 0.535.

Boundary conditions are periodic, so that the populations interact, diffuse, and reenter the domain.

The results of the evolution in the two cases are shown in Fig. 6.16, where the most important feature is the formation of patterns that depend on the value of δ.

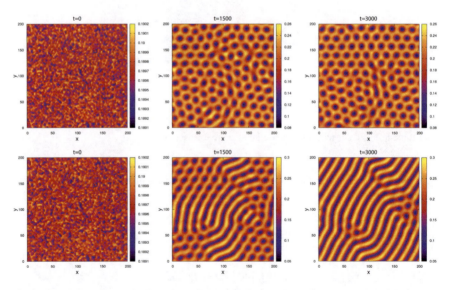

Fig. 6.16 Population u_1 at various times for parameter $\delta = 0.6$ at the top and $\delta = 0.535$ at the bottom. At initial time, the population has equilibrium value u_1^* and perturbed with random noise. Eventually, patterns are formed with time that illustrates how population u_1 accommodates. In the two cases, for comparison, we use the same initial conditions, including the same random noise, so that the pattern distribution of the population is only due to δ

The code `Diffusion2p1.f90` solves the first and second problems and is listed in Appendix B. The code `PredatorPreyDiff2p1.f90` which solves the third problem is available as Electronic Supplementary Material.

6.7 Projects

With the methods described in this chapter, it is possible to carry out the following projects:

6.7.1 Stationary Solar Wind and CMEs

The results from Sect. 5.3 related to the formation of the solar wind, and the evolution of a spherical ejection can be generalized and brought to more realistic scenarios still within the hydrodynamical regime without considering the magnetic field carried by the wind, this time by only breaking the spherical symmetry. The equations modeling this scenario are those in the IVP (6.56).

In the spherically symmetric example of Sect. 5.3, where the Sun is assumed to be at the origin, the numerical domain is chosen $[r_{min}, r_{max}]$, with r_{min} finite. In the 3D domain described with Cartesian coordinates, the implementation of such strategy is not as simple, because the sphere of radius r_{min} to be excised is in the best case a lego-sphere, and the implementation of injection boundary conditions as done in Sect. 5.3 is not straightforward. What we do is that we evolve the variables in the whole cubic domain, except within the sphere of radius r_{min}, where all the variables, primitive and conservative, are kept fixed with the values of the solar wind variables. This strategy will produce the equivalent results to those obtained with the spherically symmetric code.

The formation of the stationary SW can be simulated with conditions similar to those of the 3D blast wave of Sect. 6.5.2 but, this time, maintaining the values of the variables constant within the sphere of radius $r = r_{min}$. The initial conditions are given for density, velocity, and temperature as follows:

$$\rho_0(\mathbf{x}) = \begin{cases} \rho_{wind}, & \text{if } r \leq r_{min} \\ 0.01\rho_{wind}, & \text{if } r > r_{min} \end{cases}$$

$$\mathbf{v}_0(\mathbf{x}) = \begin{cases} (xv_{wind}/r, yv_{wind}/r, zv_{wind}/r), & \text{if } r \leq r_{min} \\ 0, & \text{if } r > r_{min} \end{cases}$$

$$T_0(\mathbf{x}) = T_{wind} \tag{6.107}$$

where $r = \sqrt{x^2 + y^2 + z^2}$ is the radial coordinate and ρ_{wind}, v_{wind}, and T_{wind} are the number density, radial velocity and temperature of the solar wind, respectively. The pressure is obtained from the temperature $p_0(\mathbf{x}) = \rho_0(\mathbf{x})T_0(\mathbf{x})$, whereas the internal energy from the ideal gas EoS $e(\mathbf{x}) = p_0(\mathbf{x})/\rho_0(\mathbf{x})/(\gamma - 1)$ and therefore the total energy density $E = \rho(\frac{1}{2}v^2 + e)$. These variables serve to start a simulation of a solar wind formation using the code in Sect. 6.5, with the condition that variables remain as initially within the sphere of radius $r_{min} = 17.18 R_\odot$. Then eventually the wind will settle as shown in Fig. 6.17 for a solar wind with number density $\rho_0 = 2100$, radial velocity $v_{wind} = 2.5 \times 10^5$ m/s, and temperature 5×10^5 K, the

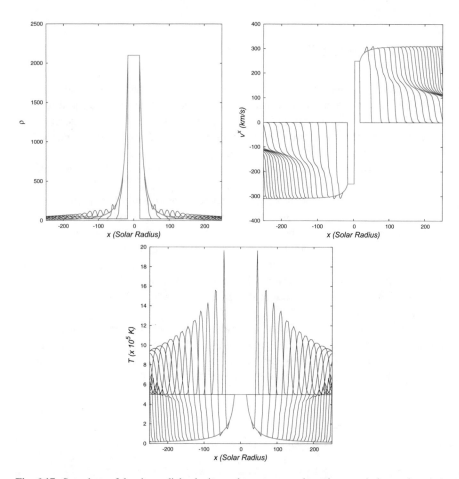

Fig. 6.17 Snapshots of density, radial velocity, and temperature along the x−axis for a solar wind pumped with properties $\rho_{wind} = 2100$, radial velocity $v_{wind} = 2.5 \times 10^5$ m/s, and temperature 5×10^5 K. Notice that within the sphere of radius r_{min}, the variables remain time-independent. These results can be compared with those obtained from the spherically symmetric simulation of Sect. 5.3, where excision was implemented inside $r = r_{min}$. In these coordinates, Earth is located on the x−axis at $214.83452 R_\odot$

6.7 Projects

same parameters used in Sect. 5.3 and of the same type as those prescribed for solar winds (see, for example, [7, 8]).

The use of a 3D code with no symmetries allows the simulation of events with no symmetries. This is the case of a coronal mass ejections (CME) that are responsible of magnetic storms all across the Solar System. Like the stationary solar wind, CME can be launched from the domain inside the sphere of radius r_{in}, with its own density, velocity, and temperature, along a particular direction with a particular opening angle. A simple way to implement this ejections is to define the direction, for example, $\phi_{0,CME}$ and $\theta_{0,CME}$ with opening angles $\delta\phi_{CME}$, $\delta\theta_{CME}$.

The injection of a CME is implemented by raising the density, velocity, and temperature from ρ_{wind}, v_{wind} and T_{wind} to ρ_{CME}, v_{CME}, and T_{CME} values within the opening angle for $r < r_{min}$ during a time window. We show an example with a gas ten times denser $\rho_{CME} = 21000$, faster $v_{CME} = 6.5 \times 10^5$ km/s, and hotter $T_{CME} = 1 \times 10^6$ K, the same conditions as in [8], injected during a time window of 3 h for the orientation $\phi_{0,CME} = 1$rad, $\theta_{0,CME} = 0$, and $\delta\phi_{CME} = \delta\theta_{CME} = 0.5$ rad, on the equatorial plane. Three snapshots appear in Fig. 6.18, where we show snapshots of density prior, during, and after the CME is ejected.

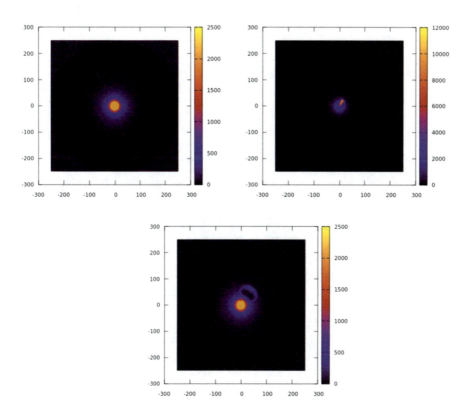

Fig. 6.18 Snapshots of density during the evolution of the CME at times prior to injection, during injection, and after injection of the CME variables. Notice the high density within a cone-shaped region within r_{min} during injection and the front shock of the CME in the third snapshot

The project consists in the generalization of the code that solves the 3D Euler equations in Sect. 6.5 and make it capable of simulating the formation of a stationary solar wind and the propagation of a CME.

The project can be brought to the next level and include detectors in the sense of Fig. 5.14, using a generalization of the code in Sect. 5.3. Real position of various satellites like Stereo A, Stereo B, Solar Orbiter, and Parker Solar Probe can be incorporated into the domain and measure the properties of the plasma of their positions. Needless to say that a detector at Earth's location would be very useful, because it would allow to know the properties of the plasma when it arrives to Earth.

Further details, for example, the conversion from physical to code units, the addition of the gravitational potential of the Sun in the source of Euler equations, as well as astronomical details like the coordinates used, can be found in [9].

6.7.2 Relativistic Hydrodynamics in 3+1 Dimensions

In Sect. 5.4, we developed the evolution equations of a relativistic inviscid fluid. Here, we rewrite Eq. (5.56):

$$\begin{aligned} \frac{\partial D}{\partial t} + \frac{\partial}{\partial x^i}(Dv^i) &= 0, \\ \frac{\partial S^i}{\partial t} + \frac{\partial}{\partial x^j}(S^i v^j + p\delta ij) &= 0, \\ \frac{\partial \tau}{\partial t} + \frac{\partial}{\partial x^i}(S^i - Dv^i) &= 0. \end{aligned} \qquad (6.108)$$

Likewise in the Newtonian case, these equations can be written, following (5.57), as a set of flux balance laws like (6.60):

$$\frac{\partial \mathbf{u}}{\partial t} + \frac{\partial (\mathbf{F}_x)}{\partial x} + \frac{\partial (\mathbf{F}_y)}{\partial y} + \frac{\partial (\mathbf{F}_z)}{\partial z} = \mathbf{S}, \qquad (6.109)$$

which are solved following the same sequence of steps needed for the Newtonian equations as described in Sect. 6.5. The main difference, a very important one, is the recovery of primitive variables, which depend on the construction of the pressure. This can be done using a generalization found for the relativistic equations in Sect. 5.4 with Eq. (5.64). The transcendental equation for the pressure needs to be slightly generalized using, instead of only $\mathcal{S}^{x2}$, the summation of the three components of $\mathcal{S}^i$, that is, $\mathcal{S}^{x2} + \mathcal{S}^{y2} + \mathcal{S}^{z2}$ in (5.64). Also, the Lorentz factor W should include instead of only $(v^x)^2$ the squared velocity $(v^x)^2 + (v^y)^2 + (v^z)^2$ within the square root of Eq. (5.63).

6.7 Projects

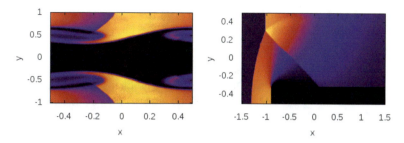

Fig. 6.19 At the left, a slice of the density during the formation of a relativistic Kelvin-Helmholtz instability solved with 3+1 relativistic hydrodynamics. At the right, a slice of the wind tunnel test for a relativistic hydrodynamical flow. The initial parameters and setup can be seen in [10]

As a sample problem, we show in Fig. 6.19 a late-time snapshot of the density of a Kelvin-Helmholtz instability, as well as the Emery wind tunnel, which are standard tests of a hydrodynamic code.

The initial conditions for these problems, other tests, and details of how to implement a code that solves the 3+1 relativistic Euler equations can be found in [10]. The project can consist in the reproduction of all the tests in such paper, which can lead to interesting problems like the evolution of astrophysical jets.

6.7.3 Magnetohydrodynamics

The relativistic equations from the previous project can be generalized to model the dynamics of relativistic plasmas with the magnetohydrodynamics (MHD) equations. The details of physics, numerical methods, tests, and subtleties of the implementation can be found also in [10].

A different generalization of relativistic Euler equations couples the gas dynamics to radiation, in which both fluid and radiation pressures interplay in the fluid dynamics. The solution of these equations has applications in high-energy astrophysics. Details of an implementation of a code that solves the relativistic radiation hydrodynamics is [11].

In the Newtonian regime, the equations from Sect. 6.5 and Project 6.7.1 can also be generalized to the MHD equations. The details of a simple implementation using the methods described in this book can be found in [12]. There are different degrees of complexity, for example, the inclusion of resistivity and thermal conduction [13]. Applications of these equations are easy to implement, for example, in [14], it is possible to find a model that describes the magnetic reconnection process as a mechanism to trigger small jets on the solar surface in two dimensions and its straightforward generalization to three dimensions [15].

Disciplined but small variants of the codes here will help to reproduce the results in the aforementioned papers.

6.8 How to Improve the Methods in This Chapter

In 3+1 problems, the use of memory becomes important. The problems solved use a discrete domain with 100^3 cells, or equivalently, one has to store a number of order 10^6 or 10^7 double precision numbers. In 1+1 problems, it is possible to cover a similar spatial domain with a resolution of order $\Delta x/10^4$. This indicates that accuracy is sacrificed when solving 3+1 problems.

One solution to this problem is parallelization, for which it is straightforward to use the MPI libraries to allow the solution of 3+1 problems in shared and distributed memory computers. This can help using higher resolution and therefore better accuracy.

A second common solution is the use of mesh refinement. With this method, resolution, therefore, memory is spent in the regions where resolution is needed. The original version of adaptive mesh refinement (AMR) was defined in [16] for hyperbolic systems of equations, with specifications to make it work for hydrodynamics [17]. However, it is recommended to program using data structures, instead of arrays of numbers.

A much simpler but inflexible alternative is fixed mesh refinement (FMR), where refined subdomains are boxes, not arbitrarily distributed refined cells like AMR. A description of this method can be found in [18] and an application for Schrödinger equation in [19].

In the particular case of *hydrodynamics*, multidimensional problems beyond the 1+1 case offer a variety of implementations. Here, we have used a semi-discrete version of Euler equations that integrates in time the contribution of the fluxes along all directions. Other methods include flux splitting along each direction, similar to the ADI one-direction-at-a-time method used in the implicit Crank-Nicolson scheme, and also Strang splitting with fractional time steps. These methods are clearly exposed in [20].

Finally, we use Euler equations of hydrodynamics as the workhorse, because they are rather difficult to solve due to the formation of discontinuities. Navier-Stokes equations conform a more general set of fluid dynamics equations because of the inclusion of viscosity, and surprisingly may not need the use of finite volume methods, because viscosity tends to smoothen functions out and prevents the formation of shocks. Instead, the use of the finite difference discretization, simpler in concept and implementation, the same used for the wave and Schrödinger equations, may work for Navier-Stokes equations straightforwardly.

References

1. Gabrielle Allen, PhD Thesis, Cardiff University (1993)
2. F. Pretorius, Evolution of binary black-hole spacetimes, Phys. Rev. Lett. **95**, 121101 (2005). arXiv:gr-qc/0507014, https://doi.org/10.48550/arXiv.gr-qc/0507014
3. https://www.astro.princeton.edu/~jstone/Athena/tests/rt/rt.html

References

4. K.R. Swanson, C. Bridge, J.D. Murray, E.C. Alvord Jr. Virtual and real brain tumors: using mathematical modeling to quantify glioma growth and invasion, J. Neurol. Sci. **216**, (2003). https://doi.org/10.1016/j.jns.2003.06.001
5. V. Chávez-Medina, J.A. González, F.S. Guzmán, Location of sources in reaction-diffusion equations using support vector machines. PLoS One **14**(12), e0225593 (2019). https://doi.org/10.1371/journal.pone.0225593
6. H. Baek, D.I. Jung, Z.W. Wang, Pattern formation in a semi-ratio-dependent predator-prey system with diffusion. Discrete Dyn. Nat. Soc. , **2013**, 657286 (2013) http://dx.doi.org/10.1155/2013/657286
7. D. Shiota, et al., Inner heliosphere MHD modeling system applicable to space weather forecasting for the other planets. Adv. Earth Space Sci. **12**, 187 (2014). https://doi.org/10.1002/2013SW000989
8. J.A. González-Esparza, A. Lara, E. Pérez-Tijerina, A. Santillán, N. Gopalswamy, A numerical study on the acceleration and transit time of coronal mass ejections in the interplanetary medium, J. Geophys. Res. **108**, 1039 (2003)
9. F.S. Guzmán, L.F. Mendoza-Mendoza, Tests of a new code that simulates the evolution of solar winds and CMEs. J. Phys. Conf. Series **2307**, 012020 (2022). https://iopscience.iop.org/article/10.1088/1742-6596/2307/1/012020/pdf
10. F.D. Lora-Clavijo, A. Cruz-Osorio, F.S. Guzmán, CAFE: a new relativistic MHD code ApJS 2015, **218**, 24–58 (2015) arXiv:1408.5846 [gr-qc], https://doi.org/10.48550/arXiv.1408.5846
11. F.J. Rivera-Paleo, F.S. Guzmán, CAFE-R: a code that solves the special relativistic radiation hydrodynamics equations. ApJS **241**, 28 (2019). arXiv:1903.04994 [astro-ph.HE], https://doi.org/10.48550/arXiv.1903.04994
12. J.J. González-Avilés, A. Cruz-Osorio, F.D. Lora-Clavijo, F.S. Guzmán, Newtonian CAFE: a new ideal MHD code to study the solar atmosphere. Mon. Not. R. Astron. Soc. **454**, 1871–1885 (2015). arXiv:1509.00225 [astro-ph.SR], https://doi.org/10.48550/arXiv.1509.00225
13. J.J. González-Avilés, F.S. Guzmán, CAFE-Q: code designed to solve the resistive MHD equations with thermal conductivity. IEEE Trans. Plasma Sci. **46**, 2378–2385 (2018). https://ieeexplore.ieee.org/document/8359445
14. J.J. González Avilés, F.S. Guzmán, V. Fedun, Jet formation in solar atmosphere due to magnetic reconnection. ApJ **836**, 24 (2017). arXiv:1609.09422 [astro-ph.SR], https://doi.org/10.48550/arXiv.1609.09422
15. J.J. González-Avilés, F.S. Guzmán, V. Fedun, G. Verth, S. Shelyag, S. Regnier, Jet formation and evolution due to 3D magnetic reconnection. ApJ **856**, 176 (2018). arXiv:1709.05066 [astro-ph.SR], https://doi.org/10.48550/arXiv.1709.05066
16. M.J. Berger, J. Oliger, Adaptive mesh refinement for hyperbolic partial differential equations. J. Comp. Phys. **53**, 484–512 (1984). https://doi.org/10.1016/0021-9991(84)90073-1
17. M.J. Berger, P. Colella, Local adaptive mesh refinement for shock hydrodynamics. J. Comp. Phys. **82**, 64–84 (1989) https://doi.org/10.1016/0021-9991(89)90035-1
18. E. Schnetter, S.H. Hawley, I. Hawke, Evolutions in 3D numerical relativity using fixed mesh refinement. Class. Quant. Grav. **21**, 1465–1488 (2004). arXiv:gr-qc/0310042, https://doi.org/10.48550/arXiv.gr-qc/0310042
19. F.S. Guzmán, F.D. Lora-Clavijo, J.J. González-Avilés, F.J. Rivera-Paleo, Rotation curves of rotating galactic BEC dark matter halos. Phys. Rev. **D 89**, 063507 (2014). arXiv:1310.3909 [astro-ph.CO], https://doi.org/10.48550/arXiv.1310.3909
20. E.F. Toro, Riemann solvers and numerical methods for fluid dynamics, 3rd edn. (Springer-Verlag, Berlin, 2009)

Chapter 7
Appendix A: Stability of Evolution Schemes

We follow the description of stability proofs in the book by J. W. Thomas [1].

Let $\mathbf{u}^n$ the solution of an evolution scheme at point (x_k, t^n) of the numerical domain in 1+1 dimensions. Assuming $\mathbf{u}^n \in \ell_{2,\Delta x}$, the space of functions with finite Riemann sum calculated with resolution Δx, then the discrete Fourier transform (DFT) of $\mathbf{u}$ is the function $\hat{u} \in L_2[-\pi, \pi]$

$$\hat{u}(\xi) = \frac{1}{\sqrt{2\pi}} \sum_{m=-\infty}^{\infty} e^{-im\xi} u_m \qquad (7.1)$$

with $\xi \in [-\pi, \pi]$. The inverse Fourier transform of $\hat{u}$ is

$$u_m = \frac{1}{\sqrt{2\pi}} \int_{-\pi}^{\pi} e^{im\xi} \hat{u}(\xi) d\xi. \qquad (7.2)$$

Parseval's relation establishes that

$$||\hat{u}||_2 = ||\mathbf{u}||_2 := \frac{||\mathbf{u}||_{2,\Delta x}}{\Delta x}. \qquad (7.3)$$

The scheme is stable if there are K and β such that the solution at time t^{n+1} is bounded:

$$||\mathbf{u}^{n+1}||_{2,\Delta x} \leq K e^{\beta(n+1)\Delta t} ||\mathbf{u}^0||_2, \qquad (7.4)$$

and in practice what is done to prove stability is to use Parseval's relation and prove that K and β exist for the norm of the DFT:

$$||\hat{u}^{n+1}||_2 \leq K e^{\beta(n+1)\Delta t} ||\hat{u}^0||_{2,\Delta x}. \qquad (7.5)$$

© The Author(s), under exclusive license to Springer Nature Switzerland AG 2023
F. S. Guzmán, *Numerical Methods for Initial Value Problems in Physics*,
https://doi.org/10.1007/978-3-031-33556-3_7

Then, if (7.5) holds, it is said that the sequence of functions $\{\hat{u}^n\}$ is stable in the space L_2. In summary, the sequence $\mathbf{u}^n$ is stable in $\ell_{2,\Delta x}$ if and only if $\{\hat{u}^n\}$ is stable in $L_2([-\pi, \pi])$ [1].

Diffusion Equation. We study the stability of two evolution schemes for the equation in problem (3.49):

$$u_t = \kappa u_{xx}. \tag{7.6}$$

The *Forward in time centered in space* (FTCS) scheme used in Sect. 3.4 is defined with forward first-order time derivative and centered space derivative around the point (x_k, t^n):

$$\frac{u_k^{n+1} - u_k^n}{\Delta t} = \kappa \frac{u_{k-1}^n - 2u_k^n + u_{k+1}^n}{\Delta x^2} \Rightarrow$$

$$u_k^{n+1} = \alpha u_{k+1}^n + (1 - 2\alpha) u_k^n + \alpha u_{k-1}^n \tag{7.7}$$

where $\alpha = \kappa \frac{\Delta t}{\Delta x^2}$. The DFT of this scheme reads

$$\hat{u}^{n+1}(\xi) = \frac{1}{\sqrt{2\pi}} \sum_{k=-\infty}^{\infty} e^{-ik\xi} u_k^{n+1}$$

$$= \frac{1}{\sqrt{2\pi}} \sum_{k=-\infty}^{\infty} e^{-ik\xi} \left[\alpha u_{k+1}^n + (1 - 2\alpha) u_k^n + \alpha u_{k-1}^n \right]$$

$$= \frac{\alpha}{\sqrt{2\pi}} \sum_{k=-\infty}^{\infty} e^{-ik\xi} u_{k+1}^n + \frac{1 - 2\alpha}{\sqrt{2\pi}} \sum_{k=-\infty}^{\infty} e^{-ik\xi} u_k^n$$

$$+ \frac{\alpha}{\sqrt{2\pi}} \sum_{k=-\infty}^{\infty} e^{-ik\xi} u_{k-1}^n, \tag{7.8}$$

shifting labels, $m = k + 1$ in the first summation and $m = k - 1$ in the third one, we obtain

$$\hat{u}^{n+1}(\xi) = \frac{\alpha}{\sqrt{2\pi}} \sum_{k=-\infty}^{\infty} e^{-i(m-1)\xi} u_{k+1}^n + \frac{1 - 2\alpha}{\sqrt{2\pi}} \sum_{k=-\infty}^{\infty} e^{-ik\xi} u_k^n$$

$$+ \frac{\alpha}{\sqrt{2\pi}} \sum_{k=-\infty}^{\infty} e^{-i(m+1)\xi} u_{k-1}^n$$

$$= \alpha e^{i\xi} \hat{u}^n + (1 - 2\alpha) \hat{u}^n + \alpha e^{-i\xi} \hat{u}^n$$

$$= \left[\alpha (e^{i\xi} + e^{-i\xi}) + (1 - 2\alpha) \right] \hat{u}^n$$

7 Appendix A: Stability of Evolution Schemes

$$= [2\alpha \cos \xi + (1 - 2\alpha)] \hat{u}^n$$

$$= \left[1 - 4\alpha \sin^2 \frac{\xi}{2}\right] \hat{u}^n, \tag{7.9}$$

where the factor in brackets is called the **symbol** $\rho(\xi)$ of the scheme (7.7). Now, the evolution from time t^0 to t^{n+1} is given by the successive application of the symbol

$$\hat{u}^{n+1} = \left[1 - 4\alpha \sin^2 \frac{\xi}{2}\right]^{n+1} \hat{u}^0, \tag{7.10}$$

and the stability is guaranteed if α is restricted such that

$$\left|1 - 4\alpha \sin^2 \frac{\xi}{2}\right| \leq 1, \tag{7.11}$$

because (7.4) is fulfilled with $K = 1$ and $\beta = 0$. Expanded, this condition reads

$$-1 \leq 1 - 4\alpha \sin^2 \frac{\xi}{2} \leq 1. \tag{7.12}$$

The first inequality reduces to $4\alpha \sin^2 \frac{\xi}{2} \leq 2$ which implies

$$\alpha = K \frac{\Delta t}{\Delta x^2} \leq \frac{1}{2}. \tag{7.13}$$

The second inequality is true for any positive α. Therefore, (7.13) is a **sufficient condition for the scheme (7.7) to be stable**.

We now analyze the stability of the *Crank-Nicolson* scheme used in Sect. 3.4.2, constructed from the discretization (3.54) around the virtual point $(x_k, t^{n+1/2})$:

$$\frac{u_k^{n+1} - u_k^n}{\Delta t} = \frac{1}{2} \kappa \left[\frac{u_{k-1}^{n+1} - 2u_k^{n+1} + u_{k+1}^{n+1}}{\Delta x^2} + \frac{u_{k-1}^n - 2u_k^n + u_{k+1}^n}{\Delta x^2} \right] \tag{7.14}$$

$$\Rightarrow -\frac{1}{2}\alpha u_{k-1}^{n+1} + (1+\alpha)u_k^{n+1} - \frac{1}{2}\alpha u_{k+1}^{n+1} = \frac{1}{2}\alpha u_{k-1}^n + (1-\alpha)u_k^n + \frac{1}{2}\alpha u_{k+1}^n \tag{7.15}$$

where $\alpha = \kappa \frac{\Delta t}{\Delta x^2}$. The DFT of this scheme reads

$$-\frac{1}{2}\alpha e^{-i\xi} \hat{u}^{n+1} + (1+\alpha)\hat{u}^{n+1} - \frac{1}{2}e^{i\xi} \hat{u}^{n+1} = \frac{1}{2}\alpha e^{-i\xi} \hat{u}^n + (1-\alpha)\hat{u}^n + \frac{1}{2}e^{i\xi} \hat{u}^n, \tag{7.16}$$

from which the symbol can be isolated:

$$\hat{u}^{n+1} = \rho(\xi)\hat{u}^n = \frac{\frac{1}{2}\alpha e^{-i\xi} + (1-\alpha) + \frac{1}{2}\alpha e^{i\xi}}{-\frac{1}{2}\alpha e^{-i\xi} + (1+\alpha) - \frac{1}{2}\alpha e^{i\xi}}\hat{u}^n = \frac{1 - 2\alpha \sin^2 \frac{\xi}{2}}{1 + 2\alpha \sin^2 \frac{\xi}{2}}\hat{u}^n. \tag{7.17}$$

The evolution from time t^0 to t^{n+1} is given by the successive application of the symbol on the initial conditions:

$$\hat{u}^{n+1} = \rho(\xi)^{n+1}\hat{u}^0, \tag{7.18}$$

and the stability condition (7.4) is satisfied if α is restricted such that the sufficient condition $|\rho(\xi)| \leq 1$ is satisfied. The symbol is more complicated than that for the FTCS scheme, and the inequalities are not as straightforward. However, looking for conditions on α where $\rho(\xi)$ has maximums or minimums suffices. Equating the derivative of (7.17) to zero, one finds that $\xi = 0, \pm\pi$ are critical points of ρ. At these points, we have that

$$\rho(0) = 1,$$
$$\rho(\pm\pi) = \frac{1 - 2\alpha}{1 + 2\alpha}.$$

This indicates that $\xi = 0$ satisfies the sufficient condition, whereas for $\xi = \pm\pi$ we have

$$-1 \leq \frac{1 - 2\alpha}{1 + 2\alpha} \leq 1$$

and the two inequalities are satisfied with no restriction on α. It is then said that the scheme is unconditionally stable.

Stability of the *centered in space centered in time (CSCT)* scheme, used for the **wave equation** (3.27) in Sect. 3.2. The scheme at point (x_k, t^n) of the numerical domain is written as

$$u_k^{n+1} = \alpha \left[u_{k+1}^n - 2u_k^n + u_{k-1}^n \right] + 2u_k^n - u_k^{n-1}, \tag{7.19}$$

with $\alpha = (\Delta t/\Delta x)^2$. The DFT of this formula is

$$\hat{u}^{n+1} = \alpha e^{i\xi}\hat{u}^n + 2(1-\alpha)\hat{u} + \alpha e^{-i\xi}\hat{u}^n - \hat{u}^{n-1},$$
$$= \left[2 - 2\alpha \sin^2 \frac{\xi}{2} \right]\hat{u}^n - \hat{u}^{n-1},$$

7 Appendix A: Stability of Evolution Schemes

where now the transform at time t^{n-1} appears, something not seen in the previous examples. As done before, we propose that $\hat{u}^{n+1} = \rho(\xi)\hat{u}^n$ and then $\hat{u}^n = \rho(\xi)\hat{u}^{n-1}$, which substituted in the expression above gives

$$\rho \hat{u}^n = \left[2 - 2\alpha \sin^2 \frac{\xi}{2}\right] \hat{u}^n - \hat{u}^{n-1}, \quad \Rightarrow$$

$$\left[\rho^2 - \rho \left(2 - 2\alpha \sin^2 \frac{\xi}{2}\right) + 1\right] \hat{u}^{n-1} = 0, \quad \Rightarrow$$

if $\beta := 2 - 2\alpha \sin^2 \frac{\xi}{2}$, then

$$\rho = \beta \pm \sqrt{\beta^2 - 1}. \tag{7.20}$$

If $\alpha > 1$, then β can be < -1. On the other hand, if $|\alpha| \leq 1$, then $|\beta| \leq 1$, which in turn implies that the modulus of the symbol $\rho = 1$, which is a sufficient condition for the scheme to be stable. Then, in summary, $\alpha = (\Delta t/\Delta x)^2 \leq 1$ is a sufficient condition for the scheme (7.19) to be stable; since both Δt and Δx are positive, the condition reduces to $\Delta t/\Delta x \leq 1$.

Chapter 8
Appendix B: Codes

The codes that reproduce the results in this book are available as Electronic Supplementary Material (ESM). In the first section of this appendix, we describe the list of codes in the ESM used to reproduce the results of each chapter, written both in Fortran 90 and C++. Also, from each chapter of the text, the Fortran 90 version of one or two codes has been selected to appear complete because they are relevant to bring the ideas in the text into practice right away. These codes are included in the second section.

8.1 Summary of Codes

A few codes selected from each chapter are printed in this appendix because they are relevant to bring the ideas in the text into practice right away. The list of codes used in each chapter are next summarized.

Chapter 2
The codes that solve Newton Cooling Law using Euler, Backward Euler, and Average Euler are listed within the text and should reproduce all the results in Sects. 2.3, 2.4, and 2.5. The results in Sect. 2.6 for the harmonic oscillator can be reproduced with the codes in the text as well.

The codes used to reproduce the results of the various applications are all available online as ESM:

[rk4_HO.f90 / rk4_HO.cpp] for the results in Sect. 2.9 and illustrates the implementation of the RK4 method.

Supplementary Information The online version contains supplementary material available at (https://doi.org/10.1007/978-3-031-33556-3_8).

© The Author(s), under exclusive license to Springer Nature Switzerland AG 2023
F. S. Guzmán, *Numerical Methods for Initial Value Problems in Physics*,
https://doi.org/10.1007/978-3-031-33556-3_8

[rk4_PP.f90 / rk4_PP.cpp] for the predator-prey model in Sect. 2.11.1.

[rk4_SIR.f90 / rk4_SIR.cpp] for the SIR model of disease dynamics in Sect. 2.11.2.

[rk4_Lorenz.f90 / rk4_Lorenz.cpp] for the Lorenz system in Sect. 2.11.3.

[rk4_Pendulum.f90 / rk4_Pendulum.cpp] for the pendulum in Sects. 2.11.4 and 2.11.5.

[rk4_PendulumBasin.f90 / rk4_PendulumBasin.cpp] used for the construction of the basin of attraction of the damped pendulum in Sect. 2.11.5.

[rk4_NewTOV.f90 / rk4_NewTOV.cpp] serves to construct the equilibrium TOV stars in Sect. 2.11.6.

[rk4_RelTOV.f90 / rk4_RelTOV.cpp] solves the construction of relativistic TOV stars in Sect. 2.11.7.

[rk4_Helmholtz.f90 / rk4_Helmholtz.cpp] uses the shooting method to solve Helmholtz equation in Sect. 2.11.8.

[rk4_SunPlanets.f90 / rk4_SunPlanets.cpp] solves the problem of test particles on a central field and is adapted to track the trajectory of planets in Sect. 2.11.9.

[rk4_TwoBodies.f90 / rk4_TwoBodies.cpp] solves the two-body problem from Sect. 2.11.10.

We **include in this appendix** the code [rk4_SIR.f90] as an example of the implementation of the solver, with a variety of initial conditions. The idea is that the implementation can be seen right away.

Also included is the code [rk4_SunPlanets.f90], because it illustrates the implementation of the alternative version of the solver, without the need of arrays.

Chapter 3

[WaveSimple.f90 / WaveSimple.cpp] solves the 1+1 wave equation using the simple discretization described in Sect. 3.2. This code appears online as ESM and can be reconstructed from the text.

[WaveImplicit.f90 / WaveImplicit.cpp] solves the 1+1 wave equation using the implicit Crank-Nicolson method from Sect. 3.3.

[DiffSimple.f90 / DiffSimple.cpp] solves the 1+1 Diffusion equation with the forward in time centered in space discretization in Sect. 3.4.1.

[DiffImplicit.f90 / DiffImplicit.cpp] solves the 1+1 diffusion equation from Sect. 3.4.2 using the implicit Crank-Nicolson method.

[SchroedingerImplicit.f90 / SchroedingerImplicit.cpp] solves the 1+1 Schrödinger equation for the particle in a box and in a harmonic trap potential from Sect. 3.5.

The code [SchroedingerImplicit.f90] is **included below in this appendix** because it contains the solver of the linear system with a tridiagonal matrix and also uses arrays of complex numbers.

Appendix B: Codes   321

Chapter 4

[MoLAdvection.f90 / MoLAdvection.cpp] solves the 1+1 advection equation and illustrates the functioning of the method of lines in Sect. 4.3.

[MoLWave.f90 / MoLWave.cpp] solves the 1+1 wave equation using the MoL in Sect. 4.4.

[MoLWaveGauge.f90 / MoLWaveGauge.cpp] solves the 1+1 wave equation on a general Minkowski space-time using the MoL in Sect. 4.5.

[MoLSchroedinger.f90 MoLSchroedinger.cpp] solves the 1+1 Schrödinger equation using the MoL in Sect. 4.6.

[MoLEFWave.f90 MoLEFWave.cpp] solves the wave equation on top of a Schwarzschild black hole in Sect. 4.7.

[EarthCrust.f90 / EarthCrust.cpp] solves the reaction-diffusion equation for the evolution of the Earth's Crust temperature in Sect. 4.8.

From this chapter, **we include** the code [MoLWave.f90] in this appendix. The reason is that it illustrates the implementation of the MoL for a system of three equations, more complex than the scalar advection equation, and also less elaborate than the wave equation on nontrivial space-times.

Chapter 5

[MoLBurgers.f90 / MoLBurgers.cpp] solves Burgers equation using the MoL and illustrates the formation of a discontinuity in Sect. 5.1.

[MoLEuler.f90 / MoLEuler.cpp] solves the 1+1 Euler equations for an ideal gas in Sect. 5.2.

[SolarWind.f90 / SolarWind.cpp] simulates the formation of the stationary solar wind and the propagation of a spherical blast wave on top of it in Sect. 5.3.

[RelativisticEuler.f90 / RelativisticEuler.cpp] solves the 1+1 relativistic Euler equations in Sect. 5.4.

[BlackHoleAccretion.f90 / BlackHoleAccretion.cpp] simulates the evolution of accretion toward Michel solution in Sect. 5.5.

In this appendix, **we include** the code [MoLEuler.f90], which is illustrative because the steps from the text concerning the finite volume method can easily be identified.

Chapter 6

[WaveSimple3p1.f90 / WaveSimple3p1.cpp] solves the 3+1 wave equation using the simple discretization in Sect. 6.2.

[WaveSimple3p1Source.f90 / WaveSimple3p1Source.cpp] solves the 3+1 wave equation with a source also in Sect. 6.2.

[MoLWave3p1.f90 / MoLWave3p1.cpp] solves the 3+1 wave equation using the MoL in Sect. 6.3.

[Schroedinger3p1.f90 / Schroedinger3p1.cpp] solves the 3+1 Schrödinger equation using the implicit Crank-Nicolson method and the ADI scheme in Sect. 6.4.

[Euler3p1.f90 / Euler3p1.cpp] solves the 3+1 Euler equations in Sect. 6.5.

[Diffusion2p1.f90 / Diffusion2p1.cpp] solves the 2+1 diffusion equation with constant and variable diffusion coefficient in Sect. 6.6.

[PredatorPreyDiff2p1.f90 / PredatorPreyDiff2p1.cpp] solves the two species model with diffusion in Sect. 6.6.

From this chapter, we show the code [Schroedinger3p1.f90], which illustrates the splitting of spatial operators, and the calculation of integrals of density and expectation values in three dimensions. **Also included** is [Diffusion2p1.f90] that contains the implementation of a 2+1 evolution code using the method of lines.

8.2 Selected Codes

Complete Code of Selected Programs by Chapter.
Chapter 2

Code [rk4_SIR.f90] for the SIR model of disease dynamics in Sect. 2.11.2:

```
! ---------------------------------------------------------
! rk4_SIR.f90
! ---------------------------------------------------------
! Author: Francisco S. Guzman
! Last modified: 17/January/2023
! Purpose: Solves the IVP for the SIR model
! Parameters to experiment with: sigma, gamma, mu
! Output file: SIR.dat
! Output:        t        S        I        R
! ---------------------------------------------------------

! -------------------------------------------
! ----->    Module with global numbers    <-----
! -------------------------------------------

module numbers

! Arrays
real(kind=8), allocatable, dimension(:)    :: t
real(kind=8), allocatable, dimension(:,:)  :: u
real(kind=8), allocatable, dimension(:)    :: k1,k2,k3,k4,rhs

! Parameters
real(kind=8) s0,i0,r0,mu,beta,gamma,sigma

! Numerical Domain and Number of Equations
real(kind=8) t0,tmax,dt
integer N,resolution_label,NE
```

Appendix B: Codes 323

```fortran
end module
! -------------------------------------------

! ---------------------------
! --->     Program starts       <---
! ---------------------------

program rk4_SIR
use numbers
implicit none

! Counters defined locally
integer i,j,k

! Define some parameter values
resolution_label = 1
NE     = 3
N      = 50000
t0     = 0.0
tmax   = 1000.0
N      = 2**(resolution_label-1)*N ! Number of cells for the discretized domain
s0     = 0.1d0
i0     = 0.9d0
r0     = 0.0d0
sigma  = 3.0d0    ! Case A = 0.5 --- Case B = 3.0
gamma  = 1.0d0/3.0d0
mu     = 1.0d0/60.0d0
beta   = sigma * ( mu + gamma )

! Allocate memory for the various arrays
! t:                time
! u(1,:):           s
! u(2,:):           i
! u(3,:):           r

allocate(t(0:N),u(1:NE,0:N),rhs(1:NE),k1(1:NE),k2(1:NE),k3(1:NE),k4(1:NE))

! ----->    PART A    <-----
dt = (tmax - t0) / dble(N)
do i=0,N
  t(i) = t0 + dt * dble(i)
end do

! ----->    PART B    <-----
! Set initial conditions
open(1,file='SIR.dat')
do k=0,8
  u(1,0) = s0 + dble(k) * 0.1d0
  u(2,0) = 1.0d0 - u(1,0)
  u(3,0) = r0
  print *, 'Scenario s0=',u(1,0),'i0=',u(2,0)
  do i=1,N
    do j=1,4
      if (j.eq.1) then
        call calcrhs( t(i-1)  ,u(:,i-1))
        k1 = rhs
      else if (j.eq.2) then
        call calcrhs( t(i-1) + 0.5d0 * dt, u(:,i-1) + 0.5d0 * k1(:) * dt)
        k2 = rhs
```

```fortran
      else if (j.eq.3) then
        call calcrhs( t(i-1) + 0.5d0 * dt, u(:,i-1) + 0.5d0 * k2 * dt)
        k3 = rhs
      else
        call calcrhs( t(i-1) + dt, u(:,i-1) + k3 * dt)
        k4 = rhs
        u(:,i) = u(:,i-1)+(1.0d0/6.0d0)*(k1 + 2.0d0 * k2 + 2.0d0 * k3 + k4 )*dt
      end if
    end do
  end do

  ! ----->   Save data    <-----
  do i=0,N,2**(resolution_label-1)
    write(1,*) t(i),u(1,i),u(2,i),u(3,i)
  end do
  write(1,*)
  write(1,*)

end do ! ends the solutions with different initial conditions
close(1)

print *, 'Program finished'

end program
! -----------------------------
! ----->   Program ends    <-----
! -----------------------------

! ---------------------------------------------------------------
! --->  Subroutine that calculates the right hand sides   <---
! ---------------------------------------------------------------
subroutine calcrhs(my_t,my_u)

  use numbers
  implicit none
  real(kind=8), intent(in) :: my_t
  real(kind=8), dimension(NE), intent(in) :: my_u

  rhs(1) = mu - mu * my_u(1) - beta * my_u(1) * my_u(2)
  rhs(2) = -mu * my_u(2) - gamma * my_u(2) + beta * my_u(1) * my_u(2)
  rhs(3) = gamma * my_u(2) - mu * my_u(3)

end subroutine calcrhs
! ---------------------------------------------------------------
```

Command on `gnuplot` that reproduce the phase-space diagrams for the SIR system in Figs. 2.18 and 2.19:

```
gnuplot> plot 'SIR.dat' u 2:3 w l
```

Time series of S, I, R use

```
gnuplot> plot 'SIR.dat' u 1:2 w l, " u 1:3 w l, " u 1:4 w l
```

Appendix B: Codes 325

Code [`rk4_SunPlanets.f90`] solves the problem of test particles on a central field and is adapted to the planets trajectory in Sect. 2.11.9:

```fortran
! --------------------------------------------------------
! rk4_SunPlanets.f90
! --------------------------------------------------------
! Author: Francisco S. Guzman
! Last modified: 17/January/2023
! Purpose: Solves the IVP for the Planets around the Sun
! Parameters to experiment with: add new planets with their own orbital parameters
! Output file: SunPlanets.dat
! Output:         t        x       y       z
!          wehere x,y,z are the Cartesian coordinates of planet's position
! --------------------------------------------------------


! -------------------------------------------
! ----->    Module with global numbers    <-----
! -------------------------------------------
module numbers

! Arrays
real(kind=8), allocatable, dimension (:) :: u,u_p
real(kind=8), allocatable, dimension (:)    :: k1,k2,k3,k4,rhs

! Parameters
real(kind=8) x0,y0,z0,vx0,vy0,vz0,MSun,G,a,e,r0,dotphi0,radian,Theta,Omega

! Numerical Domain and Number of Equations
real(kind=8) t0,tmax,dt,t
integer N,resolution_label,NE

end module
! -------------------------------------------

! ---------------------------
! --->    Program starts    <---
! ---------------------------
program rk4_CentralField

use numbers
implicit none

! Counters defined locally
integer i,j

! --->   Constants   <---
G=6.6743e-11 ! (m^3 / kg / s^2)
MSun = 1.98847e30 ! (kg)
radian = 57.2957795130823 ! Degrees

! Parameters for different objects in different blocks
! Uncomment the object you want to consider

! Earth :: with phi = 0 at initial time
!e = 0.0167086
!a = 1.49598023e11
!Theta = 0.0 ! with respect to the ecliptic
!Omega = -11.26064
```

```
! Jupiter
!e = 0.0489
!a = 7.78479e11
!Theta = 1.303/radian  ! with respect to the ecliptic
!Omega = 100.464/radian


! Pluto
e = 0.2488
a = 5.906423e12
Theta = 17.16/radian  ! with respect to the ecliptic
Omega = 110.299/radian

! Halley
!e = 0.96714
!a = 2.66792e12
!Theta = 162.26/radian  ! with respect to the ecliptic
!Omega = 58.42/radian

r0 = a * (1.0d0 - e**2) / (1.0d0 + e )
dotphi0 = (1.0d0 + e)**2 * sqrt( G * Msun / a**3 / (1.0d0 - e**2)**3 )

! Define some numerical parameter
resolution_label = 1
NE     = 6
N      = 100000000 !100000000 Pluto and Halley, smaller for other cases
t0     = 0.0
tmax   = 10000000000.  ! (sec) 10000000000. Pluto and Halley, smaller for other cases
N      = 2**(resolution_label-1) * N     ! Number of cells for the discretized domain
x0     = cos(Omega) * r0 * cos(Theta)
y0     = sin(Omega) * r0 * cos(Theta)
z0     = r0*sin(Theta)
vx0    = -sin(Omega) * r0 * dotphi0
vy0    =  cos(Omega) * r0 * dotphi0
vz0    = 0.0d0

! Allocate memory for the various arrays
! u(1):        x
! u(2):        y
! u(3):        z
! u(4):        vx
! u(5):        vy
! u(6):        vz

allocate(u(1:NE),u_p(1:NE),rhs(1:NE),k1(1:NE),k2(1:NE),k3(1:NE),k4(1:NE))

! ----->    PART A    <-----
dt = (tmax - t0) / dble(N)
print *, 'dt=',dt

! ----->    PART B    <-----
! Set initial conditions
open(1,file='SunPlanets.dat')
  u(1) = x0
  u(2) = y0
  u(3) = z0
  u(4) = vx0
  u(5) = vy0
  u(6) = vz0
  t = 0.0d0
  write(1,*) t,u(1),u(2),u(3)
! Evolve
  do i=1,N
```

Appendix B: Codes

```fortran
      u_p = u
      do j=1,4
        if (j.eq.1) then
          call calcrhs( t ,u_p(:))
          k1 = rhs


        else if (j.eq.2) then
          call calcrhs( t + 0.5d0 * dt, u_p(:) + 0.5d0 * k1(:) * dt)
          k2 = rhs
        else if (j.eq.3) then
          call calcrhs( t + 0.5d0 * dt, u_p(:) + 0.5d0 * k2 * dt)
          k3 = rhs
        else
          call calcrhs( t + dt, u_p(:) + k3 * dt)
          k4 = rhs
          u(:) = u_p(:) + (1.0d0/6.0d0)*( k1 + 2.0d0 * k2 + 2.0d0 * k3 + k4 ) * dt
        end if
      end do
      ! Saving data to a file every 10000 iterations
      t = t + dt
      if (mod(i,2**(resolution_label-1)*10000).eq.0) then
        write(1,*) t,u(1),u(2),u(3)
        print *, "iteration",i,"time=",t
      end if
    end do

close(1)
print *, 'Program finished'

end program
! -----------------------------
! ----->   Program ends    <-----
! -----------------------------


! ----------------------------------------------------------------
! --->   Subroutine that calculates the right hand sides   <---
! ----------------------------------------------------------------
subroutine calcrhs(my_t,my_u)

  use numbers
  implicit none
  real(kind=8), intent(in) :: my_t
  real(kind=8), dimension(NE), intent(in) :: my_u

  rhs(1) = my_u(4)
  rhs(2) = my_u(5)
  rhs(3) = my_u(6)
  rhs(4) = -MSun*G*my_u(1)/(my_u(1)**2 + my_u(2)**2 + my_u(3)**2)**(1.5)
  rhs(5) = -MSun*G*my_u(2)/(my_u(1)**2 + my_u(2)**2 + my_u(3)**2)**(1.5)
  rhs(6) = -MSun*G*my_u(3)/(my_u(1)**2 + my_u(2)**2 + my_u(3)**2)**(1.5)

end subroutine calcrhs
! ----------------------------------------------------------------
```

Command of gnuplot that represent time series of coordinates x, y, z:

```
gnuplot> plot 'SunPlanets.dat' w l, " u 1:3 w l, " u 1:4 w l
```

For the trajectories in three dimensions,

```
gnuplot> splot 'SunPlanets.dat' u 2:3:4 w l
```

Chapter 3

Code [`SchroedingerImplicit.f90`] solves the 1+1 Schrödinger equation for the particle in a box and in a harmonic trap potential from Sect. 3.5:

```fortran
! ------------------------------------------------------
! SchroedingerImplicit.f90
! ------------------------------------------------------
! Author: Francisco S. Guzman
! Last modified: 17/January/2023
! Purpose: Solves the IVP for the 1+1 Schroedinger Equation
!          using the CTCS Implicit Crank-Nicolson method
! Parameters to experiment with: nodes, potential
! Output file: SchroedingerImplicit.dat
! Output:      x      Re(psi)   Im(psi)    rho     exact(Re(psi))
!              error(exact(Re(psi)))
!              one data block per 8000 iterations
! Output file: SchroedingerImplicit.fancy
! Output:      t    x     Re(psi)    Im(psi)    rho
!              one data block per 8000 iterations
! Output file: SchroedingerImplicit.norms
!              t      L1(error)    L2(error)
! ------------------------------------------------------

! --------------------------
! --->   Program starts   <---
! --------------------------
program SchroedingerImplicit

implicit none

! Arrays, Global Variables and Parameters
real(kind=8), allocatable, dimension(:) :: x
complex(kind=8), allocatable, dimension(:) :: psi,psi_p
complex(kind=8), allocatable, dimension (:) :: a,b,c,d,aux
complex(kind=8) alpha,tmp
real(kind=8), allocatable, dimension(:) :: rho,V,Hermite
real(kind=8), allocatable, dimension(:) :: exact, error
real(kind=8) pi,dx,dt,xmin,xmax,tmax,CFL,t,L1,L2,En,NofP
real(kind=8) factorial
integer i,k,n,Nx,Nt,resolution_label,nodes
character(len=20) :: potential

! Define some parameter values
resolution_label = 1
nodes = 3                  ! Quantum number
CFL   = 0.125
pi    = acos(-1.0d0)
potential = 'harmonic_trap'            ! [ box, harmonic_trap ]

if (potential.eq.'box') then
   Nx   = 100
   Nt   = 1000000
   xmin = 0.0d0
   xmax = 1.0d0
```

Appendix B: Codes 329

```
else if (potential.eq.'harmonic_trap') then
  Nx     = 2000
  Nt     = 1000000
  xmin   = -10.0d0
  xmax   =  10.0d0
end if

Nx     = Nx * 2**(resolution_label-1)
Nt     = Nt * 2**(2*(resolution_label-1))


! Allocate memory for the various arrays
allocate(x(0:Nx))
allocate(psi(0:Nx),psi_p(0:Nx))
allocate(exact(0:Nx),error(0:Nx))
allocate(rho(0:Nx),V(0:Nx),Hermite(0:Nx))
allocate(a(0:Nx),b(0:Nx),c(0:Nx),d(0:Nx),aux(0:Nx))
! x:        space coordinate
! psi:      wave function at time n+1
! psi_p:    wave function at time n
! exact:    exact solution
! error:    error
! a,b,c:    arrays for the three diagonals of the matrix
! d:        right hand side of the linear system
! aux:      array for the forward-backward substitution
! Hermite:  arrays to store Hermit polynomials
! V:        potential

! Numerical Domain
dx = (xmax-xmin)/dble(Nx)
dt = CFL * dx**2
do i=0,Nx
  x(i) = xmin + dble(i) * dx
end do
alpha = cmplx( 0.0d0 , 0.25d0*dt/dx**2 )

! INITIAL DATA
if ( potential.eq.'box' ) then
   ! ---------------------------------
   ! Particle in a box
   ! ---------------------------------
   psi = cmplx( sqrt(2.0d0) * sin(dble(nodes) * pi * x) , 0.0d0 )
   V = 0.0d0                            ! Potential
   En = 0.5d0*(dble(nodes)*pi)**2       ! Energy, useful for the exact solution
   ! ---------------------------------
else if ( potential.eq.'harmonic_trap' ) then
   ! ---------------------------------
   ! Harmonic oscillator
   ! ---------------------------------
   if (nodes.eq.0) then
     Hermite = 1.0
   else if (nodes.eq.1) then
     Hermite = 2.0*x
   else if (nodes.eq.2) then
     Hermite = 4.0*x**2 - 2.0
   else if (nodes.eq.3) then
     Hermite = 8.0*x**3 - 12.0*x
   else if (nodes.eq.4) then
     Hermite = 16.0*x**4 - 48.0*x**2 + 12.0
```

```
    end if
    factorial = 1
    do i = 2, nodes
       factorial = factorial * i
    end do
    psi = 1.0d0/sqrt( sqrt(pi)*2.0**nodes*factorial ) &
          *exp(-x**2/2.0)*Hermite*cmplx(1.0d0,0.0d0)
    V   = 0.5d0 * x**2                    ! Potential
    En  = dble( nodes ) + 0.5d0           ! Energy, useful for the exact solution
    ! --------------------------------
end if

exact = real(psi)
rho   = real(psi)**2 + imag(psi)**2

open(1,file='SchroedingerImplicit.dat')
open(2,file='SchroedingerImplicit.fancy')
open(3,file='SchroedingerImplicit.norms')

! Saves data at initial time
do i=0,Nx,2**(resolution_label-1)
  write(1,*) x(i),real(psi(i)),imag(psi(i)),rho(i),exact(i),0.0
end do
write(1,*)
write(1,*)
do i=0,Nx,2**(resolution_label-1)
  write(2,*) t,x(i),real(psi(i)),imag(psi(i)),rho(i)
end do
write(2,*)

print *, 'Iteration    Time'
print *, '0              0'

! Evolution loop
do n=1,Nt

   t = t + dt     ! Updates time

   ! *****    --->    LOOP CORE    <---
   ! Recycle arrays
   psi_p = psi

   if ( potential.eq.'box' ) then
      ! ZERO boundary conditions for PARTICLE in a BOX
      b(0)  = 1.0d0 + 2.0d0*alpha+cmplx(0.0d0,0.5d0*dt*V(0))
      c(0)  = 0.0d0
      a(Nx) = 0.0d0
      b(Nx) = 1.0d0 + 2.0d0*alpha+cmplx(0.0d0,0.5d0*dt*V(Nx))
      d(0)  = psi_p(0)  * ( 1.0d0-2.0d0*alpha+cmplx(0.0d0,-0.5d0*dt*V(0)) )
      d(Nx) = psi_p(Nx) * ( 1.0d0-2.0d0*alpha +cmplx(0.0d0,-0.5d0*dt*V(Nx)) )
   else if ( potential.eq.'harmonic_trap' ) then
      ! Boundary conditions for the Harmonic Oscillator
      b(0)  = 1.0d0 + 2.0d0*alpha+cmplx(0.0d0,0.5d0*dt*V(0))
      c(0)  = -2.0d0 * alpha * psi(1)
      a(Nx) = -2.0d0*alpha*psi(Nx-1)
      b(Nx) = 1.0d0 + 2.0d0*alpha+cmplx(0.0d0,0.5d0*dt*V(Nx))
      d(0)  = psi_p(0) * ( 1.0d0-2.0d0*alpha+cmplx(0.0d0,-0.5d0*dt*V(0)) ) &
              + 2.0d0*psi_p(i)
```

Appendix B: Codes 331

```fortran
    d(Nx) = psi_p(Nx) * ( 1.0d0-2.0d0*alpha +cmplx(0.0d0,-0.5d0*dt*V(Nx)) ) &
          + 2.0d0*alpha*psi_p(Nx-1)
end if

! Entries of Matrix A and vector d for inner points
do i=1,Nx-1
  a(i) = -alpha
  b(i) = 1.0d0+2.0d0*alpha+cmplx(0.0d0,0.5d0*dt*V(i))
  c(i) = -alpha
  d(i) = psi_p(i-1)*alpha &
       + psi_p(i)*(1.0d0-2.0d0*alpha+cmplx(0.0d0,-0.5d0*dt*V(i)) ) &
       + psi_p(i+1)*alpha
end do

! Tridiagonal system
! Forward substitution
tmp = b(0)
psi(0) = d(0)/tmp
do i=1,Nx
  aux(i) = c(i-1)/tmp
  tmp = b(i)-a(i)*aux(i)
  psi(i) = (d(i)-a(i)*psi(i-1))/tmp
end do
! Backward substitution
do i = Nx-1,0,-1
  psi(i) = psi(i) - aux(i+1)*psi(i+1)
end do

! *****    --->   Ends Loop Core    <---

! Some diagnostics
rho = real(psi)**2 + imag(psi)**2

! Exact Re(psi)
if ( potential.eq.'box' ) then
  ! For the particle in the box
  exact = sqrt(2.0)*cos(En*t)*sin(dble(nodes)*pi*x)
else if ( potential.eq.'harmonic_trap' ) then
  ! For the harmonic oscillator
  exact = 1.0d0/sqrt( sqrt(pi)*2.0**nodes*factorial )*exp(-x**2/2.0) &
        * Hermite * cos( En * t )
end if

error = real(psi) - exact

! Save data during the evolution every 8000 time-steps

if (mod(n,8000*2**(2*(resolution_label-1))).eq.0) then

  if (mod(n,2**(resolution_label-1)).eq.0) then        ! Regular output
    print *, n,t       ! Send something to the screen
    write(1,*) "#_Time=",t
    do i=0,Nx,2**(resolution_label-1)
      write(1,*) x(i),real(psi(i)),imag(psi(i)),rho(i),exact(i),error(i)
    end do
    write(1,*)
    write(1,*)
  end if
```

```fortran
    ! Fancy output every 1000 time-steps
    write(2,*)
    do i=0,Nx,2**(resolution_label-1)
      write(2,*) t,x(i),real(psi(i)),imag(psi(i)),rho(i)
    end do
    write(2,*)

  end if

  ! Output of error norms and number of particles every 500 time-steps
  if (mod(n,500*2**(2*(resolution_label-1))).eq.0) then
    L1 = 0.0d0
    L2 = 0.0d0
    NofP = 0.0d0
    do i=1,Nx
      L1 = L1 + 0.5D0*(dabs(error(i-1)) + dabs(error(i)))*dx
      L2 = L2 + 0.5D0*(error(i-1)**2 + error(i)**2)*dx
      NofP = NofP + 0.5d0*(rho(i-1) + rho(i))*dx
    end do
    L2 = sqrt(L2)
    write(3,*) t,L1,L2,NofP
  end if

end do

close(3)
close(2)
close(1)

print *, 'Program finished'

end program
! -----------------------------
! -----> Program ends   <-----
! -----------------------------
```

Command of gnuplot used to produce results like those in Figs. 3.24 and 3.26. For snapshots of the real and imaginary parts of the wave function,

gnuplot> plot 'SchroImplicit.dat' u 1:2 w l, " u 1:3 w l

For snapshots of the density,

gnuplot> plot 'SchroImplicit.dat' u 1:4 w l

For norms L_1 and L_2 of the error,

gnuplot> plot 'SchroImplicit.norms' u 1:2 w l

The number of particles,

gnuplot> plot 'SchroImplicit.norms' u 1:4 w l

Appendix B: Codes   333

Chapter 4

Code [`MoLWave.f90`] solves the 1+1 wave equation using the MoL in Sect. 4.4:

```fortran
! --------------------------------------------------
! MoLWave.f90
! --------------------------------------------------
! Author: Francisco S. Guzman
! Last modified: 17/January/2023
! Purpose: Solves the IVP for the 1+1 Wave Equation using the Method of Lines and RK2
! Parameters to experiment with: amp, boundary_conditions
! Output file: MoLWave.dat
! Output:       x         phi      psi      pi       exact     error
!               one data block per 10 iterations
! Output file: MoLWave.fancy
! Output:       t         x        phi      psi      pi
!               one data block per 5 iterations
! Output file: MoLWave.norms
!               t         L1(error)       L2(error)
! --------------------------------------------------

! --------------------------
! --->   Program starts   <---
! --------------------------
program MoLWave

implicit none

! Arrays, Global Variables and Parameters
real(kind=8), allocatable, dimension(:) :: x
real(kind=8), allocatable, dimension(:) :: phi,phi_p,rhs_phi
real(kind=8), allocatable, dimension(:) :: psi,psi_p,rhs_psi
real(kind=8), allocatable, dimension(:) :: pi, pi_p, rhs_pi
real(kind=8), allocatable, dimension(:) :: exact, error
real(kind=8) dx,dt,xmin,xmax,tmax,CFL,t,amp,sigma,L1,L2
integer i,n,Nx,Nt,resolution_label,rk
character(len=20) :: boundary_conditions

! Define some parameter values
resolution_label = 1
Nx      =   200
Nx      =   2**(resolution_label-1) * Nx      ! Number of cells for the discretized domain
Nt      =   400 * 2**(resolution_label-1)
xmin    =   -1.0d0
xmax    =   1.0d0
CFL     =   0.5
amp     =   1.0d0
sigma   =   0.1
boundary_conditions = 'outgoing'           ! [ outgoing, periodic ]

! Allocate memory for the various arrays
allocate(x(0:Nx),exact(0:Nx),error(0:Nx))
allocate(phi(0:Nx),phi_p(0:Nx),rhs_phi(0:Nx))
allocate(psi(0:Nx),psi_p(0:Nx),rhs_psi(0:Nx))
allocate(pi(0:Nx), pi_p(0:Nx), rhs_pi(0:Nx) )


! x:        space coordinate
! phi:      phi at time n+1
! psi:      psi at time n+1
```

```
! pi:        pi at time n+1
! phi_p:     phi at time n
! psi_p:     psi at time n
! pi_p:      pi at time n
! rhs_phi:   rhs(phi)
! rhs_psi:   rhs(psi)
! rhs_pi:    rhs(pi)
! exact:     exact solution
! error:     error

! Numerical Domain
dx = (xmax-xmin)/dble(Nx)
dt = CFL * dx
do i=0,Nx
  x(i) = xmin + dble(i) * dx
end do

! Initialize arrays and time
t       = 0.0d0
phi     = 0.0d0
psi     = 0.0d0
pi      = 0.0d0
phi_p   = 0.0d0
psi_p   = 0.0d0
pi_p    = 0.0d0
rhs_phi = 0.0d0
rhs_psi = 0.0d0
rhs_pi  = 0.0d0
exact   = 0.0d0
error   = 0.0d0

! Initial conditions
phi = amp*exp(-x**2/sigma**2)
psi = -2.0d0*x/sigma**2 *phi
pi  = 0.0d0

exact = amp*exp(-x**2/sigma**2)

open(1,file='MoLWave.dat')
open(2,file='MoLWave.fancy')
open(3,file='MoLWave.norms')

! Saves data at initial time

  write(1,*) "#_Time=",t
  do i=0,Nx
    write(1,*) x(i),phi(i),psi(i),pi(i),exact(i),0.0
  end do
  write(1,*)
  write(1,*)
  do i=0,Nx
     write(2,*) t,x(i),phi(i),psi(i),pi(i)
  end do
  write(2,*)

  print *, 'Iteration     Time'
  print *, '0                0'

! Evolution loop
do n=1,Nt
  t = t + dt     ! Updates time
```

Appendix B: Codes

```fortran
! *****    --->    LOOP CORE    <---

! Recycle arrays
phi_p = phi
psi_p = psi
pi_p  = pi

! Heun RK2
do rk=1,2
  ! Calculate the RHS
  do i=1,Nx-1
    rhs_psi(i) = 0.5d0 * ( pi(i+1)  - pi(i-1)  ) / dx
    rhs_pi(i)  = 0.5d0 * ( psi(i+1) - psi(i-1) ) / dx
    rhs_phi(i) = pi(i)
  end do
  if ( boundary_conditions.eq.'outgoing' ) then
    ! Outgoing wave boundary conditions
    rhs_phi(0)  = - 0.5d0 * ( phi(2)    - 4.0d0*phi(1)    + 3.0d0 * phi(0)  ) / dx
    rhs_phi(Nx) = - 0.5d0 * ( phi(Nx-2) - 4.0d0*phi(Nx-1) + 3.0d0 * phi(Nx) ) / dx
  else if ( boundary_conditions.eq.'periodic' ) then
    ! Periodic boundary conditions
    rhs_psi(0)  = 0.5d0 * ( pi(1)  - pi(Nx-1)  ) / dx
    rhs_pi(0)   = 0.5d0 * ( psi(1) - psi(Nx-1) ) / dx
    rhs_phi(0)  = pi(Nx)
    rhs_psi(Nx) = 0.5d0 * ( pi(1)  - pi(Nx-1)  ) / dx
    rhs_pi(Nx)  = 0.5d0 * ( psi(1) - psi(Nx-1) ) / dx
    rhs_phi(Nx) = pi(Nx)
  end if

  if (rk.eq.1) then
    psi = psi_p + rhs_psi * dt
    pi  = pi_p  + rhs_pi  * dt
    phi = phi_p + rhs_phi * dt
  else
    psi = 0.5d0 * ( psi_p + psi + rhs_psi * dt)
    pi  = 0.5d0 * ( pi_p  + pi  + rhs_pi  * dt)
    phi = 0.5d0 * ( phi_p + phi + rhs_phi * dt)
  end if

  if ( boundary_conditions.eq.'outgoing' ) then
    ! Outgoing wave boundary conditions -- comment these 4 lines out
    ! if you use periodic BCs
    psi(0)  = - 0.5d0 * ( phi(2)    - 4.0d0*phi(1)    + 3.0d0*phi(0))  / dx
    pi(0)   = psi(0)
    psi(Nx) = 0.5d0 * ( phi(Nx-2) - 4.0d0*phi(Nx-1) + 3.0d0 * phi(Nx) ) / dx
    pi(Nx)  = -psi(Nx)
  end if

end do

! *****    --->    Ends Loop Core    <---

if ( boundary_conditions.eq.'outgoing' ) then
  ! Exact solution only for outgoing wave BCs
  exact = 0.5d0 * amp * exp(-( x - t )**2 / sigma**2) &
        + 0.5d0 * amp * exp(-( x + t )**2 / sigma**2)
else    if ( boundary_conditions.eq.'periodic' ) then
  ! An exercise is to code the exact solution for periodic BCs
end if


error = phi - exact
! Save data during the evolution every 10 time-steps
```

```fortran
   if (mod(n,10*2**(resolution_label-1)).eq.0) then       ! Regular output
     print *, n,t           ! Send something to the screen
     write(1,*) "#_Time=",t
     do i=0,Nx,2**(resolution_label-1)
       write(1,*) x(i),phi(i),psi(i),pi(i),exact(i),error(i)
     end do
     write(1,*)
     write(1,*)
   end if

   if (mod(n,5*2**(resolution_label-1)).eq.0) then
   ! Fancy output every 5 time-steps
   write(2,*)
   do i=0,Nx
     write(2,*) t,x(i),phi(i),psi(i),pi(i)
   end do
   write(2,*)
   end if

   ! Norms of the error
   L1 = 0.0d0
   L2 = 0.0d0
   do i=1,Nx
     L1 = L1 + 0.5D0*(dabs(error(i-1)) + dabs(error(i)))*dx
     L2 = L2 + 0.5D0*(error(i-1)**2 + error(i)**2)*dx
   end do
   L2 = sqrt(L2)
   write(3,*) t,L1,L2

end do

close(3)
close(2)
close(1)

print *, 'Program_finished'

end program
! -----------------------------
! ----->    Program ends    <-----
! -----------------------------
```

> Command of gnuplot used to produce the plots for ϕ, ψ, and π in Figs. 4.5 and 4.6 uses
>
> ```
> gnuplot> splot 'MoLWave.fancy' u 1:2:3 t 'phi' w l
> gnuplot> splot 'MoLWave.fancy' u 1:2:4 t 'psi' w l
> gnuplot> splot 'MoLWave.fancy' u 1:2:5 t 'pi'  w l
> ```
>
> Norms L_1 and L_2 are plotted with
>
> ```
> gnuplot> plot 'MoLWave.norms' t 'L1 norm of error' w l
> gnuplot> plot 'MoLWave.norms' u 1:3 t 'L2 norm of error' w l
> ```
>
> that will work only for outgoing wave boundary conditions, because it is the case the exact solution is calculated for. Try coding the exact solution for periodic boundary conditions. Snapshots of the wave function as in Fig. 4.7 use

(continued)

Appendix B: Codes

```
gnuplot> plot 'MoLWave.dat' i 0:10000:400 t 'phi through the
   origin' w l
```

appropriate for periodic boundary conditions.

Chapter 5

Code [`MoLEuler.f90`] solves the 1+1 Euler equations for an ideal gas in Sect. 5.2:

```fortran
! -----------------------------------------------------
! MoLEuler.f90
! -----------------------------------------------------
! Author: Francisco S. Guzman
! Last modified: 17/January/2023
! Purpose: Solves the IVP for the 1+1 Euler Equations for the shock-tube problem
!          using HLLE-minmod and RK2
! Parameters to experiment with: states L and states R
! Output file: MoLEuler.dat
! Output:      x       rho     v       E      p
!              one data block per 100 iterations
! Output file: MoLEuler.fancy
! Output:      t       x       rho     v      E       p
!              one data block per 50 iterations
! -----------------------------------------------------


! -----------------------------------------
! ----->   Module with global numbers    <-----
! -----------------------------------------
module global_numbers

! --- Numerical domain
real(kind=8), allocatable, dimension(:) :: x
real(kind=8) dx,dt,xmin,xmax,tmax,CFL,t
integer Nx,Nt,resolution_label,rk

! --- Primitive variables
real(kind=8), allocatable, dimension(:) :: rho,p,v,E,cs

! --- Conservative variables
real(kind=8), allocatable, dimension(:) :: u1,u1_p,rhs_u1
real(kind=8), allocatable, dimension(:) :: u2,u2_p,rhs_u2
real(kind=8), allocatable, dimension(:) :: u3,u3_p,rhs_u3

! --- Fluxes
real(kind=8), allocatable, dimension(:) :: flux1,flux2,flux3
real(kind=8) pL,pR,rhoL,rhoR,vL,vR,csL,csR,u1L,u1R,u2L,u2R,u3L,u3R
real(kind=8) lambda1L,lambda2L,lambda3L,lambda1R,lambda2R,lambda3R
real(kind=8) lambda1_plus,lambda2_plus,lambda3_plus,lambda1_minus,lambda2_ &
             minus,lambda3_minus
real(kind=8) f1L,f2L,f3L,f1R,f2R,f3R

! --- Hydrodynamics
real(kind=8) gamma
```

338   Appendix B: Codes

```fortran
! Shock-Tube initial data
real(kind=8) rhoL_ini,rhoR_ini,pL_ini,pR_ini,vL_ini,vR_ini

character(len=20) :: reconstructor

end module
! -------------------------------------------

! ---------------------------
! --->    Program begins     <---
! ---------------------------
program Euler

use global_numbers
implicit none

integer i,n

! Define some parameter values
resolution_label = 1
Nx      = 1000
Nx      = 2**(resolution_label-1) * Nx    ! Number of cells for the discretized domain
Nt      = 1000 * 2**(resolution_label-1)
xmin    = 0.0d0
xmax    = 1.0d0
CFL     = 0.25
rhoL_ini  = 1.0
rhoR_ini  = 0.125
vL_ini   = 0.0
vR_ini   = 0.0
pL_ini   = 1.0
pR_ini   = 0.1
gamma = 1.4
reconstructor = 'monmod'            ! [ minmod, godunov ]

! Allocate memory for the various arrays
allocate(x(0:Nx))
allocate(u1(0:Nx),u1_p(0:Nx),rhs_u1(0:Nx))
allocate(u2(0:Nx),u2_p(0:Nx),rhs_u2(0:Nx))
allocate(u3(0:Nx),u3_p(0:Nx),rhs_u3(0:Nx))
allocate(rho(0:Nx),p(0:Nx),v(0:Nx),E(0:Nx))
allocate(cs(0:Nx))
allocate(flux1(0:Nx),flux2(0:Nx),flux3(0:Nx))
! x:         space coordinate
! ui:        conservative variable i
! rho:       density
! p:         pressure
! v:         velocity
! E:         total energy
! cs:        sound speed
! fluxi:     flux i

! Numerical Domain
dx = (xmax-xmin)/dble(Nx)
dt = CFL * dx
do i=0,Nx
   x(i) = xmin + dble(i) * dx
end do
! Initial conditions for the shock tube
do i=0,Nx
   if (x(i).le.0.5d0) then
      rho(i) = rhoL_ini
      p(i)   = pL_ini
```

Appendix B: Codes   339

```fortran
      v(i)    = vL_ini
    else
      rho(i) = rhoR_ini
      p(i)    = pR_ini
      v(i)    = vR_ini
    end if
end do
E = p/rho/(gamma - 1.0D0) + 0.5D0 * rho * v**2

! Definition of Conservative Variables
u1 = rho
u2 = rho * v
u3 = p/(gamma - 1.0D0) + 0.5D0 * rho * v**2

open(1,file='MoLEuler.dat')
open(2,file='MoLEuler.fancy')

! ----- Save data at initial time

  write(1,*) "#_Time=",t
  do i=0,Nx
    write(1,*) x(i),rho(i),v(i),E(i),p(i)
  end do
  write(1,*)
  write(1,*)
  do i=0,Nx
    write(2,*) t,x(i),rho(i),v(i),E(i),p(i)
  end do
  write(2,*)

  print *, 'Iteration     Time'
  print *, '0              0'

! Evolution loop
DO n=1,Nt

  t = t + dt       ! Updates time

! *****    --->    LOOP CORE    <---

    ! Recycle arrays
    u1_p = u1
    u2_p = u2
    u3_p = u3

    ! Heun RK2
    do rk=1,2

      ! -------------
      ! ---> (a) Reconstruct variables <---
      ! -------------
      ! ----->    Calculate primitive variables
      rho = u1
      v = u2/u1
      p = (gamma - 1.0D0)*(u3 - 0.5D0*u2**2/u1)
      E = p / ( u1 * (gamma - 1.0D0) ) + 0.5D0 * rho * v**2
      cs = sqrt( p * gamma / rho )

        do i=0,Nx-1
          call reconstruct(p(i-1),p(i),p(i+1),p(i+2),dx,pL,pR)
          call reconstruct(rho(i-1),rho(i),rho(i+1),rho(i+2),dx,rhoL,rhoR)
          call reconstruct(v(i-1),v(i),v(i+1),v(i+2),dx,vL,vR)
          call reconstruct(cs(i-1),cs(i),cs(i+1),cs(i+2),dx,csL,csR)
```

```
        u1L = rhoL
        u2L = rhoL * vL
        u3L = pL/(gamma - 1.0D0) + 0.5D0 * rhoL * vL**2
        u1R = rhoR
        u2R = rhoR * vR
        u3R = pR/(gamma - 1.0D0) + 0.5D0 * rhoR * vR**2

        ! -------------
        ! ---> (b) <--- Calculate the eigenvalues at L and R
        ! -------------

        lambda1L = vL - csL
        lambda2L = vL
        lambda3L = vL + csL

        lambda1R = vR - csR
        lambda2R = vR
        lambda3R = vR + csR

        ! -------------
        ! ---> (c) <--- Calculate lambda_plus and lambda_minus
        ! -------------
        lambda1_plus  = max(0.0d0,lambda1R,lambda1L,lambda2L,lambda2R,lambda3L,lambda3R)
        lambda1_minus = min(0.0d0,lambda1R,lambda1L,lambda2L,lambda2R,lambda3L,lambda3R)
        lambda2_plus  = max(0.0d0,lambda1R,lambda1L,lambda2L,lambda2R,lambda3L,lambda3R)
        lambda2_minus = min(0.0d0,lambda1R,lambda1L,lambda2L,lambda2R,lambda3L,lambda3R)
        lambda3_plus  = max(0.0d0,lambda1R,lambda1L,lambda2L,lambda2R,lambda3L,lambda3R)
        lambda3_minus = min(0.0d0,lambda1R,lambda1L,lambda2L,lambda2R,lambda3L,lambda3R)

        ! -------------
        ! ---> (d) <--- Calculate the fluxes at L and R
        ! -------------

        f1L = u2L
        f1R = u2R

        f2L = 0.5d0*(3.0d0 - gamma)*u2L**2/u1L + (gamma - 1.0d0) * u3L
        f2R = 0.5d0*(3.0d0 - gamma)*u2R**2/u1R + (gamma - 1.0d0) * u3R

        f3L = gamma * u2L * u3L / u1L - 0.5d0 * ( gamma - 1.0d0) * u2L**3 / u1L**2
        f3R = gamma * u2R * u3R / u1R - 0.5d0 * ( gamma - 1.0d0) * u2R**3 / u1R**2

        ! -------------
        ! ---> (e) <--- Implement the formula for HLLE fluxes
        ! -------------

        flux1(i) = (  lambda1_plus * f1L - lambda1_minus * f1R     &
                    + lambda1_plus * lambda1_minus * (u1R - u1L) ) &
                    / (lambda1_plus - lambda1_minus)

        flux2(i) = (  lambda2_plus * f2L - lambda2_minus * f2R     &
                    + lambda2_plus * lambda2_minus * (u2R - u2L) ) &
                    / (lambda2_plus - lambda2_minus)

        flux3(i) = (  lambda3_plus * f3L - lambda3_minus * f3R     &
                    + lambda3_plus * lambda3_minus * (u3R - u3L) ) &
                    / (lambda3_plus - lambda3_minus)

     end do
     ! -------------
     ! ---> (f) <--- We have now the RHSs for the MoL
     ! -------------
     do i=1,Nx-1
```

Appendix B: Codes

```fortran
      rhs_u1(i) = - ( flux1(i) - flux1(i-1) ) / dx
      rhs_u2(i) = - ( flux2(i) - flux2(i-1) ) / dx
      rhs_u3(i) = - ( flux3(i) - flux3(i-1) ) / dx
    end do

    ! ------------------------------
    ! ----->    RK2 stuff    <-----
    ! ------------------------------
    if (rk.eq.1) then
      u1 = u1_p + rhs_u1 * dt
      u2 = u2_p + rhs_u2 * dt
      u3 = u3_p + rhs_u3 * dt
    else
      u1 = 0.5d0 * ( u1_p + u1 + rhs_u1 * dt)
      u2 = 0.5d0 * ( u2_p + u2 + rhs_u2 * dt)
      u3 = 0.5d0 * ( u3_p + u3 + rhs_u3 * dt)
    end if

    ! -------------
    ! ---> (g) <--- Boundary conditions
    ! -------------
    u1(0) = u1(1)
    u2(0) = u2(1)
    u3(0) = u3(1)
    u1(Nx) = u1(Nx-1)
    u2(Nx) = u2(Nx-1)
    u3(Nx) = u3(Nx-1)

    ! -------------
    ! ---> (h) <--- Recover primitive variables for output
    ! -------------
    rho = u1
    v = u2/u1
    p = (gamma - 1.0D0)*(u3 - 0.5D0*u2**2/u1)
    E = p / ( u1 * (gamma - 1.0D0) ) + 0.5D0 * rho * v**2
    cs = sqrt( p * gamma / rho )

  end do

  ! *****    --->   Ends Loop Core    <---

  ! Exact solution IF EXCERCISE SOLVED

  ! Save data during the evolution every 100 time steps

  if (mod(n,100*2**(resolution_label-1)).eq.0) then    ! Regular output
    print *, n,t            ! Send something to the screen
    write(1,*) "#_Time=",t
    do i=0,Nx,2**(resolution_label-1)
      write(1,*) x(i),rho(i),v(i),E(i),p(i)
    end do
    write(1,*)
    write(1,*)
  end if

  if (mod(n,50*2**(resolution_label-1)).eq.0) then
  ! Fancy output every 50 time steps
  write(2,*)
  do i=0,Nx
    write(2,*) t,x(i),rho(i),v(i),E(i),p(i)
  end do
  write(2,*)
  end if
```

```fortran
  END DO

close(2)
close(1)

print *, 'Program finished'

end program
! ------------------------------
! ----->   Program ends    <-----
! ------------------------------


! ----------------------------------------------------
! --->   Subroutine that reconstructs variables   <---
! ----------------------------------------------------
subroutine reconstruct(var0,var1,var2,var3,dspace,outL,outR)

use global_numbers
implicit none

real(kind=8), intent(in)  :: var0,var1,var2,var3,dspace
real(kind=8), intent(out) :: outL,outR
real(kind=8) smaxL,sminL,sigmaL,smaxR,sminR,sigmaR
real(kind=8) minmod

if (reconstructor.eq.'monmod') then
  ! ----->  IF minmod  <-----
  smaxL = ( var2 - var1 ) / dspace
  sminL = ( var1 - var0 ) / dspace
  sigmaL = minmod(sminL,smaxL)

  smaxR = ( var3 - var2 ) / dspace
  sminR = ( var2 - var1 ) / dspace
  sigmaR = minmod(sminR,smaxR)

  outL = var1 + sigmaL * 0.5d0 * dspace
  outR = var2 - sigmaR * 0.5d0 * dspace

else if (reconstructor.eq.'godunov') then
  ! ----->  IF Godunov  <-----
  outL = var1
  outR = var2

end if

end subroutine
! ----------------------------------------------------
! ------------------------------
! --->   minmod function   <---
! ------------------------------
real(kind=8) function minmod(a,b)

implicit none

real(kind=8)   a,b

  minmod = 0.5d0*(sign(1.0D0,a) + sign(1.0D0,b))*min(abs(a),abs(b))

end function minmod
! ------------------------------
```

Appendix B: Codes   343

Command of gnuplot used to produce results like those in Figs. 5.7, 5.8, 5.9, 5.10, and 5.11 for density, pressure, velocity, and internal energy:
```
gnuplot> splot 'MoLEuler.dat' u 1:2 w l
gnuplot> splot 'MoLEuler.dat' u 1:3 w l
gnuplot> splot 'MoLEuler.dat' u 1:4 w l
gnuplot> splot 'MoLEuler.dat' u 1:5 w l
```

Chapter 6

Code [Schroedinger3p1.f90] solves the 3+1 Schrödinger equation using the implicit Crank-Nicolson method and the ADI scheme in Sect. 6.4:

```fortran
! --------------------------------------------------------
! Schroedinger3p1.f90
! --------------------------------------------------------
! Author: Francisco S. Guzman
! Last modified: 3/March/2023
! Purpose: Solves the 3+1 Schroedinger equation for the particle in a box and
!          in a harmonic oscillator
!          It uses the ADI Implicit Crank-Nicolson Method
! Parameters to experiment with: nodes, potential
! Output file: psi.t
! Output:       t         N        <K>+<V>     Re(psi)(at origin)   rho(at origin)
! Output file: psi.x
! Output:       x         Re(psi)(at x-axis) Im(psi)(at x-axis) Re(psiexact)
! (at x-axis) error(at x-axis)
! Output file: psi.xy
! Output:       x    y     Re(psi)(at xy-plane)  Im(psi)(at xy-plane)  rho(at xy-plane)
! --------------------------------------------------------


! ------------------------------------------
! ----->    Module with global numbers   <-----
! ------------------------------------------
module numbers

! Arrays
real(kind=8), allocatable, dimension(:,:,:)  :: x,y,z
complex(kind=8), allocatable, dimension(:,:,:)  :: psi,psi_p,exact,error,V
real(kind=8), allocatable, dimension(:,:,:)  :: Hermitex,Hermitey,Hermitez
real(kind=8), allocatable, dimension(:,:,:)  :: rho,k_integrand,v_integrand
complex(kind=8), allocatable, dimension(:)  :: a,b,c,d,aux_cn
complex(kind=8) alpha,beta,tmp
real(kind=8) NofP,ExpectationK,ExpectationV

! Numbers
real(kind=8) dx,dy,dz,dt,xmin,xmax,ymin,ymax,zmin,zmax,tmax,CFL,t,r,pi
real(kind=8) norm_factor,nodes_x,nodes_y,nodes_z,facnx,facny,facnz,Energy
integer Nx,Ny,Nz,Nt,ifac,jfac,kfac
character(len=20) potential

end module
```

```fortran
! -------------------------------------------

! --------------------------
! --->    Program begins    <---
! --------------------------
program Schroedinger3p1

use numbers
implicit none

integer i,j,k,n

! ---------- Initial Set Up for the BOX and the HO ----------
potential = 'oscillator'              ! [ box or oscillator ]
Nx    = 100
Ny    = 100
Nz    = 100
Nt    = 100000
if (potential.eq.'box') then
  xmin = 0.0d0
  xmax = 1.0d0
  ymin = 0.0d0
  ymax = 1.0d0
  zmin = 0.0d0
  zmax = 1.0d0
  nodes_x = 3.0d0
  nodes_y = 3.0d0
  nodes_z = 3.0d0
else if (potential.eq.'oscillator') then
  xmin = -5.0
  xmax =  5.0
  ymin = -5.0
  ymax =  5.0
  zmin = -5.0
  zmax =  5.0
  nodes_x = 2.0d0
  nodes_y = 2.0d0
  nodes_z = 2.0d0
end if
CFL   = 1.0
pi    = acos(-1.0d0)

! Alocate memory of arrays
allocate(x(0:Nx,0:Ny,0:Nz),y(0:Nx,0:Ny,0:Nz),z(0:Nx,0:Ny,0:Nz))
allocate(psi(0:Nx,0:Ny,0:Nz),psi_p(0:Nx,0:Ny,0:Nz))
allocate(rho(0:Nx,0:Ny,0:Nz))
allocate(k_integrand(0:Nx,0:Ny,0:Nz),v_integrand(0:Nx,0:Ny,0:Nz))
allocate(V(0:Nx,0:Ny,0:Nz),exact(0:Nx,0:Ny,0:Nz),error(0:Nx,0:Ny,0:Nz))
allocate(Hermitex(0:Nx,0:Ny,0:Nz),Hermitey(0:Nx,0:Ny,0:Nz),Hermitez(0:Nx,0:Ny,0:Nz))
! For Nx=Ny=Nz this simplifies allocation for a,b,c,d
allocate(a(0:Nx),b(0:Nx),c(0:Nx),d(0:Nx),aux_cn(0:Nx))
! x,y,z:        space coordinates
! psi:          wave function at time n+1
! psi_p:        wave function at time n
! exact:        exact solution
! error:        error
! a,b,c:        arrays for the three diagonals of the matrix
! d:            right hand side of the linear system
! aux_cn:       array for the forward-backward substitution
! Hermitex(yz): arrays to store Hermit polynomials
! k_integrand:  integrand of the expectation value of kinetic energy
! v_integrand:  integrand of the expectation value of potential energy
```

Appendix B: Codes

```fortran
! Numerical Domain
dx = (xmax-xmin)/dble(Nx)
dy = (ymax-ymin)/dble(Ny)
dz = (zmax-zmin)/dble(Nz)
dt = CFL * min(dx**2,dy**2,dz**2)
do i=0,Nx
  x(i,:,:) = xmin + dble(i) * dx
end do
do j=0,Ny
  y(:,j,:) = ymin + dble(j) * dy
end do
do k=0,Nz
  z(:,:,k) = zmin + dble(k) * dz
end do
alpha = cmplx( 0.0d0 , 0.25d0 * CFL )
beta  = cmplx( 0.0d0 , 0.5d0 * dt)

! Initial Data
if (potential.eq.'box') then

  V   = cmplx( 0.0d0 , 0.0d0 )
  psi = cmplx(sqrt(8.0d0) * sin(nodes_x*pi*x) * sin(nodes_y*pi*y) *
  sin(nodes_z*pi*z),0.0d0)

else if (potential.eq.'oscillator') then

  V = cmplx( 0.5d0 * ( x**2 + y**2 + z**2 ) , 0.0d0 )
  ! Normalization coefficients
  facnx = 1.0d0
  facny = 1.0d0
  facnz = 1.0d0
  do ifac=2,int(nodes_x)
    facnx = facnx * dble(ifac)
  end do
  do jfac=2,int(nodes_y)
    facny = facny * dble(jfac)
  end do
  do kfac=2,int(nodes_z)
    facnz = facnz * dble(kfac)
  end do

  ! Only nx,ny,nz less than 5, written by hand
  if (nodes_x.eq.0) then
    Hermitex = 1.0d0
  else if (nodes_x.eq.1) then
    Hermitex = 2.0d0 * x
  else if (nodes_x.eq.2) then
    Hermitex = 4.0d0 * x**2 - 2.0d0
  else if (nodes_x.eq.3) then
    Hermitex = 8.0d0 * x**3 - 12.0 * x
  else if (nodes_x.eq.4) then
    Hermitex = 16. * x**4 - 48. * x**2 + 12.
  end if
  if (nodes_y.eq.0) then
    Hermitey = 1.0d0
  else if (nodes_y.eq.1) then
    Hermitey = 2.0d0 * y
  else if (nodes_y.eq.2) then
    Hermitey = 4.0d0 * y**2 - 2.0d0
  else if (nodes_y.eq.3) then
    Hermitey = 8.0d0 * y**3 -12.0 * y
  else if (nodes_y.eq.4) then
    Hermitey = 16. * y**4 - 48. * y**2 + 12.
```

```fortran
    end if
    if (nodes_z.eq.0) then
      Hermitez = 1.0d0
    else if (nodes_z.eq.1) then
      Hermitez = 2.0d0 * z
    else if (nodes_z.eq.2) then
      Hermitez = 4.0d0 * z**2 - 2.0d0
    else if (nodes_z.eq.3) then
      Hermitez = 8.0d0 * z**3 -12.0 * z
    else if (nodes_z.eq.4) then
      Hermitez = 16. * z**4 - 48.0 * z**2 + 12.
    end if

    norm_factor = 1.0d0 / sqrt( pi**(1.5) * 2.**nodes_x * 2.**nodes_y * 2.**nodes_z &
               * facnx * facny * facnz )

    psi = cmplx(norm_factor * Hermitex * Hermitey * Hermitez &
                * exp(-0.5d0*x**2) * exp(-0.5d0*y**2) * exp(-0.5d0*z**2) &
                , 0.0d0 )
end if

! At initial time the numerical equals the exact solution
exact = psi

open(1,file='psi.t',status='replace')
close(1)
open(2,file='psi.x',status='replace')
close(2)
open(3,file='psi.xy',status='replace')
close(3)

! Saves data at initial time
call save_scalars
call save_axes
call save_planes

print *,'      Iteration        Time                  N=int(rho)    Energy                  psi(0,0,0)'
print *, 0,t,NofP,ExpectationK+ExpectationV,psi(Nx/2,Ny/2,Nz/2)

! Evolution loop
do n=1,Nt

   t = t + dt     ! Updates time

   ! *****    --->    LOOP CORE    <---
   ! ADI
   psi_p = psi
   ! --->    Step 1: Solution for R    <---
   do j=1,Ny-1
   do k=1,Nz-1
     ! Boundary conditions on xmin and xmax
     b(0)  = 1.0d0 + 2.0d0*alpha
     b(Nx) = 1.0d0 + 2.0d0*alpha
     if (potential.eq.'box') then
        c(0)  = 0.0d0
        d(0)  = psi_p(0,j,k)  * ( 1.0d0 - 2.0d0*alpha )
        a(Nx) = 0.0d0
        d(Nx) = psi_p(Nx,j,k) * ( 1.0d0 - 2.0d0*alpha )
     else if (potential.eq.'oscillator') then
        c(0)  = -2.0d0*alpha
        d(0)  = psi_p(0,j,k) * ( 1.0d0 - 2.0d0*alpha ) &
              + 2.0d0*alpha*psi_p(1,j,k)
```

Appendix B: Codes

```
      a(Nx) = -2.0d0*alpha
      d(Nx) = psi_p(Nx,j,k) * ( 1.0d0 - 2.0d0*alpha ) &
            + 2.0d0*alpha*psi_p(Nx-1,j,k)
    end if
    ! All other entries of the tridiagonal matrix
    do i=1,Nx-1
      a(i) = -alpha
      b(i) = 1.0d0 + 2.0d0*alpha
      c(i) = -alpha
      d(i) = psi_p(i-1,j,k) * alpha &
           + psi_p(i,j,k)   * ( 1.0d0 - 2.0d0*alpha ) &
           + psi_p(i+1,j,k) * alpha
    end do
    ! Forward substitution
    tmp = b(0)
    psi(0,j,k) = d(0)/tmp
    do i=1,Nx
      aux_cn(i) = c(i-1)/tmp
      tmp = b(i) - a(i)*aux_cn(i)
      psi(i,j,k) = (d(i)-a(i)*psi(i-1,j,k))/tmp
    end do
    ! Backward substitution
    do i = Nx-1,0,-1
      psi(i,j,k) = psi(i,j,k) - aux_cn(i+1)*psi(i+1,j,k)
    end do
  end do
end do
! --->   Step 2: Solution for S    <---
! Refill the array for psi with the values obtained for R
psi_p = psi
do i=1,Nx-1
do k=1,Nz-1
    ! Boundary conditions on ymin and ymax
    b(0)  = 1.0d0 + 2.0d0*alpha
    b(Ny) = 1.0d0 + 2.0d0*alpha
    if (potential.eq.'box') then
      c(0)  = 0.0d0
      a(Ny) = 0.0d0
      d(0)  = psi_p(i,0,k)  * ( 1.0d0 - 2.0d0*alpha )
      d(Ny) = psi_p(i,Ny,k) * ( 1.0d0 - 2.0d0*alpha )
    else if (potential.eq.'oscillator') then
      c(0)  = -2.0d0*alpha
      a(Ny) = -2.0d0*alpha
      d(0)  = psi_p(i,0,k)  * ( 1.0d0 - 2.0d0*alpha ) &
            + 2.0d0*alpha*psi_p(i,1,k)
      d(Ny) = psi_p(i,Ny,k) * ( 1.0d0 - 2.0d0*alpha ) &
            + 2.0d0*alpha*psi_p(i,Ny-1,k)
    end if
    ! All other entries of the tridiagonal matrix
    do j=1,Ny-1
      a(j) = -alpha
      b(j) = 1.0d0 + 2.0d0*alpha
      c(j) = -alpha
      d(j) = psi(i,j-1,k) * alpha &
           + psi(i,j,k)   * ( 1.0d0 - 2.0d0*alpha ) &
           + psi(i,j+1,k) * alpha
    end do
    ! Forward substitution
    tmp = b(0)
    psi(i,0,k) = d(0)/tmp
    do j=1,Ny
      aux_cn(j) = c(j-1)/tmp
      tmp = b(j)-a(j)*aux_cn(j)
```

```
      psi(i,j,k) = (d(j)-a(j)*psi(i,j-1,k))/tmp
    end do
    ! Backward substitution
    do j = Ny-1,0,-1
      psi(i,j,k) = psi(i,j,k) - aux_cn(j+1)*psi(i,j+1,k)
    end do
  end do
end do
! --->   Step 3: Solution for T    <---
! Refill the array for psi with the values obtained for S
psi_p = psi
do i=1,Nx-1
do j=1,Ny-1
  ! Boundary conditions on zmin and zmax
  b(0)  = 1.0d0 + 2.0d0*alpha
  b(Nz) = 1.0d0 + 2.0d0*alpha
  if (potential.eq.'box') then
    c(0)  = 0.0d0
    a(Nz) = 0.0d0
    d(0)  = psi(i,j,0)  * ( 1.0d0 - 2.0d0*alpha )
    d(Nz) = psi(i,j,Nz) * ( 1.0d0 - 2.0d0*alpha )
  else if (potential.eq.'oscillator') then
    c(0)  = -2.0d0*alpha
    a(Nz) = -2.0d0*alpha
    d(0)  = psi(i,j,0)  * ( 1.0d0 - 2.0d0*alpha ) &
          + 2.0d0*alpha*psi(i,j,1)
    d(Nz) = psi(i,j,Nz) * ( 1.0d0 - 2.0d0*alpha ) &
          + 2.0d0*alpha*psi(i,j,Nz-1)
  end if
  ! All other entries of the tridiagonal matrix
  do k=1,Nz-1
    a(k) = -alpha
    b(k) = 1.0d0 + 2.0d0*alpha
    c(k) = -alpha
    d(k) = psi_p(i,j,k-1) * alpha &
         + psi_p(i,j,k)   * ( 1.0d0 - 2.0d0*alpha ) &
         + psi_p(i,j,k+1) * alpha
  end do
  ! Forward substitution
  tmp = b(0)
  psi(i,j,0) = d(0)/tmp
  do k=1,Nz
    aux_cn(k) = c(k-1)/tmp
    tmp = b(k)-a(k)*aux_cn(k)
    psi(i,j,k) = (d(k)-a(k)*psi(i,j,k-1))/tmp
  end do
  ! Backward substitution
  do k = Nz-1,0,-1
    psi(i,j,k) = psi(i,j,k) - aux_cn(k+1)*psi(i,j,k+1)
  end do
end do
end do
! --->   Step 4: Solution for Psi at time n+1   <---
! Refill the array for psi with the values obtained for T
psi_p = psi
! It is a diagonal system
psi = (1.0d0 - beta * V) * psi / (1.0d0 + beta * V )

! *****    --->    Ends Loop Core    <---

! Exact solution
if (potential.eq.'box') then
  Energy = 0.5d0 * pi**2 * ( nodes_x**2 + nodes_y**2 + nodes_z**2 )
```

Appendix B: Codes   349

```
        exact = sqrt(8.0d0)*sin(nodes_x*pi*x)*sin(nodes_y*pi*y)*sin(nodes_z*pi*z) * &
                cmplx( cos( Energy * t ) , sin( Energy * t ) )
     else if (potential.eq.'oscillator') then
        Energy = nodes_x + nodes_y + nodes_z + 1.5d0
        exact = norm_factor * Hermitex * Hermitey * Hermitez &
              * exp(-0.5d0*x**2) * exp(-0.5d0*y**2) * exp(-0.5d0*z**2) &
              * cmplx( cos( Energy * t ) , sin( Energy * t ) )
     end if
     error = psi - exact

     ! Save data during evolution every 100 time steps

     if (mod(n,100).eq.0) then       ! Output every certain # of iterations
        call save_scalars
        call save_axes
        call save_planes
        print *, n,t,NofP,ExpectationK+ExpectationV ! Send something to the screen
     end if

  end do

  print *, 'Program finished'

end program
! -------------------------
! --->   Program ends    <---
! -------------------------

! -----------------------------------------
! --->   Subroutine that saves scalars  <---
! -----------------------------------------
subroutine save_scalars

use numbers
implicit none

integer i,j,k

rho = conjg(psi)*psi ! For the integration of N
k_integrand = 0.0d0  ! For the expectation value of kinetic energy
do i=1,Nx-1
do j=1,Ny-1
do k=1,Nz-1
   k_integrand(i,j,k) = real(psi(i,j,k)) * ( &
           ( real(psi(i+1,j,k)) - 2.0d0 * real(psi(i,j,k)) + real(psi(i-1,j,k)) ) &
           / dx**2 &
         + ( real(psi(i,j+1,k)) - 2.0d0 * real(psi(i,j,k)) + real(psi(i,j-1,k)) ) &
           / dy**2 &
         + ( real(psi(i,j,k+1)) - 2.0d0 * real(psi(i,j,k)) + real(psi(i,j,k-1)) ) &
           / dz**2 ) &
                     + imag(psi(i,j,k)) * ( &
           ( imag(psi(i+1,j,k)) - 2.0d0 * imag(psi(i,j,k)) + imag(psi(i-1,j,k)) ) &
           / dx**2 &
         + ( imag(psi(i,j+1,k)) - 2.0d0 * imag(psi(i,j,k)) + imag(psi(i,j-1,k)) ) &
           / dy**2 &
         + ( imag(psi(i,j,k+1)) - 2.0d0 * imag(psi(i,j,k)) + imag(psi(i,j,k-1)) ) &
           / dz**2 )
end do
end do
end do

! Copy the value of the integral at the faces rfom neighbor points. This cam
   improve.
```

```fortran
k_integrand(0,:,:)  = k_integrand(1,:,:)
k_integrand(Nx,:,:) = k_integrand(Nx-1,:,:)
k_integrand(:,0,:)  = k_integrand(:,1,:)
k_integrand(:,Ny,:) = k_integrand(:,Ny-1,:)
k_integrand(:,:,0)  = k_integrand(:,:,1)
k_integrand(:,:,Ny) = k_integrand(:,:,Nz-1)

k_integrand = -0.5d0 * k_integrand              ! Remember the factor
             -1/2 before the Laplacian
v_integrand = V * ( real(psi)**2 + imag(psi)**2 )     ! Simply (Psi* V Psi)

! Integration of N, <K>, <V>
NofP = 0.0d0
ExpectationK = 0.0d0
ExpectationV = 0.0d0
do i=0,Nx-1
do j=0,Ny-1
do k=0,Nz-1
  NofP = NofP + 0.125*( rho(i,j  ,k  ) + rho(i+1,j  ,k  ) &
                      + rho(i,j+1,k  ) + rho(i+1,j+1,k  ) &
                      + rho(i,j  ,k+1) + rho(i+1,j  ,k  ) &
                      + rho(i,j+1,k  ) + rho(i+1,j+1,k+1) ) &
                      * dx * dy * dz
  ExpectationK = ExpectationK + 0.125*( k_integrand(i,j  ,k  ) + k_integrand(i+1,j  ,k  ) &
                      + k_integrand(i,j+1,k  ) + k_integrand(i+1,j+1,k  ) &
                      + k_integrand(i,j  ,k+1) + k_integrand(i+1,j  ,k  ) &
                      + k_integrand(i,j+1,k  ) + k_integrand(i+1,j+1,k+1) ) &
                      * dx * dy * dz
  ExpectationV = ExpectationV + 0.125*( v_integrand(i,j  ,k  ) + v_integrand(i+1,j  ,k  ) &
                      + v_integrand(i,j+1,k  ) + v_integrand(i+1,j+1,k  ) &
                      + v_integrand(i,j  ,k+1) + v_integrand(i+1,j  ,k  ) &
                      + v_integrand(i,j+1,k  ) + v_integrand(i+1,j+1,k+1) ) &
                      * dx * dy * dz
end do
end do
end do

open(1,file='psi.t', status='old',position='append')
  write(1,*) t,NofP,ExpectationK+ExpectationV,real(psi(Nx/2,Ny/2,Nz/2)), &
    rho(Nx/2,Ny/2,Nz/2)
close(1)
end subroutine save_scalars

! ----------------------------------------------------------
! --->   Subroutine that saves grid functions along x    <---
! ----------------------------------------------------------
subroutine save_axes

use numbers
implicit none

integer i,j,k

open(2,file='psi.x', status='old',position='append')
  write(2,*) '#Time=',t
  do i=0,Nx
    write(2,*) x(i,Ny/2,Nz/2),real(psi(i,Ny/2,Nz/2)),imag(psi(i,Ny/2,Nz/2)), &
               real(exact(i,Ny/2,Nz/2)), real(error(i,Ny/2,Nz/2))
  end do
  write(2,*)
  write(2,*)
```

Appendix B: Codes

```fortran
  close(2)
end subroutine save_axes

! -----------------------------------------------------------------
! --->    Subroutine that saves grid functions at the xy-plane    <---
! -----------------------------------------------------------------
subroutine save_planes

use numbers
implicit none

integer i,j,k

open(3,file='psi.xy', status='old',position='append')
  write(3,*) '#Time=',t
  do i=0,Nx
    write(3,*)
    do j=0,Ny
      write(3,*) x(i,j,Nz/2),y(i,j,Nz/2),real(psi(i,j,Nz/2)),imag(psi(i,j,Nz/2)),
        rho(i,j,Nz/2)
    end do
  end do
  write(3,*)
  write(3,*)
close(3)
end subroutine save_planes
! -----------------------------------------------------------------
```

> Command of gnuplot that show results like those in Figs. 6.7 and 6.9 for the real part of Ψ, the imaginary part of Ψ, and $\rho = \Psi^*\Psi$:
>
> ```
> gnuplot> splot 'psi.xy' i 3 u 1:2:3 ev 2:2 t 'Re(Psi)' w l
> gnuplot> splot 'psi.xy' i 3 u 1:2:4 ev 2:2 t 'Im(Psi)' w l
> gnuplot> splot 'psi.xy' i 3 u 1:2:(3**2+4**2) ev 2:2 t 'rho
>    = Re(Psi)**2 + Im(Psi)**2' w l
> ```
>
> For scalars like those in Figs. 6.8 and 6.10, for N, $\langle E \rangle$, real parts of Ψ at the origin and ρ at the origin use
>
> ```
> gnuplot> plot 'psi.t' u 1:2 t 'Number of Particles' w l
> gnuplot> plot 'psi.t' u 1:3 t 'Expectation value of E' w l
> gnuplot> plot 'psi.t' u 1:4 t 'Central value of the real
>    part of Psi' w l
> gnuplot> plot 'psi.t' u 1:5 t 'Central density' w l
> ```

> **Code** [Diffusion2p1.f90] solves the 2+1 diffusion equation with constant and variable diffusion coefficient in Sect. 6.6.

```
! -----------------------------------------------------------------
! Diffusion2p1.f90
! -----------------------------------------------------------------
```

```
! Author: Francisco S. Guzman
! Last modified: 17/february/2023
! Purpose: Solves the IVP for the 2+1 Diffusion Equation
!                using the Method of Lines and RK3
!         Uses zero Boundary Conditions
!         Case constant kappa for initial data of the exact solution
!         Case space-dependent kappa for an initial gaussian
! Parameters to experiment with: change profile of kappa
! Output file: Diff2p1.t
! Output:       t       max(u)
!
! Output file: Diff2p1.xy
! Output:       x       y       u
!               one data block per 0.01/dt iterations
! -------------------------------------------------------

! --------------------------------------------
! ----->   Module with global numbers   <-----
! --------------------------------------------
module global_numbers

implicit none

! --- Numerical domain
real(kind=8), allocatable, dimension(:,:) :: x,y
real(kind=8)  dx,dy,dz,dt,xmin,xmax,ymin,ymax,tmax,CFL,t,pi
integer Nx,Ny,Nz,Nt,rk,ghost

! --- Primitive variables
real(kind=8), allocatable, dimension(:,:) :: u,u_p,rhs_u,kappa

! -- Some parameters of the equation
real(kind=8)  nodesx,nodesy,amp

character(len=20) :: kappa_case

end module
! --------------------------------------------

! ---------------------------
! --->   Program begins   <---
! ---------------------------
program Diffusion2p1

use global_numbers
implicit none

integer i,j,k,n

! Define some parameter values
Nx    = 100
Ny    = 100
Nt    = 100000
```

Appendix B: Codes

```
xmin    = 0.0d0
xmax    = 1.0d0
ymin    = 0.0d0
ymax    = 1.0d0
CFL     = 0.125/4.0d0
ghost   = 0          ! 1 for periodic boundary conditions, 0 for Dirichlet
pi      = acos(-1.0d0)
amp     = 1.0d0
nodesx  = 2.0d0
nodesy  = 3.0d0
kappa_case = 'constant'         ! [ constant, tanh ]

call allocate_memory      ! Alocates memory

! Numerical Domain
dx = (xmax - xmin) / dble(Nx)
dy = (ymax - ymin) / dble(Ny)
dt = CFL * dx**2
do i=-ghost,Nx+ghost
  x(i,:) = xmin + dble(i) * dx
end do
do j=-ghost,Ny+ghost
  y(:,j) = ymin + dble(j) * dy
end do
print *, 'dt=',dt

! ------------------------
! Initial conditions
! ------------------------
t = 0.0d0

if (kappa_case.eq.'constant') then
  ! For the Exact Solution with K=constant
  u = amp * sin(nodesx*pi*x) * sin(nodesy*pi*y)
  kappa = 1.0d0
else if (kappa_case.eq.'tanh') then
  ! For a Gaussian and kappa = tanh function
  u= exp(-((x-0.5)**2+(y-0.5)**2)/0.05)
  kappa=(2.0d0*tanh((x-0.5)/0.05)+3)
end if

! ----- Save data at initial time
open(1,file='Diff2p1.t')
open(2,file='Diff2p1.xy',status='replace')
call savedata       ! Save data at initial time

! Evolution loop
print *, 'Iteration    time'   ! Send something to the screen
print *, '0',t  ! Send something to the screen

do n=1,Nt
```

```
  t = t + dt       ! Updates time
  ! *****    --->    LOOP CORE    <---

  ! Recycle arrays
  u_p   = u

  DO rk=1,3
    ! Calculate the RHS
    do i=1,Nx-1
    do j=1,Ny-1
      rhs_u(i,j) =     0.5d0 * ( kappa(i+1,j) - kappa(i-1,j) ) / dx &
                     * 0.5d0 * (     u(i+1,j) -     u(i-1,j) ) / dx &
                     + 0.5d0 * ( kappa(i,j+1) - kappa(i,j-1) ) / dx &
                     * 0.5d0 * (     u(i,j-1) -     u(i,j-1) ) / dx &
           + kappa(i,j) * (( u(i+1,j)-2.0d0*u(i,j)+u(i-1,j) ) / dx**2 &
                          +( u(i,j+1)-2.0d0*u(i,j)+u(i,j-1) ) / dy**2 )

    end do
    end do
    if (rk.eq.1) then
      u = u_p + rhs_u * dt
    else if (rk.eq.2) then
      u = 0.75d0 * u_p + 0.25d0 * u + 0.25d0 * rhs_u * dt
    else
      u = u_p/3.0d0 + 2.0d0 * u/3.0d0 + 2.0d0 * rhs_u * dt/3.0d0
    end if
  END DO


  call Dirichlet_BCs

  ! *****    --->    Ends Loop Core    <---

  ! Save data every 10 iterations
  if (mod(n,int(0.01/dt)+1).eq.0) then
    call savedata
    print *, n,t   ! Send something to the screen
  end if

end do

close(1)
close(2)

print *, 'Program finished'

end program
! -------------------------
! --->   Program ends    <---
! -------------------------


! ----------------------------------------
! --->   Subroutine that saves data    <---
! ----------------------------------------
```

Appendix B: Codes

```fortran
subroutine savedata

use global_numbers
implicit none

integer i,j
real(kind=8) max

do i=0,Nx
  do j=0,Ny
    if (u(i,j).gt.max) max = u(i,j)
  end do
end do
write(1,*) t, max

! 2D output
write(2,*) '#Time=',t
do i=0,Nx
do j=0,Ny
  write(2,*) x(i,j),y(i,j),u(i,j)
end do
write(2,*)
end do
write(2,*)
write(2,*)

end subroutine

! ---------------------------------------------------------
! --->    Subroutine that implements Dirichlet BCs    <---
! ---------------------------------------------------------
subroutine Dirichlet_BCs

use global_numbers
implicit none

  u(0,:)  = 0.0d0
  u(Nx,:) = 0.0d0

  u(:,0)  = 0.0d0
  u(:,Ny) = 0.0d0

end subroutine

! --------       Allocate memory subroutine       ---------
! ---------------------------------------------------------
! --->    Subroutine that allocates memory     <---
! ---------------------------------------------------------
subroutine allocate_memory

use global_numbers
implicit none
```

```
allocate(x(-ghost:Nx+ghost,-ghost:Ny+ghost))
allocate(y(-ghost:Nx+ghost,-ghost:Ny+ghost))
allocate(u(-ghost:Nx+ghost,-ghost:Ny+ghost))
allocate(kappa(-ghost:Nx+ghost,-ghost:Ny+ghost))
allocate(rhs_u(-ghost:Nx+ghost,-ghost:Ny+ghost))

end subroutine
! ---------------------------------------------
```

Instruction of gnuplot to display results like those in Figs. 6.14 and 6.15 for u:

```
gnuplot> set pm3d map
gnuplot> splot 'Diff2p1.xy' i 3
```

Index

B
Basic methods, 151

C
Convergence, 3, 7, 8, 13–15, 23, 26, 28–31, 38, 40, 42, 53, 105–108, 115–118, 121, 131, 134, 137, 147, 148, 150, 151, 161, 162, 185, 187, 188, 225, 274

D
Diffusion equation, 5, 99, 121, 132–139, 144, 147, 169, 196, 259, 300–305, 314–317, 351

E
Error theory, 29–31, 105–107

F
Finite volumes, viii, 5, 6, 205–257, 259, 286, 310, 321

H
Hydrodynamics, 2, 5, 64, 174, 226–233, 238, 239, 241, 252, 256, 285–300, 305, 308–310, 337

M
Method of lines (MoL), 5, 121, 155–202, 209, 210, 215, 216, 221, 238, 263, 271–277, 287, 295, 300, 302, 304, 321, 333, 340, 352

O
Ordinary differential equations (ODEs), vii, 1, 5, 7–96, 100, 105, 106, 156–158, 199, 256, 300

P
3+1 partial differential equations, vii
Partial differential equations (PDEs), vii, 1, 2, 5, 52, 64, 95, 99–152, 155–202, 205, 271

S
Schrödinger equation, vii, 2, 4, 79, 100, 139–152, 155, 183–188, 202, 259, 277–285, 310, 320, 321, 328, 343

W
Wave equation, 4, 99, 108–132, 135, 139, 155, 161, 163–183, 188–196, 200–201, 205, 207, 208, 249, 259, 263–277, 316, 320, 321, 333

Printed in the United States
by Baker & Taylor Publisher Services